# Tutorials in Chemoinformatics

# Tutorials in Chemoinformatics

*Edited by Alexandre Varnek*

*University of Strasbourg, Strasbourg*
*France*

*Registered Office(s)*
John Wiley & Sons, Inc., 111 River Street, Hoboken, NJ 07030, USA
John Wiley & Sons Ltd, The Atrium, Southern Gate, Chichester, West Sussex, PO19 8SQ, UK

*Editorial Office*
9600 Garsington Road, Oxford, OX4 2DQ, UK

For details of our global editorial offices, customer services, and more information about Wiley products visit us at www.wiley.com.

Wiley also publishes its books in a variety of electronic formats and by print-on-demand. Some content that appears in standard print versions of this book may not be available in other formats.

*Limit of Liability/Disclaimer of Warranty*
In view of ongoing research, equipment modifications, changes in governmental regulations, and the constant flow of information relating to the use of experimental reagents, equipment, and devices, the reader is urged to review and evaluate the information provided in the package insert or instructions for each chemical, piece of equipment, reagent, or device for, among other things, any changes in the instructions or indication of usage and for added warnings and precautions. While the publisher and authors have used their best efforts in preparing this work, they make no representations or warranties with respect to the accuracy or completeness of the contents of this work and specifically disclaim all warranties, including without limitation any implied warranties of merchantability or fitness for a particular purpose. No warranty may be created or extended by sales representatives, written sales materials or promotional statements for this work. The fact that an organization, website, or product is referred to in this work as a citation and/or potential source of further information does not mean that the publisher and authors endorse the information or services the organization, website, or product may provide or recommendations it may make. This work is sold with the understanding that the publisher is not engaged in rendering professional services. The advice and strategies contained herein may not be suitable for your situation. You should consult with a specialist where appropriate. Further, readers should be aware that websites listed in this work may have changed or disappeared between when this work was written and when it is read. Neither the publisher nor authors shall be liable for any loss of profit or any other commercial damages, including but not limited to special, incidental, consequential, or other damages.

*Library of Congress Cataloging-in-Publication data applied for*

ISBN: 9781119137962

Cover image: (molecules) © alice-photo/Gettyimages; (computers) © loops7/iStockphoto
Cover design by Wiley

Set in 10/12pt Warnock by SPi Global, Pondicherry, India

10  9  8  7  6  5  4  3  2  1

# Contents

# List of Contributors

**João Montargil Aires de Sousa**
LAQV-REQUIMTE and Departamento
de Quimica
Universidade Nova de Lisboa
Portugal

**Jürgen Bajorath**
Department of Life Science Informatics
Rheinische Friedrich-Wilhelms-
Universität Bonn
Bonn
Germany

**Jenny Balfer**
Department of Life Science Informatics
Rheinische Friedrich-Wilhelms-
Universität Bonn
Bonn
Germany

**Igor I. Baskin**
Department of Physics
Lomonosov Moscow State University
Moscow
Russia

**Guillaume Bret**
Medalis Drug Discovery Center
University of Strasbourg
Illkirch
France

**Sharon D. Bryant**
InteLigand GmbH
Vienna
Austria

**Antonio de la Vega de Leon**
Department of Life Science Informatics
Rheinische Friedrich-Wilhelms-
Universität Bonn
Bonn
Germany

**Jérémy Desaphy**
Medalis Drug Discovery Center
University of Strasbourg
Illkirch
France

**Dragos Horvath**
Laboratory of Chemoinformatics (UMR
7140 CNRS/UniStra)
University of Strasbourg
Strasbourg
France

**Gökhan Ibis**
InteLigand GmbH
Vienna
Austria

**Esther Kellenberger**
Medalis Drug Discovery Center
University of Strasbourg
Illkirch
France

**Thierry Langer**
University of Vienna, Department of
Pharmaceutical Chemistry
Vienna
Austria

*Eugen Lounkine*
Novartis Institutes for Biomedical
Research In Silico Lead Discovery
Cambridge, Massachusetts
USA

*Timur Madzhidov*
Institute of Chemistry
Kazan Federal University
Kazan
Russia

*Gilles Marcou*
Laboratory of Chemoinformatics (UMR
7140 CNRS/UniStra)
University of Strasbourg
Strasbourg
France

*Ramil Nugmanov*
Institute of Chemistry
Kazan Federal University
Kazan
Russia

*Giulio Poli*
Department of Pharmacy
University of Pisa
Pisa
Italy

*Didier Rognan*
Medalis Drug Discovery Center
University of Strasbourg
Illkirch
France

*Thomas Seidel*
University of Vienna, Department of
Pharmaceutical Chemistry
Vienna
Austria

*Inna Slynko*
Medalis Drug Discovery Center
University of Strasbourg
Illkirch
France

*Vitaly Solov'ev*
A.N. Frumkin Institute of Physical
Chemistry and Electrochemistry RAS
Moscow
Russia

*Alexandre Varnek*
Laboratory of Chemoinformatics (UMR
7140 CNRS/UniStra)
University of Strasbourg
Strasbourg
France

*Martin Vogt*
Department of Life Science Informatics
Rheinische Friedrich-Wilhelms-
Universitat Bonn
Bonn
Germany

# Preface

Chemoinformatics methods are widely used both in academic research and in industrial applications. They represent an important part of curricula of numerous chemistry and pharmacology MSc and PhD programs running practically in all European countries, in United States, Australia, Japan, Korea, and Canada. Since early 2000, several dozens of books devoted to various applications of chemoinformatics have been published. Surprisingly, among them, there were very few textbooks.[1–3] To our knowledge, no books describing practical exercises in chemoinformatics have been published so far. Here, we fill this gap by presenting 30 tutorials assembling more than 100 exercises developed either for the master programs running at the University of Strasbourg (France), Rheinische Friedrich-Wilhelms University of Bonn (Germany), New University of Lisbon (Portugal), University of Vienna (Austria), Lomonosov Moscow State University (Russia), and Kazan Federal University (Russia), or for five international Strasbourg Summer Schools in Chemoinformatics (CS3) in the period 2008–2016. These tutorials cover all main areas in chemoinformatics: chemical databases, library design, chemical data analysis, structure—property/activity modeling, pharmacophore modeling, and docking. Each tutorial contains a short theoretical part, algorithm description (if necessary), and software instructions.

The book is divided into 10 parts devoted to different applications of chemoinformatics. Part 1 is devoted to chemical databases with four tutorials related to data curation and standardization, creation of a local database, manipulations with the Markush structures, and structure encoding by text and bit strings. Part 2 concerns design of focused and diverse compounds libraries, whereas Part 3 describes data visualization and analysis using hierarchical clustering and self-organizing maps. Parts 4 and 5 are devoted to structure-property/activity models (QSAR/QSPR). Beginners should start with the exercises in Part 4 demonstrating general workflow of building and validation of individual regression and classification models. Part 5 describing various ensemble modeling approaches (bagging, boosting, stacking, random subspaces) is addressed to more advanced readers. Parts 6 through 9 consider different approaches explicitly accounting for structure of biological targets (proteins, nucleic acids). Thus, Part 6 describes 3D pharmacophores modeling—a popular method of chemical structures encoding and virtual screening. Part 7 provides a description of a protein preparation followed by the virtual screening using pharmacophore and docking approaches. Various aspects of ligand-to-protein docking are discussed in Part 8. Pharmacological profiling using shape analysis of the protein binding sites is described in Part 9. Part 10 is particularly useful for developers of chemoinformatics tools because it focuses on the implementation of basic chemoinformatic tasks and fundamental algorithms in a high-level programming language. Since the tutorials were initially prepared for different programs taught in specifically developed courses for a given university, certain heterogeneity of the material presentation is inevitable. Hopefully, this won't pose a problem for readers.

Overall, three types of software tools are used in the exercises: in-house programs (ISIDA, VolSite, and Shaper), open-source programs (KNIME, WEKA, CDK, RDKit), and commercial programs which are either free (ChemAxon, OpenEye) or relatively inexpensive (MOE, LigandScoute, LeadIT) for academics. Note that the graphical interfaces of some programs (e.g., WEKA) vary from one version to another. Therefore, the reader may find tiny differences between the current version of sofware and some screenshots given in the book. Some exercises use a Python code. In-house software as well as the data sets needed for the exercises are available for the readers from the dedicated website www.wiley.com/go/varnek/chemoinformatics.

We believe that this book can be a valuable support in teaching chemoinformatics and molecular modeling disciplines at BSc and MSc level. It doesn't replace traditional textbooks (see references list) certainly needed for deep understanding of chemoinformatics approaches.

The authors would like to thank Dr Olga Klimchuk for the help with the manuscript preparation and Dr Stefano Pieracchini for the fruitful discussion.

*Alexandre Varnek*

## References

**1** J. Gasteiger and T. Engel, *Chemoinformatics: A Textbook*, Wiley, 2004.

**2** A.R. Leach and V.J. Gillet, *An Introduction to Chemoinformatics*, Springer, 2007.

**3** T. Madzhidov, I. Baskin, and A. Varnek, *An Introduction to Chemoinformatics*, vol. 1–4, University of Kazan, 2013-2016.

## About the Companion Website

Don't forget to visit the companion website for this book:

www.wiley.com/go/varnek/chemoinformatics

There you will find valuable material designed to enhance your learning, including datasets and in-house software to support the exercises in this book.

Scan this QR code to visit the companion website.

Part 1

Chemical Databases

# 1

## Data Curation

*Gilles Marcou and Alexandre Varnek*

*Goal*: Identify and curate problematic chemical information from a data collection. The raw dataset is processed so that it will be ready to feed a relational database dedicated to the organoleptic properties of small organic molecules. Information is interpreted and re-encoded as categories or bit vectors when relevant.

*Software*: KNIME 3.0, ChemAxon

*Data*: The following files are provided in the tutorial:

- `thegoodscent_dup.csv` – The raw data formatted in a semicolon separated file extracted from the web site of The Good Scent Company. The data is prepared and the most visible errors and discrepancies are already corrected.
- `thegoodscent_dup.raw` – The raw data without any processing related to the tutorial.
- `MissingOdorTypes.csv` – Manually curated Odor Types provided for some difficult cases.
- `StructureCuration.csv` – File containing the curation rules for some deficient SMILES of the input.
- `TutoDataCuration.zip` – The final KNIME workflow. Unzip the archive in the KNIME workspace and it will appear in your LOCAL workflows.
- `Slurp.pl` – A Perl script exploring the website of The Good Scents Company in search of some chemical information.

The Good Scent Company is an online shop providing cosmetic, flavor, and fragrance ingredients. It provides information for the flavor, food, and fragrance industry since 1994, and sales ingredients since 1980.

## Theoretical Background

Chemical datasets can be collected from literature, compendiums, web sites, lab-books, databases, and so on. Aggregation and automatic treatment of data represent additional sources of errors. Therefore, verification of quality and accuracy of chemical information is a crucial step of data valorization.[1]

*Tutorials in Chemoinformatics*, First Edition. Edited by Alexandre Varnek.
© 2017 John Wiley & Sons Ltd. Published 2017 by John Wiley & Sons Ltd.
Companion website: www.wiley.com/go/varnek/chemoinformatics

The problem of the quality of publicly available chemical data can be illustrated on the searching the Web for the chemical structure of antibacterial compound Vancomycine, for which stereochemistry information is essential. One can suggest two possible queries using InChIKey notations:[2,3]

*Query 1*: "MYPYJXKWCTUITO" "Vancomycine"
*Query 2*: "MYPYJXKWCTUITO-LYRMYLQWSA-N" "Vancomycine"

*Query 1* corresponds to the first layer of the InChI code of Vancomycine; it encodes only elemental constitution and atoms connectivity, whereas *Query 2* includes detailed stereochemistry information.

A search on Google (29/01/2016) retrieves 82 and 71 entries for *Queries 1* and *2*, respectively. Entries found with *Query 2* correspond to the correct chemical structure of Vancomycine, whereas all 11 additional entries retrieved with *Query 1* refer to its different enantiomers, see example on Scheme 1.1.

From this example, one can see that an estimate of the erroneous data associating Vancomycine to the wrong chemical structure is about 13%. Analysis of some 6800 publications in drug discovery[4] show that the average error rate of reported chemical structures is about 8% and, it seems, nothing has changed so far. Numerous examples and alerts about data curation problems, especially in public databases, can be found in the literature.[4–8]

In this tutorial, a dataset regarding organoleptic properties of cosmetic related chemicals was collected from the website http://thegoodscentscompany.com/(January 2016). The dataset contains eight records: the name of the chemical substance, the CAS number, an odor category and description, the source of the odor description, a taste description, the literature for the source of the taste description, and the SMILES encoding the chemical structure of the substance. The data were retrieved automatically

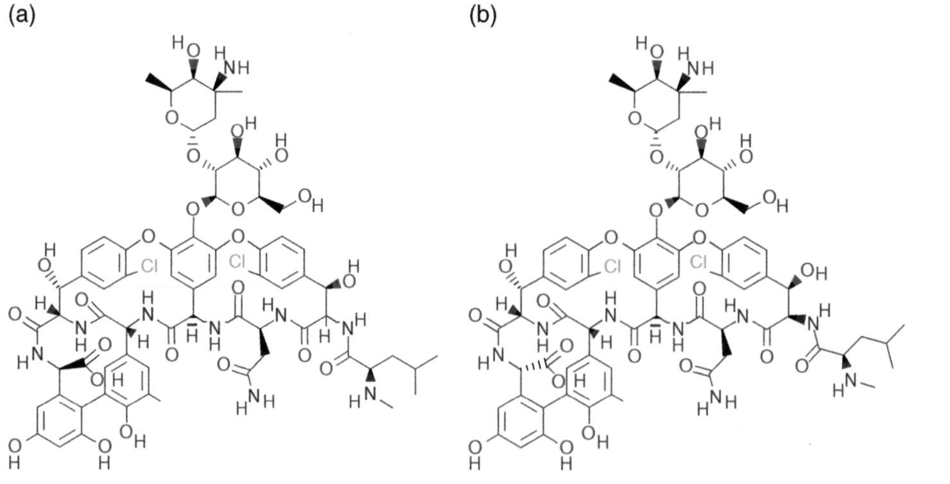

(a)　　　　　　　　　　　　　　(b)

**Scheme 1.1** Chemical structures of Vancomycine from PubChem. (a) PubChem CID 441141, InChIKey: MYPYJXKWCTUITO-UTHKAUQRSA-N. (b) PubChem CID 14969, InChIKey: MYPYJXKWCTUITO-LYRMYLQWSA-N. Notice that Vancomycine corresponds to structure (b), whereas structure (a) is, in fact, its enantiomer.

using a script provided with the tutorial (however, the script might need changes to work properly if the website has changed its structure in the meantime).

Each substance should be associated to exactly one organoleptic category, its odor type. Besides, some additional descriptions of the odor and tastes can be present. These textual descriptions are interpreted in terms of a dictionary of concepts used to describe the odors and tastes: the organoleptic semantic. With the help of this semantic each substance can be represented as a bit vector: each bit is related to an organoleptic descriptor. A bit is "on" if a particular description is relevant for the substance and it is "off" otherwise. Similarly, the chemical structures are interpreted in terms of MACCS fingerprints. In such a vector, a bit is "on" if the chemical structure of the substance possesses some feature (includes some element or chemical function for instance). Binary descriptions are suitable for further analysis, to compute distances or association rules.

Chemical structures and organoleptic descriptions, organoleptic category and bibliographic references are split into different files that can be loaded into separate tables and then merged into a relational database.

KNIME is an Integrated Development Environment (IDE) and a workflow-programing language. Processing units, called nodes, are connected to each other. Data is directed from one node to another following the connections between them. By default, KNIME is divided into eight zones (Figure 1.1). The first one (**1**) is the *toolbar* of buttons for quick shortcuts. These buttons include creating a new project, saving the current projects, zooming and automatic cleanup of the workbench, running and managing the workflow. The second (**2**) area is the *workbench*, the place to drag and drop the nodes and to connect them in order to design a workflow. A miniature of the workbench is provided inside the sixth area (**6**), the *Outline*, in order to help navigating the workflow. The third area (**3**) it the *KNIME Explorer*, a storage area for workflows: it is divided by default into LOCAL and EXAMPLES. The EXAMPLES require an Internet connection to connect with a public KNIME server (login as guest, no password) where is found useful KNIME examples implementing solutions for many basic and advanced operations. The fourth (**4**) area is the *Node Repository*; this is the place where all nodes, representing data processing operations, are stored. Nodes are organized in a tree and a navigation bar provides a node search tool. The most frequently used nodes and the annotated ones are available inside the seventh area (**7**), the *Favorite Nodes*. The fifth area (**5**) is the *Node Description*. When a node is selected, it displays the help text describing the purpose of the node, its parameters, and the format of input and output. The eighth area (**8**) is the *Console* where errors and warning messages are displayed.

When KNIME is activated the first time, it requests a directory to use as workspace. This workspace is used to store temporary files and the workflows. The location and name of the workspace is up to the choice of the user. This choice can be changed later in the **Preferences** menu of KNIME.

Using KNIME consists in manipulating the following basic concepts:

- Drag and drop a node from the node repository into the workbench to use it.
- A node (Figure 1.2) has a main title describing its purpose, a traffic light describing the state of a node, and a custom name. On the side of a node are located handles. The left handles are input and right handles are output.

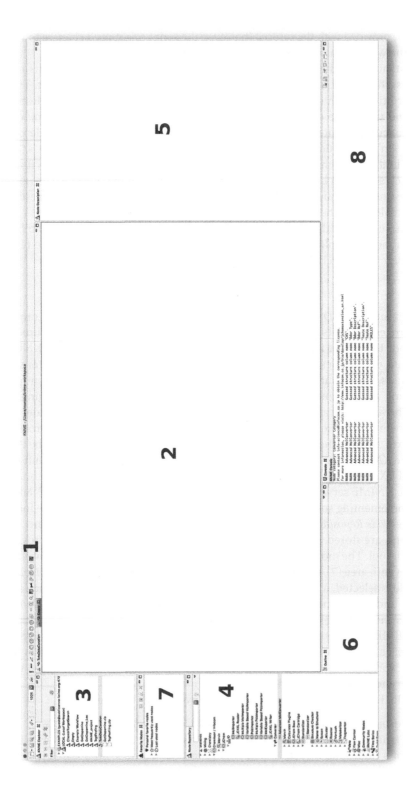

Figure 1.1 KNIME Overview. The interface is organized as follows: (1) the toolbar, (2) the workbench, (3) the KNIME Explorer, (4) the Node Repository, (5) the Node Description, (6) the Outline, (7) the Favorite Nodes, (8) the Console.

- The traffic light is red if the node is not configured, orange if the node is ready, green if the node was successful in processing the data. It is modified if the node generated an error or a warning.
- Click on a right handle of a node, pull and release the mouse button on a left handle triangle of another node to connect the two nodes. The connection represents the dataflow. The output of a node (right handed triangle) is the input (left handed triangle) of the next node.
- Right click on a node to open a popup menu. The main action of this menu is to configure the node. Other common actions are to execute the node, edit the tooltip message, or to get a preview of the data processing by the node.
- Lay the mouse over an in or out triangle of a node to get a snippet of the state of the data at this location of the workflow.
- Right click to an edge connecting two nodes to edit or delete it.
- It is recommended to find a particular node using the search tool of the *Node Repository.*

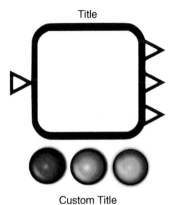

Title

Custom Title

Figure 1.2  Schematic view of a KNIME node. The main title describes the data processing. To the left and right, the handles represent the input and output respectively. The traffic light indicates the node status and below is a custom title for the node.

## Step-by-Step Instructions

The tutorial requires the installation of the ChemAxon nodes for KNIME (version 2.12 or later) called JChemExtension (version 2.8 or later). To use them, a ChemAxon[1] and a JChemExtension[2] license are needed. To install new nodes, proceed as follows:

1) In the **help** menu of KNIME chose **Install new software...**
2) In the pop-up window click the button **Add...**
3) Give a meaningful name (for instance JChemExtension) and the relevant URL (currently https://www.infocom.co.jp/bio/knime/update/3.1).
4) Click the **Next** button and continue through the installation wizard packages for verification and license agreement.

The JChemExtension license file is loaded through the **Preferences** menu of KNIME, then into the topic **KNIME, JChem**. ChemAxon license files are installed using the dedicated ChemAxon **License Manager** software.

At any time in the tutorial, it is recommended to use the **New Workflow Annotation** in a right-click popup menu in the workbench to add comments and colors to the workbench. This helps to order and clarify the workflow. It is also recommended to customize the names of the nodes so that the workflow becomes more self-explaining. A main objective of a workflow-programing environment such as KNIME is to provide a clear picture of the process.

---

1 https://www.chemaxon.com/products/
2 http://infocom-science.jp/product/detail/jchemextension_update_en.html

| Instructions | Comments |
|---|---|
| • Launch KNIME and open a new workbench. | This starts a new project. The user is invited to give a meaningful name for their project. |
| • Use the node **CSV Reader** and configure it so that the *Column Delimiter* is a semicolon (;) and the *Quote Char* is a double quote ("). Uncheck the option *Has Column Header* and the option *Has Row Header*. Set the file source as `thegoodscent_dup.csv` | The input file was roughly formatted so that the column delimiter is a semicolon and all data element is delimited by double-quotes. This is usually a good choice, since these characters are usually not present in linear chemical notations like SMILES of InChI. There is no column title line in this file. |
| • Add the **Column Rename** node and connect its input to the output of the **CSV Reader**. | The columns of the raw CSV file need to be identified. |
| • Configure the **Column Rename** node to rename each column as follows <br> ○ Col0: *Name* <br> ○ Col1: *CAS* <br> ○ Col2: *Odor Type* <br> ○ Col3: *Odor Description* <br> ○ Col4: *Odor Ref* <br> ○ Col5: *Taste Description* <br> ○ Col6: *Taste Ref* <br> ○ Col7: *SMILES* | The configuration interface shall look like the Figure 1.3. |
| • Execute the workflow by clicking on the green lecture buttons of the tool bar. Check the *Console* window for errors and warnings. | Upon execution of the workflow, at least one error will occur. On line 11860 of the CSV file, the compound is named `6"-O-malonyl genistin`. The character " is taken for a word delimiter. <br><br> The remedy is to use a text editor to replace inside the line 14003, the " by a '. In fact, many inconsistencies were present in the raw extract of the website: misinterpreted characters, fused words, confusion between l, I and 1, and O and 0. |

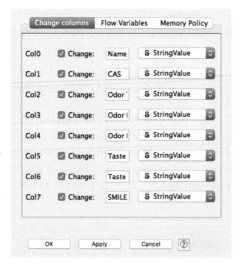

Figure 1.3 Configuration interface of the Column Rename node.

| Instructions | Comments |
|---|---|

- Right click on the node **Column Rename** and select the item *Renamed/Retyped table.*
- Look at lines 7688-7798 to identify, by their name, some undesired entries.
- Add a **Row Filter** node and configure it.
- In the configuration interface, select *exclude rows by attribute value*, chose the *select the column to test* as *Name* and as *matching criteria* select the *use pattern matching* option. Then use as *pattern* the string:

  peg-.*|ppg-.*

- Tick the box *regular expression.*

Browsing through the entries, for instance at lines 7688-7798, polyethylene glycol (PEG) and polypropylene glycol (PPG) can be found. Since the objective of the database is to collect organoleptic properties of small organic compounds, such substances should be excluded.

Those compounds are found by their names, containing the substring "peg-" or "ppg-". In regular expression semantics supported by KNIME, a simple query "peg-.*" matches any line containing the substring "peg-" followed by any number of characters ".*" The "|" character is a logical OR. Thus, the query "peg-.*|ppg-.*" searches simultaneously any substring containing either "peg-" or "ppg-" (Figure 1.4).

Many other PEG/PPG containing substances were present in the database (steareths, octoxynols, ceteths, oleths, isoceteths, poloxamines, trideceths, pareths, ceteareths) as well as lauryl ether sulfates. They are widely used as surfactant, emulsifying agents, cleansing agents, and solubilizing agents for cosmetics. They were removed from the dataset in this tutorial, but they illustrate the particularities of chemistry that must be taken care of when processing data.

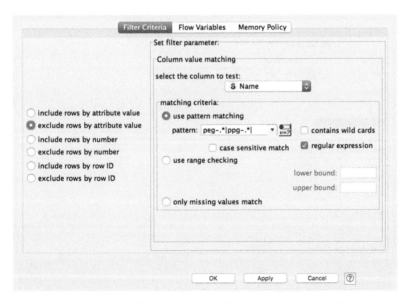

**Figure 1.4** Configuration of the Row Filter node used to remove polyethylene glycol and polypropylene glycol.

*(Continued)*

| Instructions | Comments |
|---|---|

- Right click on the last node **Row Filter** and select the item *Filtered*.
- Look at lines 2666-2668 to identify by their name, some undesired entries.
- Add a **Row Filter** node and configure it.
- In the configuration interface, select *exclude rows by attribute value*, chose the *select the column to test* as *Name* and as *matching criteria* select the *use pattern matching* option. Then use as *pattern* the string:

```
.*\bextract\b.*|.*\boil\b.*
```

- Tick the box *regular expression*.
- Note that the node can be configured temporarily to *include rows by attribute value*. After execution, it allows to check those lines that are matched by the node.

- Add a **Rule-based Row Filter** node and connect its input to the output of the last **Row Filter** node of the workflow.
- Configure the **Rule-base Row Filter** node.
- In the Expression field use the following rule:

```
($Odor Type$ = "") AND ($Odor
Description$ = "") AND ($Taste
Description$ = "") => TRUE
```

- Select the *Exclude TRUE matches* option.
- Execute the workflow

Another specificity is the use of chemically ill-defined substances. For instance, looking at lines 2666-2668 some extracts are present.

This node is dedicated to search and remove lines mentioning the words oil or extract. The regular expression is:

$$.*\bextract\b.*|.*\boil\b.*$$

The \b are word delimiters, in order to avoid matching accidentally the substring "oil" with words such as "foil," "boil," or "soiled" for instance.

Perfumery and cosmetics uses natural extract, therefore, the raw data must be investigated for such keywords as "absolute," "resin," "concrete," "root," "wood," "leaf," "flower," "fruit," "juice," "seed," "bean," "tincture," "straw," or "water" for instance.

Keyword filtering is often insufficient. It is also necessary to check what are the data matched by the keyword and expert decision and perhaps to include or exclude a data point.

Entries with no recorded odor or taste information are not needed. This requires a test on several columns. Therefore, a more complex request must be build, motivating the use of the **Rule-based Row Filter** node. This node manages complicated logical rules (involving any combination of logical operators NOT, AND, OR, XOR) based on any columns. Columns are referred to by their name surrounded by $.

The configuration of the node should look like Figure 1.5. So the rule matches lines with empty "Odor Type," "Odor Description," and "Taste Description" fields.

**Figure 1.5** Configuration of the Rule-based Row Filter node to remove instances providing no information on the organoleptic properties.

| Instructions | Comments |
|---|---|

- Connect the input of a **Rule Engine** node to the output of the last **Rule-based Row Filter** node.
- Configure the **Rule Engine** as shown in Figure 1.6.
  - Add the following lines into the *Expression* text-edit region:

```
($Odor Type$ = "") AND
($Odor Description$ MATCHES
"\b\w*\b") => $Odor
Description$
($Odor Type$ = "") AND
($Taste Description$ MATCHES
"\b\w*\b") => $Taste
Description$
(NOT $Odor Type$ = "") =>
$Odor Type$
```

  - You can use the *Column List, Category* and *Description* elements of the configuration interface to help you write the above three rules.
  - Invalid syntaxes are underlined in red and the node cannot be executed.
  - Select the *Replace Column* option and choose the column *Odor Type*.
- Validate and execute the node.

The *Odor Type* field defines the organoleptic category to which the substance belongs. However, for a few hundreds of compounds, this information is missing.

The first attempt to recover this information is to use the *Odor Description* or the *Taste Description* fields if they contain only one word.

This manipulation requires the use of the **Rule Engine**. The node is programmed using three rules to compute a value, in decreasing order of priority:

1) If the *Odor Type* is empty and the *Odor Description* is composed of one word, use the *Odor Description* as the *Odor Type*.

2) If the *Odor Type* is empty and the *Taste Description* is composed of one word, use the *Taste Description* as the *Odor Type*.

3) If the *Odor Type* is not empty, use the value of *Odor Type* as the *Odor Type*.

The *Odor Description* and *Taste Description* fields are compared to the following regular expression:

$$\verb|\b\w*\b|$$

It matches those fields composed of word characters (\w) in any number (*), as long as they are in one same unique word (enclosed by \b).

Once a rule is applied the others are ignored.

The third rule is important because the result overwrites the *Odor Type* column. If no rule applies to an instance the result of the operation is *undefined* (annotated as "?"), and the *undefined* value will overwrite the initial content of the *Odor Type* field. Using the third rule, the *Odor Type* value is kept unchanged by default.

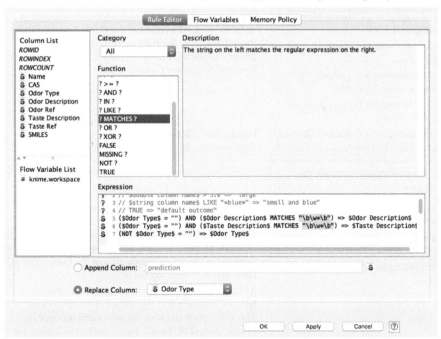

**Figure 1.6** Configuration of Rule Engine node using the Odor Description and Taste Description fields to fill in the missing values of the Odor Type field.

*(Continued)*

| Instructions | Comments |
|---|---|
| • Include and configure a **CSV Reader** node (Figure 1.7). It will read the file `MissingOdorTypes.csv`. The *Column Delimiter* is a semicolon (;) and the *Quote Char* is a double quote ("). Check the option *Has Column Header* and uncheck the option *Has Row Header*. | Yet, for about 300 substances, the *Odor Description* and *Taste Description* are too complicated to be used as an *Odor Type*. An expert had to review these items and manually corrected them. The corrections are collected into the file *MissingOdorTypes.csv*. This file is loaded with a **CSV Reader** node and generates a second data flow. |
| • Include a **Joiner** node. Connect the **CSV Reader** node output to the *Right table* input of the **Joiner** (the bottom left handle) and the **Rule Engine** output to the *Left table* input of the **Joiner** (the top left handle). | The field *SMILES*, which is common to the main dataflow (or left table) and the new dataflow (right table), is used to merge them via a **Joiner node**. The columns that come from the right table are filled up with missing values if no matching row exists in the right table. All columns from the left table are included while only the *Odor Type* field of the right table is added to the output. |
| • Configure the **Joiner** node. | |
| ○ In the *Joiner Settings* tab (Figure 1.8) select the *Left Outer Join* option as the *Join mode*. Then in the *Joining Columns* menu, select the match all of the following option and select the column *SMILES* for both left and right table. | |
| ○ In the *Column Selection* tab (Figure 1.9), uncheck the *Always include all columns* option of the *Right table*. In the column selection menu interface of the *Right table* exclude all columns except the *Odor Type*. | |
| • Use and configure a **Column Merger** node (Figure 1.10). | Then, a new **Column Merger** node is configured in order to merge the values of the *Odor Type* field from the left and from the right table. If the value is missing inside the *Primary Column*, the value is taken from the *Secondary Column*. |
| ○ Select as *Primary Column* the column *Odor Type*. | |
| ○ Select as *Secondary Column* the column *Odor Type (#1)*. | |
| ○ Select the option *Replace primary column*. | |
| • Add and configure a **Column Filter** node (Figure 1.11). The column *Odor Type (#1)* should be inside the *Exclude* area while all other columns should be inside the *Include* area. | Finally, the **Column Filter** node discards the *Odor Type* from the left table, as it is no longer needed. |
| • Add to the workflow a **Domain Calculator** field (Figure 1.12). | Although the KNIME system does not have a Nominal type to describe the data (they are manipulated as normal strings), some data integrity checking can be enforced. The **Domain Calculator** node is used to list all words used in the *Odor Type*, *Odor Ref*, and *Taste Ref* fields. |
| • Add the *Include* area, the columns *Odor Type*, *Odor Ref* and *Taste Ref*. | |
| • Set the *Restrict number of possible values* to 200. | |
| | After execution, it can be investigated into the *Spec – Columns* tab of the output *Data table* view. This is the place to check for errors and misspelling ("floal" instead of "floral," "cirus" instead of "citrus" for instance) that can occur during manual data entry. |

Figure 1.7 Configuration of the CSV Reader node, loading corrected Odor Type fields from the file MissingOdorTypes.csv.

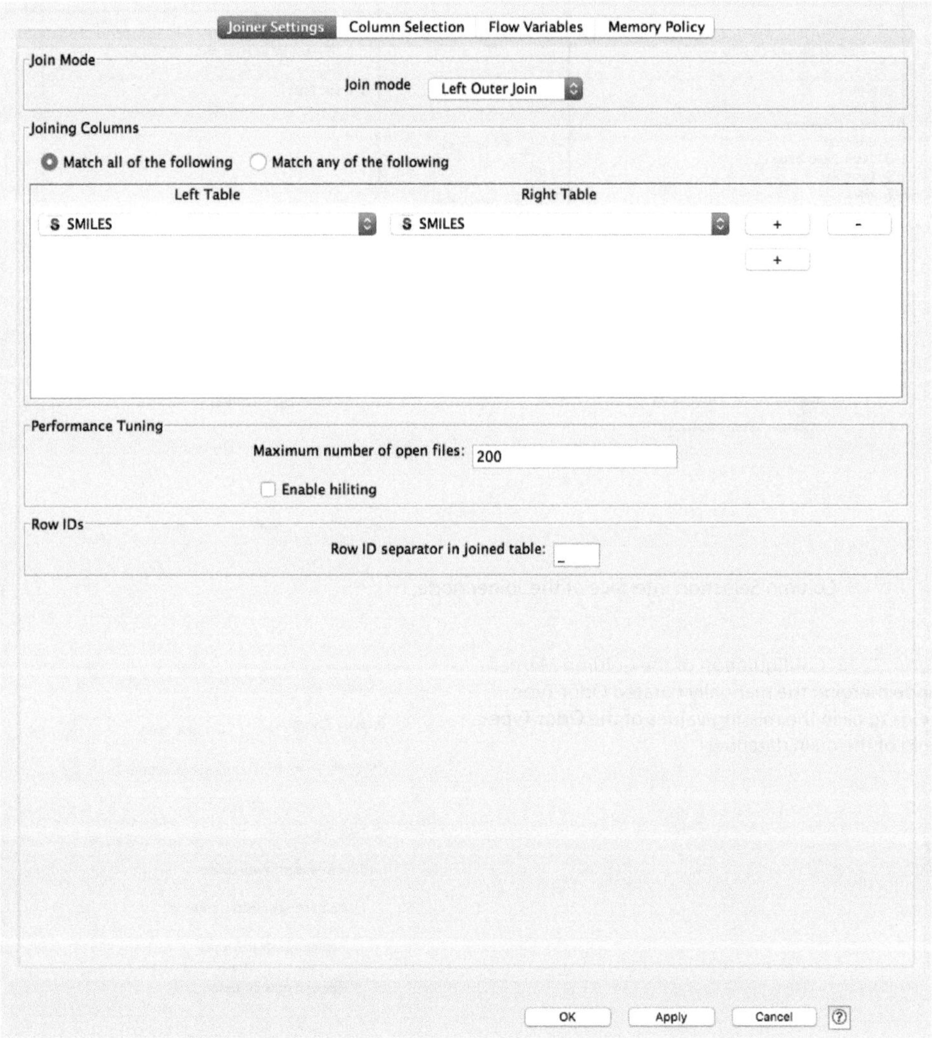

Figure 1.8 Joiner Settings of the Joiner node interface.

(*Continued*)

Figure 1.9 Column Selection interface of the Joiner node.

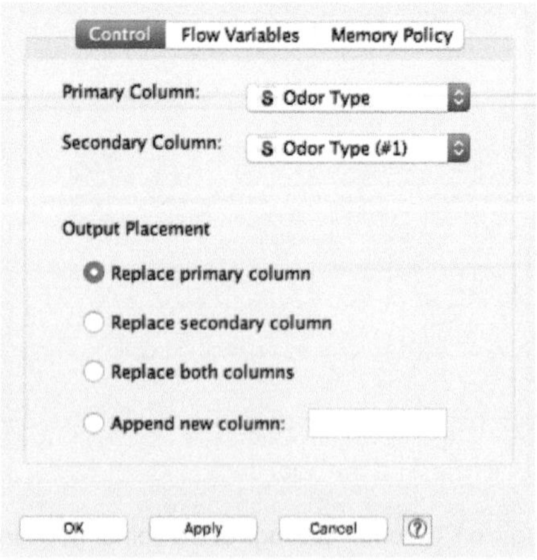

Figure 1.10 Configuration of the Column Merger node merging the manually curated Odor Type fields to fill in the missing values of the Odor Type field of the main dataflow.

| Instructions | Comments |
|---|---|
| • Connect a **Column Combiner** node to the last node of the workflow.<br>• Configure the node so that the *Include* area contains only the *Odor Type, Odor Description*, and *Taste Description*.<br>• Set the *Delimiter* to "," and choose the *Replace Delimiter by* option. Use the space character inside the text box of this option.<br><br>Set the *Name of appended column* to *Raw Organoleptic*. | Another part of the workflow shall be dedicated to the aggregation of *Odor Type, Odor Description*, and *Taste Description* in a general description of the organoleptic description of the substance.<br><br>The first step is to merge the columns *Odor Type, Odor Description* and *Taste Description* (Figure 1.13). |

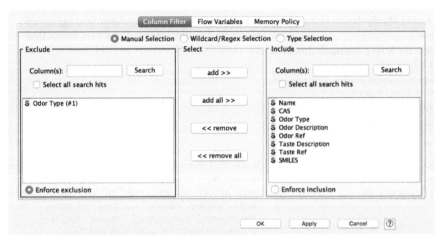

Figure 1.11  The Column Filter interface, used to discard the manually curated Odor Type field. It is now useless since the information is already merged into the main dataflow.

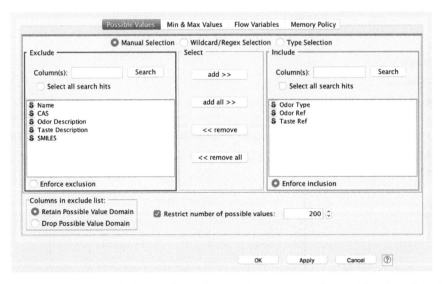

Figure 1.12  Domain Calculator node interface, used to enumerate the possible values of the Odor Type field, and the odor and taste descriptions references.

(Continued)

| Instructions | Comments |
|---|---|

- Connect a **String Replacer** node to the output of the **Column Combiner**.
- Configure the **String Replacer** node (Figure 1.14).
- Set the *Target Column* to *Raw Organoleptic*.
- Choose the *Regular Expression* option.
- Type in the following string as the *Pattern*:

  , | \ . | - | / | & | \d+

- Enter a white space as the *Replacement text*.
- Choose the *all occurrences* option.

The new field, *Raw Organoleptic*, added to the workflow contains points, comma, ampersand, hyphens, and digits. These characters do not participate directly to the semantic of organoleptic properties. So, they are removed.

In the regular expression, the dot character is represented by a "\." because otherwise the dot is a wild card for *any* character. The notation "\d+" matches a digit (\d) appearing one or more times (+).

- Add a **String Manipulation** node to the workflow.
- Configure the **String Manipulation** node (Figure 1.15).
  - Select the option *Replace Column* and choose *Raw Organoleptic*.
  - Inside the *Expression* text edit region, type the command:

lowerCase ($Raw Organoleptic$)

Additionally, the *Raw Organoleptic* field aggregates sentences describing an odor or a taste. So the field uses a mixture of lowercase and uppercase.

- Connect the output of the **String Manipulation** node to the input of a **String Replacer** node.

It also uses words that are not related to an odor description. These words are: "with," "a," "an," "the," "and," "of," "for," "in," "low," "high," "it," "like," "likes," "likely," "slight," "slightly," "nuance," "nuances," "enhanced," "reminiscent," "penetrating," "intense," "has," "all," "minutes," "breath," "mouth," "feel," and "feels."

Therefore, one node is used to lowercase all characters, and a second node is used to remove the undesired words from the organoleptic description.

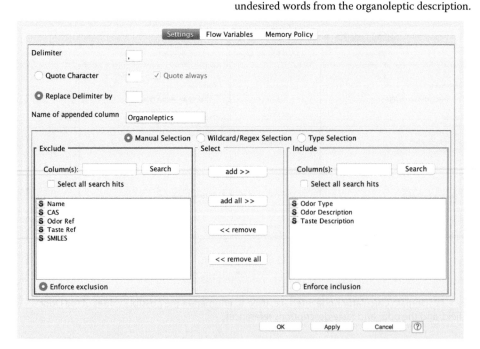

Figure 1.13 Configuration of the Column Combiner Node, used to agregate odor and taste description into a global organoleptic description.

Figure 1.14  Configuration of the String Replacer node removing punctuation, special characters, and numbers from the Raw Organoleptic field.

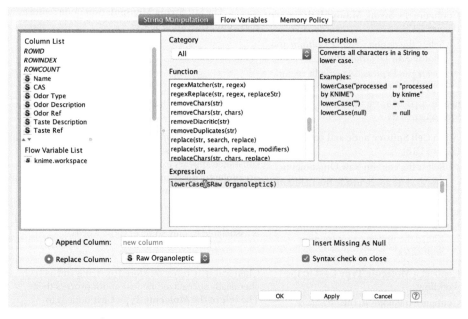

Figure 1.15  String Manipulation node, used to lowercase all characters of the Raw Organoleptic field.

*(Continued)*

| Instructions | Comments |
|---|---|

- Configure the **String Replacer** node (Figure 1.16).
  - ○ Set the *Target column* to *Raw Organoleptic*. As *Pattern Type* select *Regular expression*. As *Replace* option select *all occurrences*.
  - ○ Inside the *Pattern* text edit region, type in the following command:

```
\bwith\b|\bthe\b|\band\b|\
ball\b|\ban\b|\ba\b|\bof\b|\
blike\b|\blikes\b|\bslight\b|\
bslight.+\b|\breminiscent\b|\
benhanced\b|\bnote\b|\bnote.+\
b|\bin\b|\bbackground\b|\
blasts\b|\bbreath\b|\bfor\b|\
bminutes\b|\bmouth\b|\bfeel\
b|\bfeels\b|\bpenetrating\b|\
bnuance\b|\bnuance.+\b|\bit\
b|\bhas\b|\bintense\b|\blow\
b|\bhigh\b
```

- Add a **String Manipulation** node.
- Configure the **String Manipulation** node (Figure 1.17).
  - ○ Select the option *Replace Column* and choose *Raw Organoleptic*.
  - ○ Inside the *Expression* text edit region, type the command:

```
removeDuplicates($Raw
Organoleptic$)
```

These operations generate successions of white spaces. The last **String Manipulation** node removes these duplicate white spaces.

- Add a **Cell Splitter** node and configure it (Figure 1.18).
  - ○ Select the column Raw Organoleptic.
  - ○ Use a white space inside the field *Enter a delimiter*.
  - ○ Tick the box *remove leading and trailing white space chars (trim)*. Chose the option *as set (remove duplicates)*.

The *Raw Organoleptic* field needs to be converted to a set. Each word describing the organoleptic of a substance appears exactly once within a set. The organoleptic description of the substance reduces to the organoleptic semantic.

The set of all these words constitute a compendium of the description of chemical substances. A chemical is associated to only some of these words.

- Connect a **Molecule Type Cast** to the workflow.
- Configure the node so the *Structure Column* is *SMILES* and the *Structure Type* is *smiles*.

Chemical structures are stored as string and as such chemically specialized nodes cannot process them. The role of the **Molecule Type Cast** node is to annotate and interpret the part of the dataflow concerning chemical structure.

This node is very permissive. No quality check of the chemical structure is done.

- Connect the output of the **Molecule Type Cast** node to the input of the **Structure Checker** node.

This is the role of the **Structure Checker** node. The actions of this node are performed in the same order as they are listed in the configuration interface.

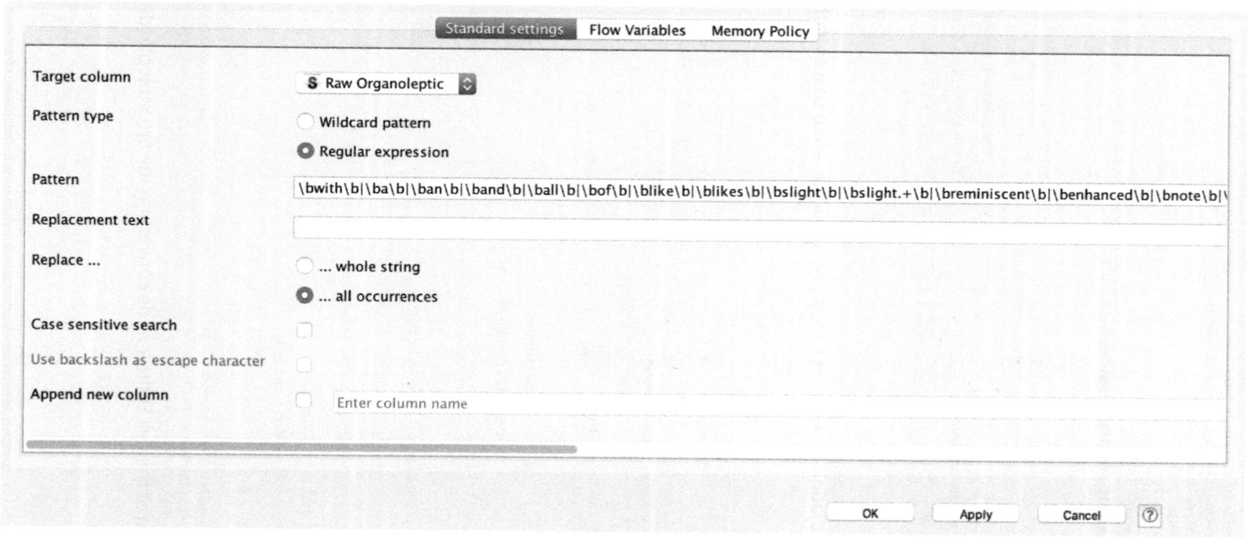

Figure 1.16 The String Replacer node, used to remove the undesired words from the Raw Organoleptic field using regular expressions.

(Continued)

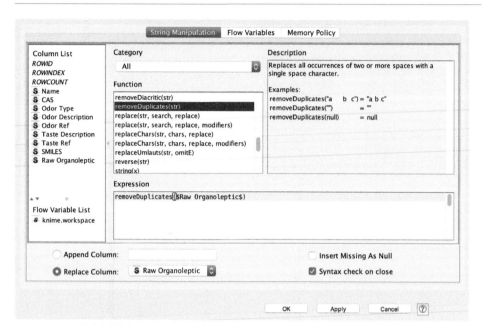

Figure 1.17  String Manipulation node, used to remove duplicate white spaces in the Raw Organoleptic field.

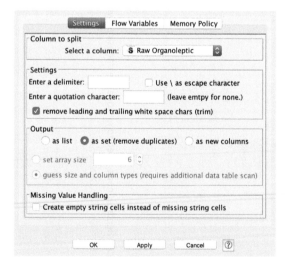

Figure 1.18  Configuration of the Cell Splitter node converting the organoleptic description to a set of descriptive words.

| Instructions | Comments |
|---|---|

- Configure the **Structure Checker** (Figure 1.19). Add the following actions to the right-hand window:
  - ○ *Empty Structure Checker*
  - ○ *Rare Element Checker*
  - ○ *Covalent Counterion Checker*
  - ○ *Valence Error Checker*
  - ○ *Ring Strain Error Checker*
  - ○ *Aromaticity Error Checker*
- It is usually possible to customize an action by clicking on it into the right-hand window. Customize the *Aromaticity Checker* so that it uses a *Basic* aromatization scheme and set the *Fix* option to *Dearomatize*.
- Connect the discarded structure output (the bottom right handle of the node) a **Table Writer** node. Configure the node so the output table file is called *Discarded.table*. Check to *Overwrite OK* box.

It starts by checking that the chemical structure field contains a chemical structure.

Then, the atoms of the structure are checked. Rare atoms are those not in the following list: H, Li, Na, K, Mg, Ca, B, C, N, O, F, Cl, Br, I, Al, P, S, Cr, Mn, Fe, Co, Ni, Cu, Zn.

The third step is to check if an alkali metal or alkaline earth metal is covalently connected to an N, O, or S. If this is the case, the covalent bond is deleted and the atoms are charged.

The fourth step searches for valence overflows (forbidden by the octet rule). Unless the problem is an additional H (which is then removed), the chemical structure cannot be automatically repaired.

In the fifth step, impossible intracyclic bonds are searched for. These bonds are trans double bonds, successive double bonds and triple bonds. These bonds cannot exist within small organic molecule rings.

Finally, the sixth step considers the aromatic representation of rings. The *Basic* style uses Hückel's rule.

The node has two output handles. The bottom one is the dataflow containing chemical structures that are discarded by the node, while the top one can be safely processed.

The discarded structures cannot be further processed. They should be fixed separately and this is why the **Table Write** node is connected immediately to the discarded structure handle.

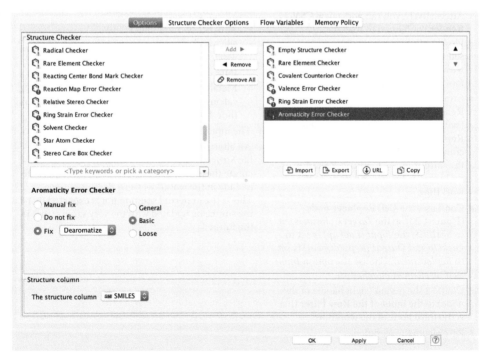

**Figure 1.19** Configuration of the Structure Checker focused on atoms and valences of the chemical structures.

(*Continued*)

| Instructions | Comments |
|---|---|
| • Edit the file *thegoodscent_dup.csv* to correct some errors. | The procedure above should identify the following problems. |
| ○ Line 7607, replace "H2S" by "S" | • Some SMILES structures are wrong. |
| ○ Line 1673, replace "InChI = 1/C7H8S/ c1-6-4-2-3-5-7(6)8/h2-5,8H,1H3" by "Cc1ccccc1S" | ○ Line 7607, hydrogen sulfide is coded "H2S" instead of "S." |
| ○ Line 6711, replace "SJWFXCIHNDVPSH-MRVPVSSYSA-N" by "CCCCCC[C@@H](C)O" | ○ Line 1673, ortho-thiocresol is encoded with the InChI code. |
| ○ Line 6045, replace "CC[S-]C(=O) CNC(=O)C" by "CCSC(=O) CNC(=O)C" | ○ Line 6711, the (R)-(-)-2-octanol is encoded by its InChIKey. |
| ○ Line 7645, replace "O = [Bi].Cl" by "O = [Bi]Cl" | • The name and structure are misspelled. |
| ○ Line 7164, replace "CC(=O)c1ccc(C) n1" by "CC(=O)c1ccc(C)[nH]1" | ○ Line 6045, "S-ethyl 2-acetyl aminoethane thioate" should be understood as "S-ethyl 2-acetamidoethanethioate." |
| ○ Line 15308 and 15896, replace "Cc1c(sc2c1c(nc(=O)[nH]2)N)C" by "Cc1sc2NC(=O)N = C(N)c2c1C". | ○ Line 7645, "bismuth(III) oxychloride" is simply "bismuth oxychloride." |
| • If you edited the CSV file directly, reset the first **CSV Reader** node of the workflow and re-execute the whole workflow. | • Warnings about loose definition of aromatic rings |
| • Alternatively, you can add a new **CSV Reader** configured to read the file *StructureCuration.csv*, using a semicolon (";") as *Column Delimiter* and with *Has Column Header* ticked and *Has Row Header* not ticked. | ○ Line 7164, the nitrogen of 2-acetyl-5-methyl pyrrole has a proton that should be included into the SMILES code. |
| | ○ Line 15308 and 15896, pyrimidin-2-one in 4-Amino-5,6-dimethylthieno(2,3-d) pyrimidin-2(1H)-one should not be aromatic. |
| • Connect the output of this new **CSV Reader** to the dictionary table input (bottom left handle) of a new **Cell Replacer** node. Connect Table containing column (top left handle) to the output of the **Column Rename** node that is the second node of the workflow. | • Structures that attract attention. |
| | ○ Line 2651, cyclohexasiloxane. |
| | ○ Line 43778, vanadium pentoxide. |
| | ○ silica and silicates at lines 7429, 15231, 4414, 4469, 4471, 4476. |
| | ○ Pure elements, such as helium (line 7335). |
| • Configure the **Cell Replacer** node (Figure 1.20) so the *Target Column* is set to *SMILES*, the *Input (Lookup)* is set to *LookUp* and *Output (Replacement)* is set to *Replacement*. Choose the option *Input as If no element matches use.* | ○ Metal oxides such as titanium oxide (line 8348), calcium oxide (line 4353), and magnesium oxide (line 4360). |
| | • The input file can be edited manually. |
| | • An alternative is to use the prepared file *StructureCuration.csv* containing the rules to curate these deficient SMILES codes. It requires to include in the workflow some dedicated nodes. This is the preferred solution if it is expected that the source be updated with the very same errors in the future. |
| • Connect the output (right handle of the node) to the input of the **Row Filter** that is the third node of the workflow. | |
| • Re-execute the workflow. | |

| Instructions | Comments |
|---|---|

**Instructions**

- Add a **Standardizer** node to the workflow.
- Configure the **Standardizer** node (Figure 1.21):
  - Add a *Dearomatize* action
  - Add a *Mesomerize* action
  - Add a *Tautomerize* action
  - Add an *Aromatize* action
  - Add a *Transform Nitro* action
  - Add a *Clean 2D* action
- As before, an action can be customized by clicking on the corresponding line in the right-hand space of the interface.
- Configure the *Aromatize* action in order to use a *Basic style aromatization*.
- Tick the box *Append column*.

**Comments**

The chemical structures in the dataset are still heterogeneous. For instance, some aromatic rings are in a Kekulé form while others are in aromatic form, which would bias substructure search, similarity searching or any attempt to model the dataset based on the chemical structure.

The goal of the **Standardizer** node is to define a set of rules according to which a chemical structure shall be drawn. The rules are applied to each chemical structure of the workflow following the order of the list defined inside the right-hand area of the configuration interface of the node.

In the present situation, the first step is to dearomatize a structure. The dearomatized structure is used as a reference state to compute a standard mesomer and tautomer state of the compound. Then the structure is aromatized, using the *Basic* style. This implements an homogeneous representation of the chemical structure.

Yet, some chemical functions can pose a challenge to any automatic procedure to cure the chemical structures. For instance, the PubChem database has enumerated about 60 different representation of a nitro group [9]. The rule integrated into the **Standardizer** node concerns only one of them: if the pentavalent nitrogen is found in a nitro group pattern, it is replaced by positively charged tetravalent nitrogen and one oxygen becomes negatively charged.

The pattern matching and replacement strategy is becoming obsolete. Already in 2005, an early attempt to create a chemical ontology was referencing 231 chemical functions [10]. More recently [9], PubChem was analyzed in terms of atom environments (atoms plus the neighbor atoms with chemical bonds). Considering only an atom and the chemical bonds around, about 1600 unique atom environments were enumerated. With up to one neighbor atom, about 110000 unique atom environments were found. Based on these numbers, the number of rules to exhaustively standardize chemical functions would be between 10000 to millions.

Figure 1.20  Configuration of the Cell Replacer node used to curate some SMILES of chemical structures.

(*Continued*)

| Instructions | Comments |
|---|---|
| • Add a **MACCS Fingerprint** node. In the configuration interface (Figure 1.22), select *Binary* as *Output type* and *SMILES (Standardized)* as *The structure column*.<br>• Add a **Naming** node (Figure 1.23). Choose the option *Traditional Name* and select *SMILES (Standardized)* as *The structure column*. | Once the chemical structures are better validated, some calculations can be done a priori to help the management of data.<br><br>The first one is to compute molecular fingerprints. Yet, the chemical data to process refer to substances, not molecules. A possible relevant choice for such a situation are MACCS fingerprints [11]. These fingerprints are very qualitative, encoding for elemental analysis, presence of particular substructure and annotation about the nature of the chemical (molecule, substance, mixture, monomer...). Although in the present implementation, only a subset of 166 bits is used, they are robust enough to be relevant for organoleptic substances.<br><br>Another useful computed property of the chemical structures is the name of the compound. As the compounds are fairly common, the common name is preferred. |
| • Add a **Category to Number** node and connect the *Data* handle to the output of the previous node of the workflow.<br>• Configure the node (Figure 1.24).<br>• Add to the *Include* section, the columns *Odor Type, Odor Ref*, and *Taste Ref*. | It is time to export the dataflow, using a format suitable for management into a relational database system.<br>The export shall include at least four tables:<br>• The odors references<br>• The tastes references<br>• The odor types<br>• The substances. |

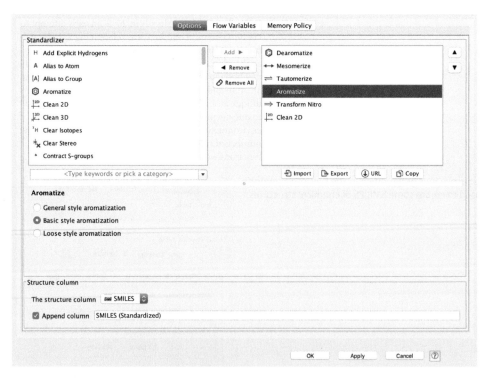

Figure 1.21  Standardizer node configuration interface setting rules for aromatic ring representation, taking into account mesomers and tautomers.

**Figure 1.22** Configuration interface of the MACCS Fingerprint node.

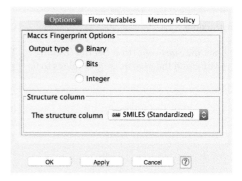

**Figure 1.23** The configuration interface of the Naming node.

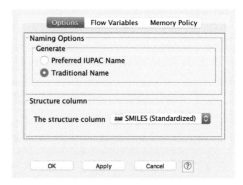

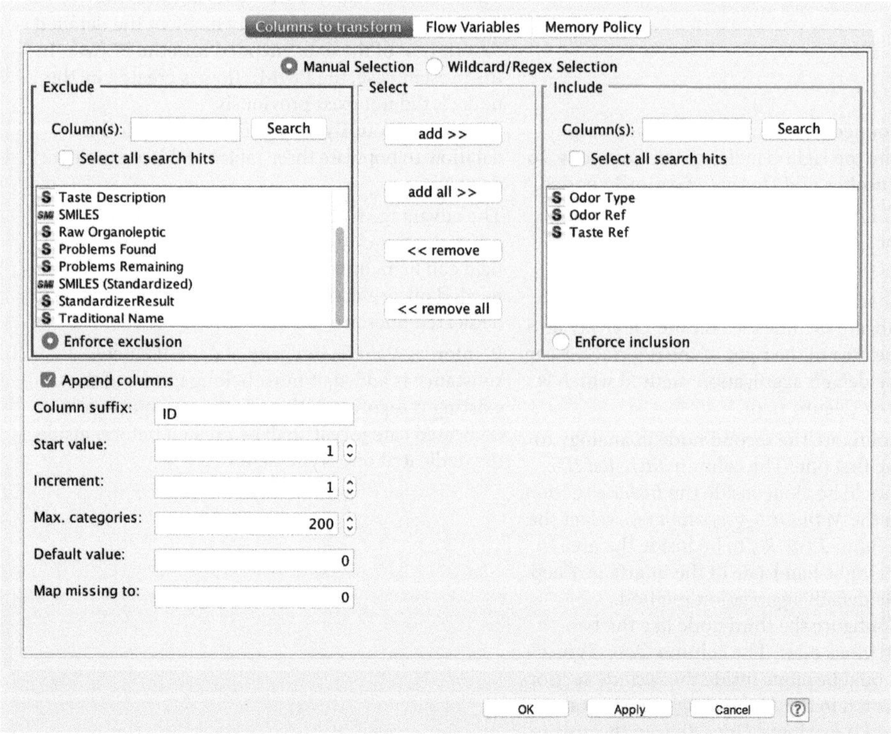

**Figure 1.24** Configuration interface of the Category To Number node used to generate foreign keys.

*(Continued)*

| Instructions | Comments |
|---|---|
| • Set the *Start value* to 1. Set the *Default value* and the *Map missing to* values to 0. | The table shall be connected through foreign keys.

The goal of the **Category To Number** node is to map each category of nominal data to an integer. Here, each reference for the description of an odor (respectively as taste) is one category. The generated integer becomes the foreign key.

The *Default value* and the *Map missing to* values are set so that in any case, the absence of information will also generate a valid foreign key for the future relational database.

The bottom handles of the **Category To Number** node are dedicated to access to a PMML (Predictive Model Markup Language[12]) dataflow. The PMML is a general format to encode models. Since, KNIME manages dataflow using basing type and data structures, it is sometimes necessary to add a PMML dataflow that helps define a datatype more strictly. There are no intrinsic categorical data in KNIME (they are treated as general string).

For instance, a closer look at the *Transformed PMML Input* (bottom left handle of the node), shows that the columns to which the **Domain Calculator** node was applied (*Odor Type*, *Odor Ref*, and *Taste Ref*) are described into the PMML dataflow using all the enumerated values.

Those PMML handles are there, because the operation of this node is likely to affect the detailed description of the data encoded into the PMML. In the present case, the PMML flow is created by the node: it did not exist previously. |
| • Connect the output of the *Data* output (the top right handle) of the **Category To Number** node to three **GroupBy** nodes. | The three **GroupBy** nodes are used to split the dataflow to populate three tables of the organoleptic database. |
| • Configure the first node by adding only the columns *Odor Ref ID* to the *Include* section (Figure 1.25). | The advantage is that the references and the organoleptics categories defined by the *Odor Type* field can be maintainded separately. If a correction is needed on a reference, it is done only once inside the dedicated reference table.

It enforces also the integrity of data. If a new substance is added, it must belong to one of the existing categories. If the substance requires a dedicated category it shall be created before, inside the dedicated odor type table. |
| • Then click on the *Manual Aggregation* tab and include the column *Odor Ref* into the area to the right of the interface. Keep the default aggregation method which is *First* (Figure 1.26). | |
| • Configure the second node in analogy to the first one. The column *Taste Ref ID* should be alone inside the *Include* section. In the *Manual Aggregation* tab, select the column *Taste Ref* to be inside the area to the right-hand side of the interface. Keep the default aggregation method. Configure the third node like the two previous ones. The column *Odor Type ID* should be alone inside the *Include* section. Switch to the *Manual Aggregation* tab. Add the column *Odor Type* in the area to the right-hand side of the interface. Keep the default aggregation method. | |

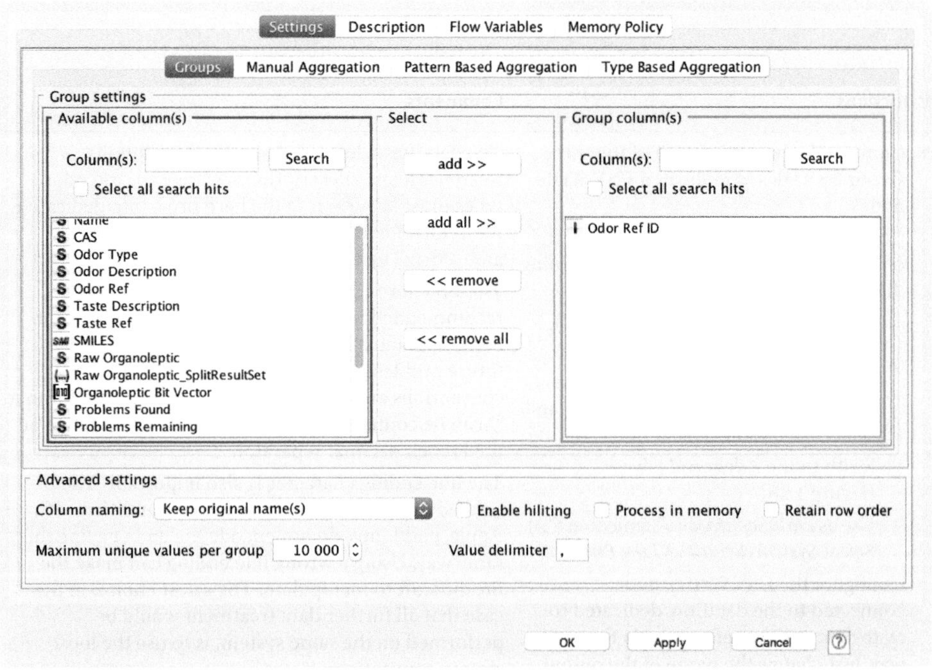

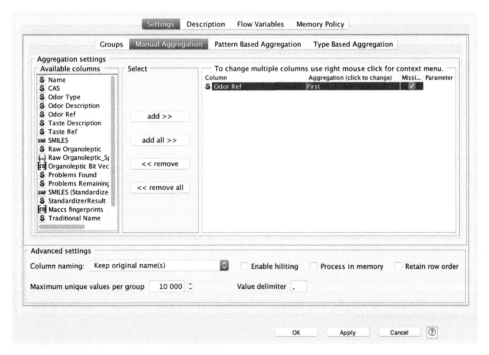

Figure 1.25 The main tab of the configuration interface of the GroupBy node using the Odor Ref ID for aggregation.

Figure 1.26 The Manual Aggregation tab of the configuration interface of the GroupBy node aggregating the Odor Ref column by keeping the first value.

*(Continued)*

| Instructions | Comments |
|---|---|
| • Connect the output of each of the three **GroupBy** nodes to dedicated **CSVWrite** nodes.<br><br>• Configure the first **CSVWriter** node, the one connected to the dataflow concerning the odor references.<br><br>    ○ Inside the *Settings* tab (Figure 1.27), give a name for the output file: `OdorRefTable.csv`<br><br>    ○ Tick the box *Write column header* and select the option *Overwrite*.<br><br>    ○ Swith to the *Advanced* tab (Figure 1.28).<br><br>    ○ Use as *DataSeparator* a semicolon (";"). Select *System default* as *Line Endings*.<br><br>• Configure the **CSVWriter** node connected to the dataflow dedicated to taste descriptions references in the same way. Just change the name of the output file, it shall be: `TasteRefTable.csv`.<br><br>• Direct the dataflow dedicated to the odor type to the third **CSVWriter** node. Set the output file name to be: `OdorTypeTable.csv`. All other parameters should be identical to the other to **CSVWriter** nodes. | The dataflows dedicated to odor descriptions references, taste descriptions references, and odor types must be writen to the hard drive into dedicated files: `OdorRefTable.csv`, `TasteRefTable.csv` and `OdorTypeTable.csv`<br><br>The CSV format can be customized. It is recommended to use a semicolon as data separator. Some care must be given to this choice to avoid future problems with the generated file due to conventions and localization. For instance the comma "," can be confused with punctuation in a sentence or the French decimal separator.<br><br>The line ending character is also important. The created data files are text files. However, Microsoft, Apple, and Linux do not encode the end of a line the same way. Using a wrong line ending can make the file difficult to manipulate. The safest choice, in the case that all further data treatment would be performed on the same system, is to use the local system conventions. |

Figure 1.27 The Settings tab of the configuration interface of the CSVWriter node for output of the odor descriptions references.

| Instructions | Comments |
|---|---|
| <ul><li>Add a **Column Filter** node and connect its input to the data output of the **Category To Number** node (the upper right handle).</li><li>Configure the **Column Filter** node (Figure 1.29).</li></ul>The *Include* section should contain all the columns except the columns *Odor Type, Odor Ref, Taste Ref, Raw Organoleptic, Problem Founds, Problems Remaining,* and *StandardizerResult.* | The remaining columns of the data stream are not yet saved to a file. They should be dedicated to the substance table only. Therefore all information that is stored into separated tables (odor and taste descriptions references and odor types) must be discarded. |
| <ul><li>Add a new **GroupBy** node to the workflow.</li><li>Select the field *SMILES (Standardized)* as the *Group column(s)* (Figure 1.30).</li><li>Select the *Keep original names* option as *Column naming.*</li><li>Switch to the *Manual Aggregation* tab (Figure 1.31).</li><li>Add the following fields into area to the right of the *Aggregation settings* with *List* option as *Aggregation*:</li></ul> | A closer examination of the standardized structures reveals that some chemical structures are duplicates. This happens because some substances have been added to the database at various concentrations of the perceptive substance, or substances got duplicated in the CAS. Other reasons can be errors of the chemical structure, essentially due to incorrect stereochemistry or stereoisomers that were confused with the racemic mixture. |

**Figure 1.28** The Advanced tab of the configuration interface of the CSVWriter node for output of the odor descriptions references.

(*Continued*)

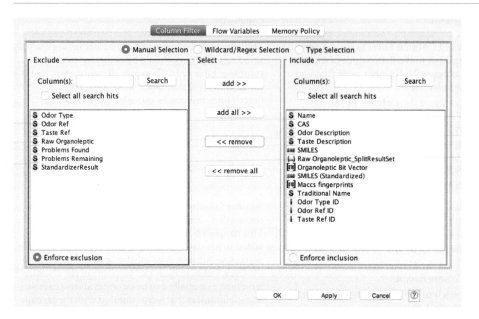

Figure 1.29 Configuration interface of the Column Filter used to process the dataflow that is planed to populate the chemical substance table of a database.

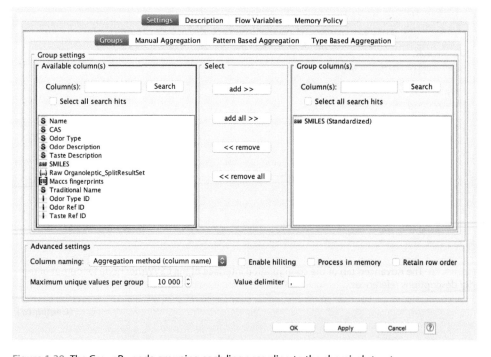

Figure 1.30 The GroupBy node grouping each line according to the chemical structures.

| Instructions | Comments |
|---|---|
|     ○ *Name* <br>     ○ *CAS* <br>     ○ *Odor Description* <br>     ○ *Taste Description* <br>     ○ *SMILES* <br>     ○ *Odor Ref ID* <br>     ○ *Taste Ref ID* <br> • Add the following fields into area to the right of the *Aggregation settings* with *First* option as *Aggregation*: <br>     ○ *Odor Type* <br>     ○ *Traditional Name* <br>     ○ *Odor Type ID* <br>     ○ *MACCS fingerprints* <br> • Add the following fields into area to the right of the *Aggregation settings* with *Union* option as *Aggregation*: <br> • *Raw Organoleptic_SplitResult* | If two items of the dataflow share the same chemical structure, the related information has to be merged. For fields such as *Name, CAS, Odor Decription, Taste Description, SMILES, Odor Ref ID*, and *Taste Ref ID* it is interesting to keep all information. Therefore, the data are aggregated into collections. <br><br> For data specific to the chemical structure, the *Traditional Name, the MACCS fingerprints* the information is redundant and only the first encountered value is sufficient. <br><br> For the *Odor Type* and *Odor Type ID* there should be some discrepancies. A duplicated chemical structure might have been categorized in different ways. So at this point, each discrepancy shall be analyzed and a decision must be taken by the user whether to discard a given chemical structure or to edit the *Odor Type* value. <br><br> Finally the *Raw Organoleptic_SplitResult* may differ for the duplicates. However, any organoleptic description is equally interesting. Therefore, the lists of organoleptic terms are merged into one sole list using the *Union* action. |

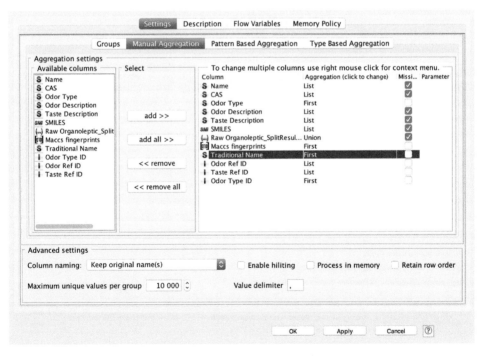

Figure 1.31 The GroupBy node Manual Aggregation interface managing the duplicate chemical structures in the dataset.

*(Continued)*

| Instructions | Comments |
|---|---|
| • Add and configure a **Create Bit Vector** node (Figure 1.32).<br><br>　○ Choose the option *Create bit vectors from a single collection column.*<br><br>　○ Choose the column: *Raw Organoleptic_SplitResultSet.*<br><br>　○ As *Output column* name it *Organoleptic Bit Vector.* | The *Raw Organoleptic_SplitResultSet* contains a synthetic description of the odors and tastes of the substance. If a substance appeared more than once, all descriptions have been merged into this field. Each word describing the organoleptic of a substance appears exactly once within a set.<br><br>The set of all these words constitute a compendium of the description of chemical substances. A chemical substance is associated to a subset of these words. This relation can be efficiently encoded as a bit-vector. The bit-vectors generated by the node **Create Bit Vectors** are an efficient way to represent each chemical substance by its organoleptic properties. This is an alternative description to molecular fingerprint that are computed from the chemical structures. |

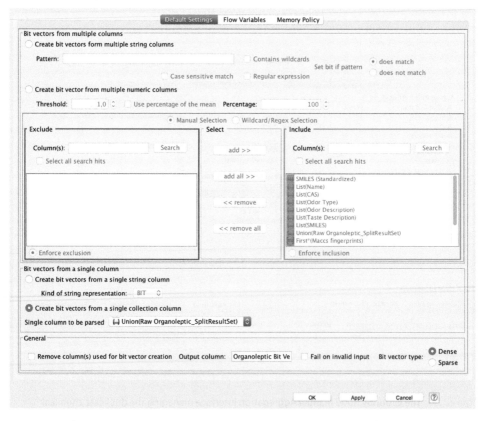

Figure 1.32 The Create Bit Vector node, providing an organoleptic description of the chemical substances as bitstrings.

| Instructions | Comments |
|---|---|
| • Add a **Column Rename** to the workflow and configure it (Figure 1.33). <br><br> • Double click on the following column titles, inside the Column Search area, to add them in the right-hand area of the interface: *SMILES (Standardized)*, | The dataflow contains fields that are formatted so they cannot be readily saved in a text file, such as a CSV. <br><br> In particular, there are fingerprints represented as bit vectors. To save them, the value of each bit (on/off) must be translated into a 8 bit encoded character ("1"/"0"). |

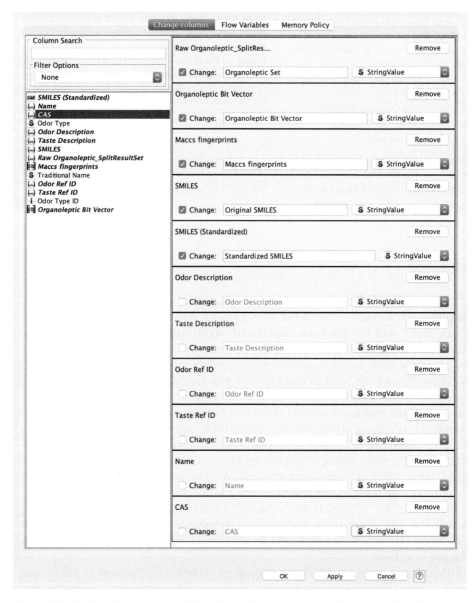

Figure 1.33 Configuration interface of the Column Rename node used to type cast the bit vectors and collections to standard strings.

(*Continued*)

| Instructions | Comments |
|---|---|
| *Name, CAS, Odor Description, Taste Description, SMILES, Raw Organic_ SplitResultSet, MACCS fingerprints, Odor Ref ID, Taste Ref ID, Organoleptic Bit Vector.*<br>• Tick the Change box in each of the added lines.<br>• Change the output type of each line to *StringValue.*<br>• Rename each line so the dataflow fields are more meaningful:<br>   ○ *Raw Organoleptic_SplitResultSet* as *Organoleptic Set*<br>   ○ *SMILES* as *Original SMILES*<br>   ○ *SMILES (Standardized)* as *Standardized SMILES.* | Another difficult type are Collections. An example is the column *Raw Organoleptic_SplitResult.* A collection is a safe way to aggregate into one column lots of information. A collection encodes also the rules to split this information into separated columns. Yet in the present case, the collection content should be stored as a simple text.<br><br>Here, the **Column Rename** is used to type cast those fields of the dataflow into standard strings that can be saved into a text file.<br><br>Doing so, the data representation is less efficient and some information is lost. For instance, after the operation, it is no longer possible to safely split the initial collection into separated fields. |
| • Add a **CSV Writer** node to the workflow.<br>• Configure the **CSV Writer** node like the others (Figure 1.27 and Figure 1.28).<br>• Set the name of the output file as `SubstanceTable.csv.` | The dataflow dedicated to substance specific information is saved into the file `SubstanceTable.csv.` |

## Conclusion

The final workflow is shown in Figure 1.34. The process is very tedious but it is a prerequisite to chemical information management: databases, modeling, and so on. The raw input data file has been interpreted, scrutinized for errors, and particular caution was paid to the chemical structures. Finally, the substances information, odor types, odor and taste description references populate different CSV files with foreign keys. These files will be used to build a relational database dedicated to organoleptic properties of small organic compounds.

It is remarkable that many aspects of the dataflow are specific to this particular task of assembling a database of organoleptic properties of organic substances. It is extremely difficult to generalize the curation process. Each problem necessitates specific treatments. This is the main reason why solutions like flow programming languages, such as KNIME, are so efficient to deal with this task. A clear picture of the process alleviates the complexity of the data treatment. Many routine actions are repeatedly needed, although always in a different manner, during a curation process: splitting files, joining files, merging columns or rows, manipulating strings, and checking and standardizing the chemical structures. These benefit from abstract representation as configurable nodes in a workflow.

Finally, it is important to note that management of chemical structures is still an open question today. There is no satisfactory solution to this problem. The many valid possible representations of a compound as a graph still constitute a challenge today and there are no rules to decide which one is the more relevant in a particular context.

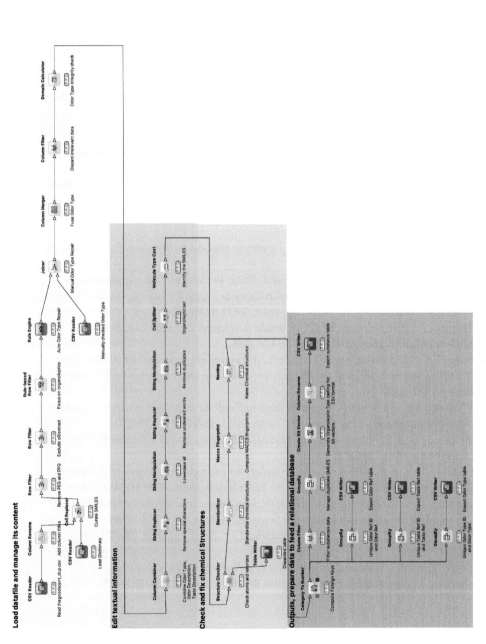

Figure 1.34 Overview of the KNIME workflow for the treatment of an organoleptic dataset aiming at populating a relational database.

Almost all software editors in chemoinformatics provide some automated solution for chemical structure curation. There are also some publications providing an easy to follow workflow [5]. It can give a false feeling of security and the problem is often disregarded. Checking and fixing chemical structures is a process that is an intrinsic part of a scientific process: it technically translates the opinion of an expert on a domain of chemistry. For instance, aggregating data on binders of a particular receptor protein, using the very same public sources, will lead to very different datasets depending on the expert working on it, with very different results concerning the outcome of projects based on these datasets.

Beyond the automatic processing of data, other solutions may be envisaged. An example of such alternative is the crowd sourced curation procedure that is used to validate the data in some main public databases [13].

## References

1 Curry, E., A. Freitas, and S. O'Riáin, *The role of community-driven data curation for enterprises*, in *Linking enterprise data*. 2010, Springer. 25–47.

2 McNaught, A., *The IUPAC international chemical identifier*. Chemistry International, 2006, 12–14.

3 Heller, S.R. and A.D. McNaught, *The IUPAC international chemical identifier (InChI)*. Chemistry International, 2009, **31**(1), 7.

4 Oprea, T., et al., *On the propagation of errors in the QSAR literature*. EuroQSAR, 2002, 314–315.

5 Fourches, D., E. Muratov, and A. Tropsha, *Trust, but verify: on the importance of chemical structure curation in cheminformatics and QSAR modeling research*. Journal of Chemical Information and Modeling, 2010, **50**(7), 1189–1204.

6 Tropsha, A., *Best practices for QSAR model development, validation, and exploitation*. Molecular Informatics, 2010, **29**(6–7), 476–488.

7 Boser, B.E., I.M. Guyon, and V.N. Vapnik. *A training algorithm for optimal margin classifiers*. in *Proceedings of the fifth annual workshop on Computational learning theory*. 1992, ACM.

8 Williams, A.J. and S. Ekins, *A quality alert and call for improved curation of public chemistry databases*. Drug Discovery Today, 2011, **16**(17), 747–750.

9 Hähnke, V.D., E.E. Bolton, and S.H. Bryant, *PubChem atom environments*. Journal of Cheminformatics, 2015, **7**(1), 1–37.

10 Feldman, H.J., et al., *CO: A chemical ontology for identification of functional groups and semantic comparison of small molecules*. FEBS letters, 2005, **579**(21), 4685–4691.

11 Durant, J.L., et al., *Reoptimization of MDL keys for use in drug discovery*. Journal of Chemical Information and Computer Sciences, 2002, **42**(6), 1273–1280.

12 Guazzelli, A., et al., *PMML: An open standard for sharing models*. The R Journal, 2009, **1**(1), 60–65.

13 Ekins, S. and A.J. Williams, *Reaching out to collaborators: crowdsourcing for pharmaceutical research*. Pharmaceutical Research, 2010, **27**(3), 393–395.

2

## Relational Chemical Databases: Creation, Management, and Usage

*Gilles Marcou and Alexandre Varnek*

*Goal*: The tutorial focuses on the creation, management, and usage of a relational database managements system (RDBMS) in the context of chemical information.
*Software*: InstantJChem[1]
Data: The data concerns a set of substances with organoleptic properties. The following files are used in the tutorial:

- `SubstanceTable.csv` – the chemical structures of the organoleptic substances.
- `OdorRefTable.csv` – bibliographic references about the odor description of substances.
- `TasteRefTable.csv` – bibliographic references about the taste description of substances.
- `OdorTypeTable.csv` – enumerated keywords list describing the odor types of substances.
- `JoinSubstanceOdorRef.csv` and `JoinSubstanceTasteRef.csv` – Join tables needed to build many-to-many relationships.

The dataset is adapted from the result of the data treatment described in the data curation tutorial. It is a compilation of data extracted from the web site of The Good Scent Company, http://thegoodscentscompany.com/(January 2016).

## Theoretical Background

A relational database is constituted of tables and relationships. A table is a way to store information analogously to a spreadsheet: each item is associated to a number of fields. This may lead to creation of data duplicates. In our case, each publication describes several substances, therefore a unique table would require duplicating the bibliographic references.

*Tutorials in Chemoinformatics*, First Edition. Edited by Alexandre Varnek.
© 2017 John Wiley & Sons Ltd. Published 2017 by John Wiley & Sons Ltd.
Companion website: www.wiley.com/go/varnek/chemoinformatics

Data duplication poses several problems. First, redundancies in the database uselessly inflate its size and affect the efficiency of information retrieval and storage. Second, duplicated data cannot be controlled in an efficient way, therefore chances of introducing errors into the database are increased with duplicated information. Finally, if an error is found in a duplicate data, it is needed to search and update each item of the dataset, sharing this same information.

Relational database management system is a common way to avoid this problem.[2] Data entries sharing the same piece of information are pointing to the same item in a dedicated table. In the present case, a dedicated table will contain the information about the bibliographic references of the odor of the substances. Several substances, described by the same publication will point to the same item in this table. If the citation has to be updated, it is updated only once, in the dedicated table.

The structure of the database to be constructed is represented in Figure 2.1. The bibliographic references for both odor and taste descriptions are located into dedicated tables. They are connected to the organoleptic substances table by Many-To-Many relationships. Such relationship models the fact that a bibliographic source may describe several substances and a substance can be described in several publications. The list of odor type, mutually exclusive terms describing the odor of the substance, is located in a dedicated table in One-To-Many relationship with the organoleptic substance table. Many substances belong to the same odor classification. This kind of relationship models the nominal character of the odor type as a field of the database: the odor type table can be viewed as dropdown list and each substance pick one and only one value into this list.

As a rule, the more separated the data are, the more simple is it to mutualize and manage. For instance, the list of journal names could be a topic of another table to which would point Odor Reference and Taste Reference tables. However, the goal is to keep it as simple as possible in the frame of this tutorial.

There is a critical point distinguishing chemical databases from other databases. The data types supported by RDBMS are described in Table 2.1 of the **Standard table** section. From this table, it is clear that a chemical structure can be stored in a database system only in an encoded form and there are no built in interfaces from the code to the chemical structure and the other way around. This is why the software InstantJChem[1] is needed. It is build on top of a mainstream RDBMS (Derby[3], MySQL[4], or Oracle[5]) and extend its capabilities to support chemically meaningful data. In InstanstJChem, it is refered to as **JChem table** (Table 2.1).

There are other solutions to add a chemically aware interface to a standard RDBMS. They are sometimes called chemical cartridge, using the term promoted by Oracle—the current leader of RDBMS market.[5] A cartridge defines the manner in which the server interprets, stores, retrieves, and indexes the data. They are software components plugged into a mainstream RDBMS to extend its capabilities in chemistry. Examples of such tools are MyChem (http://mychem.sourceforge.net/), pgChem:Tigress (http://pgfoundry.org/projects/pgchem/), OrChem (http://orchem.sourceforge.net/).[6] Some are fully build-in solutions in commercial packages such as MOE[7] from CCG, Maestro[8] from Schrodinger, or ChemFinder[9] from CambridgeSoft. They do not extend an existing RDBMS but rather implement their own system.

In InstantJChem, the implemented solution is to store in the database a code for the chemical structure itself, and to also store a fingerprint description of it.

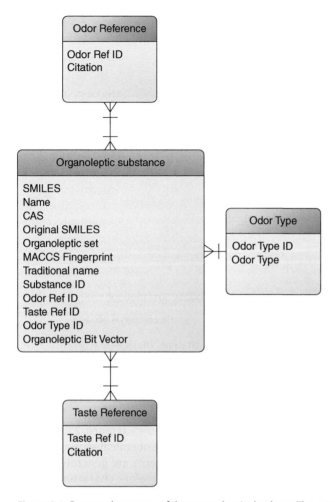

**Figure 2.1** Proposed structure of the organoleptic database. The organoleptic substances table is the endpoint of Many-To-Many relationships with the bibliographic citations in the Odor Reference and Taste reference tables. The organoleptic substance table is also in a One-To-Many relationship with the Odor Type table.

**Table 2.1** Description of data types commonly supported by RDBMS and by InstantJChem.

| Type | Description |
|---|---|
| **Standard table** | |
| Integer | Long integer: $232 = 4294967296$ |
| Text | User can specify widths of text fields as large as needed. |
| Real | Real double-precision. |
| Date | Allows storing dates. |
| Boolean | Value is True or False. |
| List (Standard) | To store a list of database items. |
| **JChem table** | |
| Chemical terms | A list of functions evaluated on chemical structures: logD, pKa, tautomers. |
| Structure | Chemical structure, automatically created with a Jchem table. |

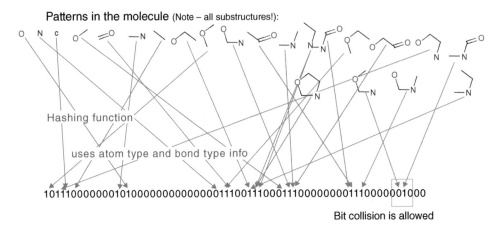

**Figure 2.2** Chemical Hashed Fingerprints are represented by a bit string. The detection of substructure in the chemical structure sets on a pattern of bits in the bit string. A bit collision occurs when two patterns overlay on one or more bits. Source: Chemical Hashed Fingerprint 2015.[10] Reproduced with permission from ChemAxon Technical Support Forums..

InstantJChem uses Chemical Hashed Fingerprints[10] (CHFP) (Figure 2.2). A bit string represents them. The presence of a substructure in the chemical structure sets a pattern of bits on in the bit string. The substructures patterns are generated from the chemical structure using a hash function. A CHFP is therefore characterized by the size of the substructure patterns detected, controlling the number of different patterns. The size of the bit string and the size of the bit patterns used to encode a substructure pattern are the other two parameters of CHFP. Thus, the larger the substructures, the more precise is the encoding for a given chemical structure, but it requires larger bit strings to encode. Thus more memory is used to store data and database searching is slowed down. Yet, each chemical structure contains a small number of substructures compared to the total number of chemical substructure, thus each bit string is mainly composed of bits off. A possibility is therefore to use shorter bit strings with larger bit patterns in order to store the information more efficiently, but accepting a larger number of bit collisions. A bit collision occurs when the bit encoding of two patterns overlay on one or more bits. Thus, the bit string becomes ambiguous and ultimately reduces the accuracy of chemical structure search in the database.

CHFP are computed and stored by InstantJChem in a dedicated table, the JChem table (Table 2.1). They are used to compute indexes and for efficient chemical structure-based queries. The full chemical structure is used only as an end step to get more precise answers on those structures retrieved using the fingerprints. Therefore, the management of the fingerprint is critical for chemical structures retrieval from the database.

The full structure and CHFP may also be used to compute many properties for each chemical structure of the database. These computed properties include the molecular weight, the brute formula, the number of hydrogen donors and acceptors, the logP, and so on. They are collectively designated as Chemical terms (Table 2.1).

Of course, most if not all the chemically aware RDBMS are working on analogous solutions. The differences are in the nature of fingerprints used and details of algorithms used to treat chemical structure graphs.

## Step-by-Step Instructions

This part of the tutorial requires the installation of InstantJChem. The web site of the editor, Chemaxon (www.chemaxon.com), provides versions of the software for Windows, Mac, and Linux. Trial licenses are available for a limited amount of time, and special free licenses are proposed for teaching and academic researchers.[11] To install a license, start from the initial installation of the software, and then open the *Licenses...* in the *Help* menu (Figure 2.3).

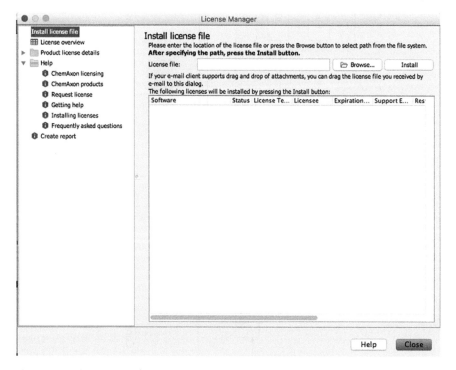

Figure 2.3 Select the interface Install License File on the left hand frame, then choose a valid license file by clicking on the Browse button, then click the button Install.

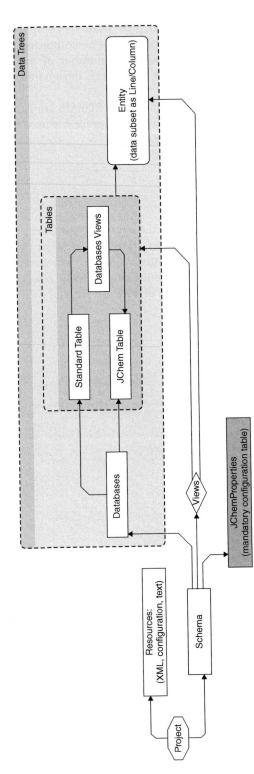

Figure 2.4 Concepts introduced by InstantJChem to manipulate in a transparent way the chemical information along with standard information.

It opens the **License Manager** software. Select the *Install license file* interface on the left hand frame; click the button *Browse* to select a valid license file and validate the process by clicking the *Install* button. Click the button *Close* to close the interface.

The InstantJChem software provides a transparent access to all its components by introducing a number of additional concepts (Figure 2.4). Data are organized in *Projects*, which can contain both generic files and *Schema(s)*. All files placed to the project folder will appear in the graphical user interface as belonging to this project. These could be (*i*) configuration files for specific applications, (*ii*) molecular files used for the data exchange between different applications, and (*iii*) text files for comments. *Schema* could be considered as a layer between the graphical interface and an RDBMS. Technically, it primarily contains the procedures to connect to an RDBMS (URL, identifiers, ...), encode user queries in a meaningful way for the RDBMS, and decode answers in a meaningful way for the user. It represents an ensemble of related tables, each of which can be manipulated and displayed in *InstantJChem*. Two types of tables are used: *Standard* tables containing only standard data types and *Structure* tables containing molecular information as structures, calculated properties, and so on. *Structure* tables are organized around a molecular graph representing a compound. The *Schema* also contains *Entities*: these are organized data so that they could be understood as a spreadsheet. A standard table can be promoted to an entity trivially. However, the most common entity is the one promoted from a JChem table: technical details are hidden and only the chemical structure interface is visible. Entities themselves are organized into a *Data Tree*. Finally, the *Schema* contains the different specialized *Views* designed by the user to access information.

| Instructions | Comments |
| --- | --- |
| • Start *InstantJChem*. | Start the software *InstantJChem* (Figure 2.5). |
| • Click on the main *File- > New Project...* or alternatively click on the ⊞ icon indicated on Figure 2.5. | Start a new project to open the **New Project** wizard. |
| • Select the **IJC Poject (empty)** option, and then click on **Next** (Figure 2.6). | These operations create a new directory named *Organoleptics* that will contain the database. However, the database itself is not yet created. |
| • In the next step, name your project as *Organoleptics* in the field **Project Name**. | Note that the *Organoleptics* directory is supposed to contain all files needed to manage a local database. Therefore, it is necessary to share this directory in order to make the database available to a third party. |
| • Choose a proper folder in the field **Project Location** (Figure 2.7). A good choice can be the folder containing the tutorial files. | |
| • The field **Project Folder** is adequately set automatically. | |
| • Click the **Finish** button. | |

*(Continued)*

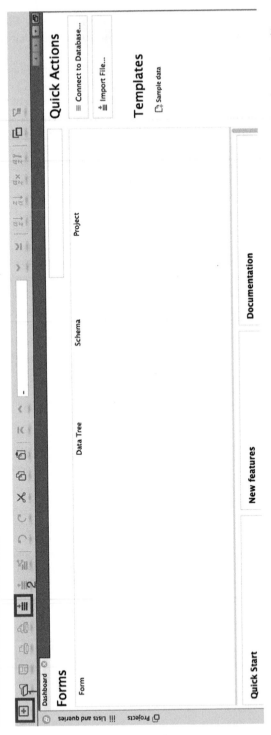

Figure 2.5 The InstantJChem software starts with a Dashboard summarizing user's projects. To start a new project, click on the framed icon 1 in red in this picture. To import a dataset click the framed icon 2.

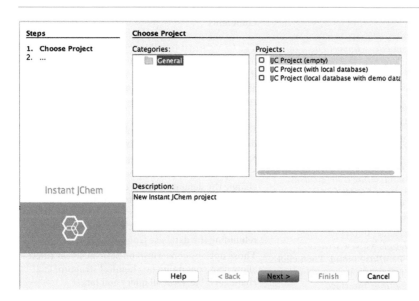

Figure 2.6 First step of the project creation wizard. The first option creates a project, the second creates a project and sets up a local database in one step and the third is a demo mode.

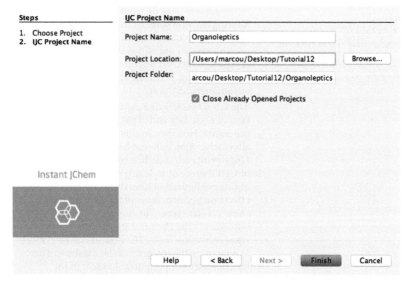

Figure 2.7 Second step of the project creation wizard. Once a project name and location is set, the project folder is set automatically.

(*Continued*)

| Instructions | Comments |
|---|---|
| • Click on the main *File- > Import File...* or alternatively click on the ⬥ icon indicated (see Figure 2.5). | The CSV format is ambiguous: in some systems it is a Comma Separated Value format, in others it is a Semicolon Separated Value format. Besides, as a text file that can contain accentuated and special characters, it is subject to character encoding hell [12]. |
| • Choose the file *SubstanceTable.csv*. | |
| • As *File type* chose *Delineated text file*. Then click on the button to the right of this field (button 1 in Figure 2.8). It opens the *Parsers option* interface.[12] | For these reasons, it is essential to give more details about decoding the input CSV file. |
| • In the *Parsers option* interface: | Note that the first field of the file contains chemical structure. This is why it is needed to change the default table set up to configure it as a JChem table, supporting this information. The configuration of this table is critical: many choices at this point cannot be changed without rebuilding the database from scratch. |
| ○ Tick the option *First line is field names*. | |
| ○ Set the *Separator* as *SEMICOLON*. | |
| ○ Tick options *Strip Quotes* and *Trim Whitespaces*. | |
| ○ Set the *Characterset* to *UTF-8*. | |
| • Set the *Table details* as *New structure entity (using JChemBase table)*. Then click the button 2 (Figure 2.8). It opens the *Table creation options* interface (Figure 2.9). | Those options concern some basic business rules: allow data containing no chemical structure in a chemical table; treat all query and target structures as absolute stereo (if off, a missing chiral flag means that a chemical structure can be queried as drawn or as the enantiomer); check for duplicate structures and enumerate tautomers for duplicate filtering. |
| • In the *Table creation options* interface: | |
| ○ Set the *Display Name* to *Substance* | In the present case we need to avoid duplicate filtering because some of the input structure appear as duplicates: the structure can be present with an undefined stereo center to represent a racemic mixture or with fully specified stereo when the substance is enantiomerically pure. |
| ○ Tick the option *Assume absolute stereo*, but not *Duplicate filtering* and *Tautomer duplicate checking*. | |
| • Take a look at other tabs *ID generation*, *Fingerprints*, *Standardizer*. Then click the button *Close*. | The tab *ID Generation* controls how the primary keys of each data added to the database are generated. However, in this tutorial, no alternative choice should be available. The *Fingerprints* tab is dedicated to the definition of CHFP used for indexing the database. Effects of this configuration have deep and unintuitive effects on performances of the system to query chemical structure. The *Standardizer* implements the chemical structures standardization rules. These rules can be critical to guarantee the integrity of the database. This functionality is an important aspect of the specification of any chemical database management system. |
| • Click the button *Next*. | |

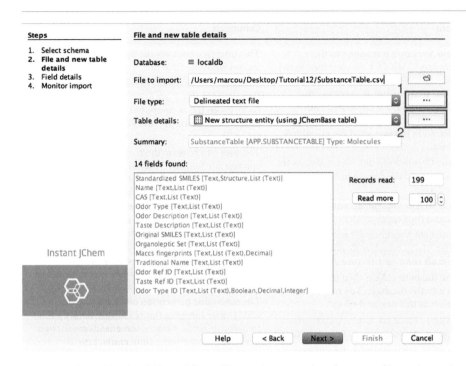

**Figure 2.8** Open File wizard. The red-framed button 1 opens an interface to set file parsing options. The framed button 2 opens a configuration interface for the chemically aware table (JChem table).

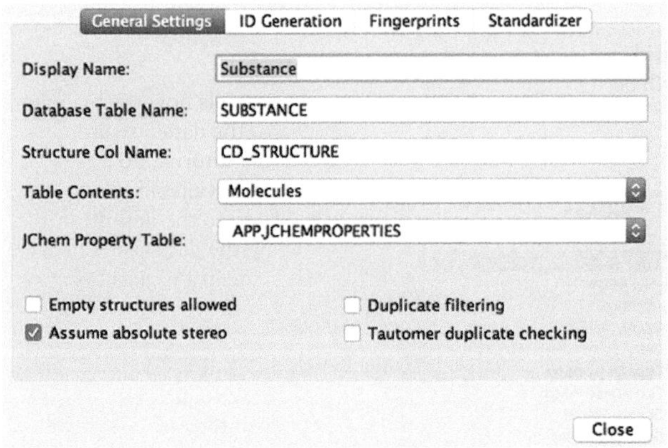

**Figure 2.9** The Table creation options interface. The choices in this interface control the input of chemical structures and their encoding. Many of these choices here cannot be modified later without fully rebuilding the database.

*(Continued)*

| Instructions | Comments |
|---|---|
| • The field *Structure* is mapped to the *Standardized SMILES* field identified in the imported file. A primary key (*CdId*), the molecular weight (*Mol Weight*) and the formula (*Formula*) are computed and automatically added to the database. Other fields found in the imported file are added by default. | This interface controls the data type of each field loaded from the CSV file. |
| | Fields are automatically detected and configured (Figure 2.10). They are annotated with a color code: green if it is computed and orange if the field cannot be empty (the *Required* parameter is set to *True*). |
| • Click the *Organoleptic Bit Vector* and set the type to *Text* and the *Length* to 1116 (Figure 2.11). | The user can remove them or add them using the appropriate buttons. Additionally, if the data are to be added to an existing table, fields can be merged or added. When added, the new data are added to the table. When merged, some existing fields are used to pair a row of the database table with a row of the imported file; then additional fields of the imported file are added to existing fields of the database and the content of the existing field is overwritten by the information in the imported file for those fields that are mapped. |
| • Click the *Maccs fingerprints* and set the type to *Text* and the *Length* to 195. | |
| • Click the field *Substance ID* and set the type to *Integer*, the parameters *Required* to *TRUE* and *Default value* to 1 (Figure 2.12). | |
| • Click on the button *Next* then the data are loaded into the database. You may receive an error such as this one in the log: | |
| `Warning: Failed to import row 0: Empty structures not allowed` | The automatic perception of data type from the CSV can be inappropriate as for the binary fingerprint fields. Besides some fields are foreign keys that impose some constraints: keys in relationship should be of the same type (by convention they are integers) and, using a local database, they must be defined as requiring a value. |
| But it can be safely ignored: the software may try to read the title line of the CSV file. This behavior can be assimilated to a bug of the software. However, errors during loading of data are stored in files named after the input file name terminated by `_errors`. | |
| • Click the *Finish* button to open a grid overview of the created entity. It will be similar to Figure 2.13. | |

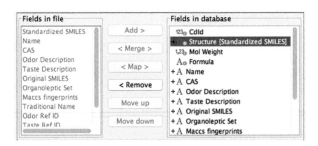

Figure 2.10  Field configuration interface. The configuration acts at the entity level.

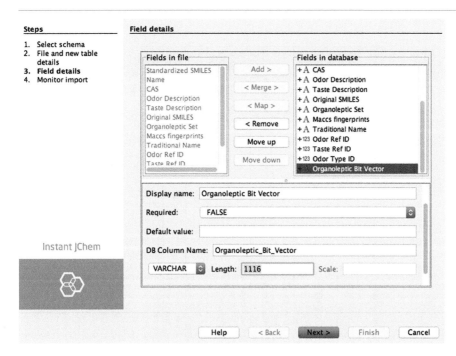

Figure 2.11 Configuration of the field Organoleptic Bit Vector. Some attention must be paid to the default value of the Length of the text that is too short to store the whole chain containing 1116 bit values.

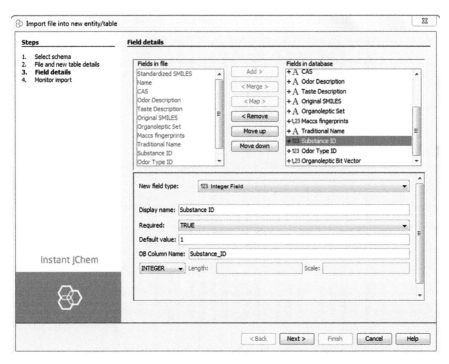

Figure 2.12 Configure the Substance ID as in Integer field. As it will be used as a foreign key, it is needed to set the parameters Required to TRUE and Default value to 1.

(Continued)

| CAS | Odor Description | Taste Description | Original SMILES | Organoleptic Set | Maccs fingerprints | Traditional Name | Substance ID | Odor Type ID | Organoleptic Bit Vector |
|---|---|---|---|---|---|---|---|---|---|
| 103-64-0, 1340-14-3 | strong leafy green hyacinth, | | [BrC=CC1=CC=CC | green, strong, leafy, hyacinth | 0000000000000000 0000000000000000 | $\beta$-bromostyrene | | 1 | 13 0000000000000000 0000000000000000 |
| 28217-92-7 | green musty nutty | | [C(C1CCCC1)C1=NC=CN=C1] | nutty, green, musty | 0000000000000000 0000000000000000 | 2-(cyclohexylmethyl) | 2 | | 28 0000000000000000 0000000000000000 |
| 101-49-5 | green waxy sweet honey hyacinth, | | [C(C1OCCO1)C1=CC=CC=C1] | green, waxy, sweet, honey, | 0000000000000000 0000000000000000 | 2-benzyl-1,3-dioxolane | 3 | | 13 0000000000000000 0000000000000000 |
| 3208-40-0 | sweet fruity green | | [C(C1CCCO1)CC1=CC=CC=C1] | fruity, sweet, green | 0000000000000000 0000000000000000 | 2-(3-phenylpropyl)oxolane | 4 | | 10 0000000000000000 0000000000000000 |
| 2110-18-1 | Fresh green pepper-like with a | Green, vegetative, green and jalapeno | [CC1=CC=CC=C1) | green, fresh, pepper, | 0000000000000000 0000000000000000 | 2-(3-phenylpropyl)pyridine | 5 | | 13 0000000000000000 0000000000000000 |
| 80858-47-5, 80858-47-5 | floral hyacinth green metallic | | [CC1=CC=CC=C1) | floral, hyacinth, green, metallic, | 0000000000000000 0000000000000000 | [2-(cyclohexyloxy) ethyl]benzene | 6 | | 24 0000000000000000 0000000000000000 |
| 103-50-4 | sweet fruity cherry earthy mushroom | fruity spicy nutty | [C(OCC1=CC=CC=C1 | earthy, sweet, fruity, cherry, | 0000000000000000 0000000000000000 | dibenzyl ether | 7 | | 26 0000000000000000 0000000000000000 |
| 4437-22-3 | coffee nutty earthy | coffee mushroom-like | [C(OCC1=CC=CO1) | coffee, nutty, earthy, mushroom | 0000000000000000 0000000000000000 | 2-[(furan-2-ylmethoxy)methyl] | 8 | | 25 0000000000000000 0000000000000000 |
| 35250-78-3 | fruity, woody | | [C(OCC1=CC=CS1)C1=CC=CS1] | fruity, woody | 0000000000000000 0000000000000000 | 2-[(thiophen-2-ylmethoxy)methyl] | 9 | | 10 0000000000000000 0000000000000000 |
| 538-74-9 | | bitter roasted | [C(SCC1=CC=CC=C1) | bitter, roasted | 0000000000000000 0000000000000000 | benzyl sulfide | 10 | | 8 0000000000000000 0000000000000000 |

Figure 2.13  The final aspect of the table containing chemical substances.

| Instructions | Comments |
|---|---|

- If the *Projects* window (Figure 2.14) is not present, you can open it using the option *Window->Projects* from the main menu bar.
- Right click on the entity *Substance*. Then click on *Edit Data Tree*. Browse the database administration interface (Figure 2.15) to retrieve the architecture of concepts illustrated in Figure 2.4.

There are alternative ways to open the database configuration interface. For instance, it is possible to click the database (*localdb*) line of the *Projects* window and click *Edit Schema*.

This interface allows modifying some parameters of the connection (*Schema*). It is also possible to access the underlying standard table created in the menu *Database Tables*. For instance, it is possible to track down the fingerprints used to index chemical structures of the database. They are stored into fields with FP starting names.

**Figure 2.14** The Projects window of the Organoleptics project after loading the first table named Substance. The root level displays the schema (the database), the first level of the hierarchy displays entities (the database tables), and the third level the view (a spreadsheet-like interface in this case).

The *Database Views* interface can be confusing. A view in this interface is a slice of the database organized as it could fit in a spreadsheet. The information accessible through a view can come from different tables and result from queries. It must not be confused with data browser interfaces that are also designated as "views."

Then tables and views are treated in a homogeneous way as entities. And several entities can be organized in a hierarchical fashion into a data tree.

The higher levels of abstraction (entities and data tree) are the favored way to manage the database: adding, deleting, and editing fields.

- Click on the main *File->Import File...* or alternatively click on the icon indicated (see Figure 2.5).
- Choose the file *OdorTypeTable.csv*.
- As *File type* chose *Delineated text file*. Then click on the button to the right of this field (button 1 in Figure 2.8). It opens the *Parsers option* interface.
- In the *Parsers option* interface:
  - ○ Tick the option *First line is field names*.
  - ○ Set the *Separator* as SEMICOLON.
  - ○ Tick options *Strip Quotes* and *Trim Whitespaces*.
  - ○ Set the *Characterset* to *US-ASCII*.
- Set the *Table details* as *New database table*. Then click the button 2 (Figure 2.8). It opens the *Table creation options* interface (Figure 2.9).
- In the *Table creation options* interface:
  - ○ Set the *Display Name* to *Odor Type*
- Click the button *Next*.
- In the *Field details* interface, set the imported field *Odor Type ID* as *Integer*. Set the parameters *Required* to *True* and *Default Value* to 1.
- Click the button *Next*.
- Click the button *Finish*.
- If in the resulting table, there are two lines with the *Odor Type ID* equal to 1, select the first one then click on *Data->Delete Row(s)...* or alternatively on the delete row icon ▦.

During this step, the table containing the definition of odor types is loaded. Odor types can be understood as categories.

Each substance is associated to one odor type to the exclusion of all others.

*Substance* and *Odor Type* tables share a foreign key in the field *Odor Type ID*. Hence, it imposes on this field the constraint to be integer and to be defined for each instance, with a default value.

It is possible that the first line of the CSV imported file is loaded while it contains the field names. This produces in the final table an additional entry with the ID 1, which must be deleted.

(Continued)

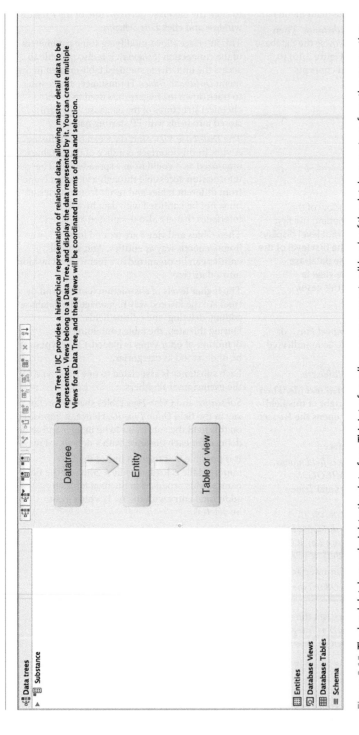

Figure 2.15 The local database administration interface. This interface allows easy access to all levels of the database system, from the connection parameters (Schema) to the most abstract organization of data as Data Tree.

| Instructions | Comments |
|---|---|
| • If needed, reopen the database administration. (From the *Projects* window, right click on the entity *Substance*. Then click on *Edit Data Tree*). | In this step a One-to-Many relationship is created between the tables *Substance* and *Odor Type*. The substances are on the *many* side and the odors are on the *one* side. This means that many substances can be related to the same odor. The field selected from the *many* side defines the foreign key column while the field from the *one* side defines the primary key column. |
| • Open the *Entity* interface (Figure 2.16) and click the *New relationship* icon ⤙. | |
| • Select *Create Many-to-One relationship* option. | |
| • Set the *From* parameter to *Substance* and the *To* parameter to *Odor Type*. | The interface creates relation between the tables. Ticking *Create DB constraints* allows to implement database foreign key constraints. This helps to avoid any action on the foreign key that could break the link between the tables and prevent entering a foreign key value outside of the range of primary key values. |
| • Set the value of both *Field* parameters to *Odor Type ID*. | |
| • The interface should look as Figure 2.17. Click the button *Next*. | |
| • Select the entity *Substance* and tick the box *Data trees child entities* (Figure 2.18). | Finally, the higher abstraction user interface manipulate and display items at the *entity* abstraction level. Therefore, relationships need to be promoted at a higher abstraction level (data trees), to be used in formular for instance. |
| • Select the entity *Odor Type* and tick the box *Data trees child entities*. | |
| • Click the button *Finish*. | |
| • Check the created relationship inside the *Data trees* interface. | |
| • Click on the main *File- > Import File...* or alternatively click on the ⬆ icon indicated (see Figure 2.5). | Loads the table of bibliographic references of substances odors. The procedure is analogous to the one to load the *Odor Type* table. |
| • Choose the file *OdorRefTable.csv*. | |
| • As *File type* chose *Delineated text file*. Open the *Parsers option* interface (button 1 in Figure 2.8). | |
| • In the *Parsers option* interface: | |
| Tick the option *First line is field names*. Set the *Separator* as *SEMICOLON*. Set the *Characterset* to *US-ASCII*. | |
| • Set the *Table details* as *New database table*. Open the *Table creation options* interface (button 2, Figure 2.8). | |
| • In the *Table creation options* interface: Set the *Display Name* to *Odor Reference* | |
| • Click the button *Next*. | |
| • In the *Field details* interface, set the imported field *Odor Ref ID* as *Integer*. Set parameters *Required* to *True* and *Default Value* to 1. | |
| • Click the button *Next*. | |
| • Click the button *Finish*. | |
| • If the resulting table has an additional first row (because of the title row), select the first one then click on *Data- >Delete Row(s)...* or alternatively on the delete row icon ▦. | |

*(Continued)*

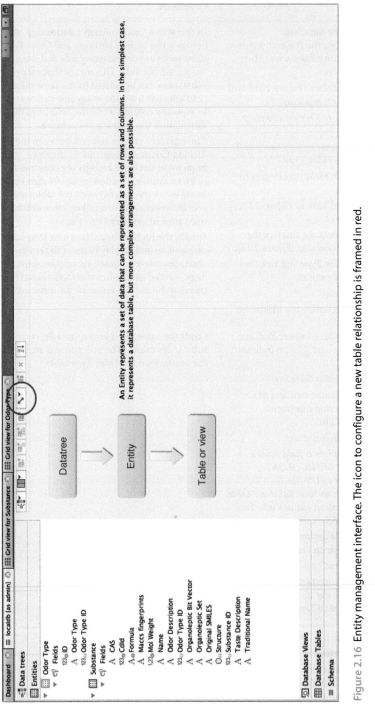

An Entity represents a set of data that can be represented as a set of rows and columns. In the simplest case, it represents a database table, but more complex arrangements are also possible.

Figure 2.16 Entity management interface. The icon to configure a new table relationship is framed in red.

Figure 2.17 Configuring a Many-to-One relationship between the table Substance and the table OdorType using the Odor Type ID as foreign key.

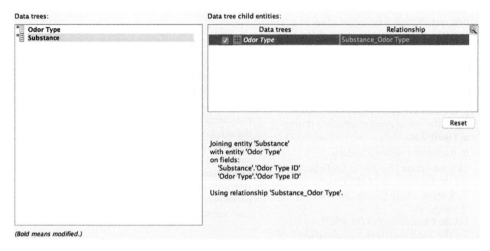

Figure 2.18 Promoting the relationship between tables Odor Type and Substance to the Data trees level. There is a tick box for each Entity (table).

- Use the *File->Import File...* tool or alternatively click on the ≜ icon indicated.
- Choose the file *TasteRefTable.csv*.
- As *File type* chose *Delineated text file*. Open the *Parsers option* interface (button 1 in Figure 2.8).
- In the *Parsers option* interface:
  Tick the option *First line is field names*.
  Set the *Separator* as *SEMICOLON*.
  Tick the options *Strip Quotes* and *Trim Whitespaces*.
  Set the *Characterset* to *US-ASCII*.
- Set the *Table details* as *New database table*. Open the *Table creation options* interface (button 2, Figure 2.8).
- In the *Table creation options* interface:
  Set the *Display Name* to *Taste Reference*

Loads the table of bibliographic references regarding tastes of substances. As for the preceding step, the procedure is analogous to the one to load the *Odor Type* table.

(Continued)

| Instructions | Comments |
|---|---|
| | |

- Click the button *Next*.
- In the *Field details* interface, set the imported field *Taste Ref ID* as *Integer*. Set parameters *Required* to *True* and *Default Value* to 1.
- Click the button *Next*.
- Click the button *Finish*.
- If in the resulting table, has an additional first row (because of the title row), select the first one then click on *Data-> Delete Row(s)...* or alternatively on the delete row icon ⊞.
- Use the *File-> Import File...* tool or alternatively click on the ⬆ icon indicated.
- Choose the file *JoinSubstanceOdorRef.csv*.
- Import the file as *Delineated text file*. Open the *Parsers option* interface (button 1 in Figure 2.8).
- In the *Parsers option* interface:

    Tick the option *First line is field names*.
    Set the *Separator* as *SEMICOLON*.
    Tick options *Strip Quotes* and *Trim Whitespaces*.
    Set the *Characterset* to *US-ASCII*.
- Set the *Table details* as *New database table*. Open the *Table creation options* interface (button 2, Figure 2.8).
- In the *Table creation options* interface: Set the *Display Name* to *Join Substance Odor Reference*
- Click the button *Next*.
- In the *Field details* interface, set the imported field *FK_Substance_ID* as *Integer*. Set parameters *Required* to *True* and *Default Value* to 1.
- In the *Field details* interface, set the imported field *FK_Odor_Ref_ID* as *Integer*. Set the parameters *Required* to *True* and *Default Value* to 1.
- Click the button *Next*.
- Click the button *Finish*.
- If in the resulting table, there are two lines with the *FK_Odor_Ref_ID* equal to 1, select the first one then click on *Data-> Delete Row(s)...* or alternatively on the delete row icon ⊞.

Comments:

This step loads a join table to build the *many-to-many* relationship between the *Odor Reference* table and the *Substance* table.

If a substance is described in two references, then the join table will contain two lines: one pairing the substance ID with the first bibliographic reference ID and the second pairing the same substance ID with the second bibliographic ID. The behavior is symetrical if a bibliographic reference describes several substances: the corresponding bibliographic ID is repeated with each substance ID.

| Instructions | Comments |
|---|---|
| • Use the *File-> Import File...* tool or alternatively click on the ≡ icon indicated. | This step loads a join table to build the *many-to-many* relationship between the *Taste Reference* table and the *Substance* table. |
| • Choose the file *JoinSubstanceTasteRef.csv*. | |
| • Import the file as *Delineated text file*. Open the *Parsers option* interface (button 1 in Figure 2.8). | |
| • In the *Parsers option* interface: | |
| Tick the option *First line is field names*. | |
| Set the *Separator* as *SEMICOLON*. | |
| Tick options *Strip Quotes* and *Trim Whitespaces*. | |
| Set the *Characterset* to *US-ASCII*. | |
| • Set the *Table details* as *New database table*. Open the *Table creation options* interface (button 2, Figure 2.8). | |
| • In the *Table creation options* interface:Set the *Display Name* to *Join Substance Taste Reference* | |
| • Click the button *Next*. | |
| • In the *Field details* interface, set the imported field *FK_Substance_ID* as *Integer*. Set parameters *Required* to *True* and *Default Value* to 1. | |
| • In the *Field details* interface, set the imported field *FK_Taste_Ref_ID* as *Integer*. Set the parameters *Required* to *True* and *Default Value* to 1. | |
| • Click the button *Next*. | |
| • Click the button *Finish*. | |
| • If in the resulting table, there are two lines with the *Odor Type ID* equal to 1, select the first one then click on *Data-> Delete Row(s)...* or alternatively on the delete row icon ▦. | |
| • If needed, reopen the database administration. (From the *Projects* window, right click on the entity *Substance*. Then click on *Edit Data Tree*). | This will create two *many-to-one* relationships. On the *many* side is the *Join Substance Odor Reference* table. On the *one* sides are the *Substance* and *Odor Reference* tables. |
| • Open the *Entity* interface (Figure 2.16) and click the *New relationship* icon ⌄·. | Thus from a substance, one can point to all references citing it. On the other end, one can use the reference table and retrieve all substances it describes. |
| • Select *Create Many-to-One relationship* option. | |
| • Set the *From* parameter to *Join Substance Odor Reference* and the corresponding *Field* parameter to *FK_Substance_ID*. | There is subtle point in promoting relationship to data trees: it must be done in two steps. The first one promote the relation between the join table to each of primary tables (*Substance* or *Odor Reference*). It creates a nested relationship, that can be promoted also at the level of data trees. |
| • Set the *To* parameter to *Substance* and the corresponding *Field* parameter to *Substance_ID*. | |

*(Continued)*

| Instructions | Comments |
|---|---|
| • The interface should look as Figure 2.19. Click the button *Next.*<br>• Select the entity *Substance* and tick boxes inside the *Data trees child entities* frame.<br>• Select the entity *Join Substance Odor Reference* and tick boxes inside the *Data trees child entities* frame.<br>• Click the button *Finish.*<br>• Open the *Entity* interface (Figure 2.16) and click the *New relationship* icon.<br>• Select *Create Many-to-One relationship* option.<br>• Set the *From* parameter to *Join Substance Odor Reference* and the corresponding *Field* parameter to *FK_Odor_Ref_ID.*<br>• Set the *To* parameter to *Odor Reference* and the corresponding *Field* parameter to *Odor_Ref_ID.*<br>• The interface should look as in Figure 2.20. Click the button *Next.*<br>• Select the entity *Substance* and tick the boxes inside the *Data trees child entities* frame.<br>• Select the entity *Join Substance Odor Reference* and tick boxes inside the *Data trees child entities* frame (Figure 2.21).<br>• Click the button *Finish.* | |

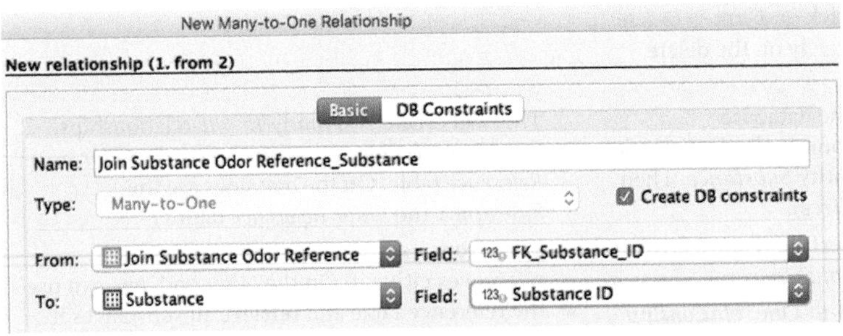


Figure 2.19 Configuration of the Many-to-One relationship between the Join Substance Odor Reference table and the Substance table.

| Instructions | Comments |
|---|---|

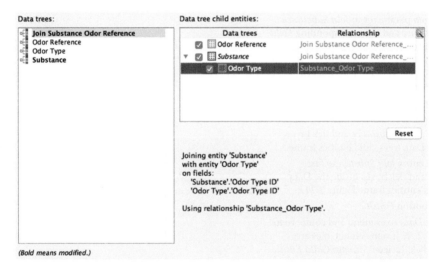

**Figure 2.20** Configuration of the Many-to-One relationship between the Join Substance Odor Reference table and the Odor Reference table.

Data trees:

Join Substance Odor Reference
Odor Reference
Odor Type
Substance

Data tree child entities:

| Data trees | Relationship |
|---|---|
| ☑ Odor Reference | Join Substance Odor Reference_... |
| ▼ ☑ Substance | Join Substance Odor Reference_... |
| ☑ Odor Type | Substance_Odor Type |

Reset

Joining entity 'Substance'
with entity 'Odor Type'
on fields:
    'Substance'.'Odor Type ID'
    'Odor Type'.'Odor Type ID'

Using relationship 'Substance_Odor Type'.

*(Bold means modified.)*

**Figure 2.21** Promote all relationship to data tree child entities. Note that the child entities of the Join Substance Odor Reference are updated by the promotion of child entities of related tables.

- If needed, reopen the database administration. (From the *Projects* window, right click on the entity *Substance*. Then click on *Edit Data Tree*).
- Open the *Entity* interface (Figure 2.16) and click the *New relationship* icon.
- Select *Create Many-to-One relationship* option.
- Set the *From* parameter to *Join Substance Taste Reference* and the corresponding *Field* parameter to *FK_Substance_ID*.

This step, achive the architecture of the database. As with the bibliographic references to odors, a *many-to-many* relationship is built between the table dedicated to substances and the one dedicated to bilbiographic references regarding tastes of substances.

The architecture visible on the *Data trees* interface is the used to manipulate the database interface.

*(Continued)*

| Instructions | Comments |
|---|---|
| • Set the *To* parameter to *Substance* and the corresponding *Field* parameter to *Substance_ID*. | |
| • The interface should look as the Figure 2.22. Click the button *Next*. | |
| • Select the entity *Substance* and tick boxes inside the *Data trees child entities* frame. | |
| • Select the entity *Join Substance Taste Reference* and tick boxes inside the Data trees child entities frame. | |
| • Click the button *Finish*. | |
| • Open the *Entity* interface (Figure 2.16) and click the *New relationship* icon ⤙. | |
| • Select *Create Many-to-One relationship* option. | |
| • Set the *From* parameter to *Join Substance Taste Reference* and the corresponding *Field* parameter to *FK_Taste_Ref_ID*. | |
| • Set the *To* parameter to *Taste Reference* and the corresponding *Field* parameter to *Taste_Ref_ID*. | |
| • The interface should look as the Figure 2.23. Click the button *Next*. | |
| • Select the entity *Substance* and tick boxes inside the Data trees child entities frame. | |
| • Select the entity *Join Substance Taste Reference* and tick boxes inside the Data trees child entities frame Figure 2.24. | |
| • Click the button *Finish*. | |
| • Check the *Data trees* menu and compare to the Figure 2.25. If some child tables are missing, click the icon *Manage Child Entities* icon (⬚) to open a tool for promoting the missing entities to the data trees level. | |
| • From the *Projects* window, right click on the *Substance* entity. Then click on *New View...* | A form view is a powerful way to access the database. A view is attached to an entity. In this case it is dedicated to the *Substance* table. Each field of the database that is attached to this entity can be binded to a *widget*. A widget is a display for the field. It can be single or multi-line text field, a structure pane, a table, and so on. |
| • Keep default parameters (Figure 2.26) and click the button *Finish*. It opens the *Design* mode of the view interface, which is currently empty. | |
| • An interactive object can be selected among those of the *Palette*. Then click anywhere on the surface of the *View* to position the element. A specific configuration menu will bind the object to one or more fields of entities linked by any relationship. | Note that the interface can be programmaticaly controled. This means that the database administrator can write short scripts which can be triggered by a user by double-clicking on a widget. So the interface is highly customizable. |
| Use the *Palette* on the right side of the interface to reproduce the example view proposed in Figure 2.27. | When the interface is ready, it can be used for instance to look at the entry 151 (Figure 2.28). The odor of this compound is described in two bibliographic references. Thus is illustrate how the many-to-many relationship is working. |

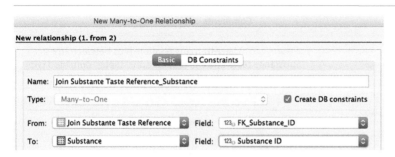

Figure 2.22 Configuration of the many-to-one relationship between the Join Substance Taste Reference and the Substance tables.

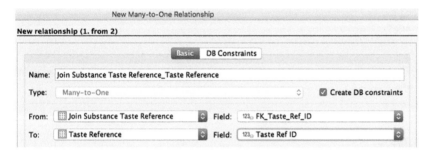

Figure 2.23 Configuration of the many-to-one relationship between the Join Substance Taste Reference and the Taste Reference tables.

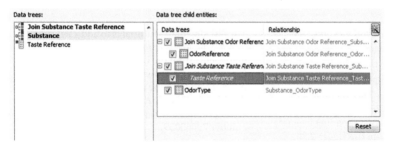

Figure 2.24 Promote all relationship to data tree child entities. Note that child entities of the Join Substance Taste Reference are updated by the promotion of child entities of related tables. There is a tick box for each Entity (table).

*(Continued)*

| Instructions | Comments |
|---|---|

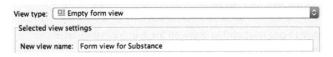

Figure 2.25 Many-to-many relationships between the Substance table and the bibliographic tables, emulated with many-to-one relationships.

**Data trees**
- ▸ Join Substance Odor Reference
- ▸ Join Substance Taste Reference
- ▸ Odor Reference
- ▸ Odor Type
- ▾ Substance
  - ▾ Join Substance Odor Reference
    - ▾ Odor Reference
      - 123 ID
      - A Odor Ref
      - 123 Odor Ref ID
      - 123 FK_Odor_Ref_ID
      - 123 FK_Substance_ID
      - 123 ID
  - ▾ Join Substance Taste Reference
    - ▾ Taste Reference
      - 123 ID
      - A Taste Ref
      - 123 Taste Ref ID
      - 123 FK_Substance_ID
      - 123 FK_Taste_Ref_ID
      - 123 ID
    - ▸ Odor Type
    - A CAS
    - 123 CdId
    - A Formula
    - A Maccs fingerprints
    - 1,23 Mol Weight

View type: Empty form view

Selected view settings

New view name: Form view for Substance

Figure 2.26 Interface to create a new form for the table Substance.

- Click the *Form View for Substance* tab.
- Click the *Data->New row...* from the main menu or click the corresponding icon (🖰).
- Fill the form as proposed in the Figure 2.29.
- Click the *Grid View for Odor Reference*. Click the *Data->New row* from the main menu or click the corresponding icon (🖰) and fill the form as in the Figure 2.30.
- Click the *Grid View for Join Substance Odor Reference*. Click the *Data->New row...* from the main menu or click the corresponding icon (🖰) and fill the form as in the Figure 2.31.
- Browse the new structure in the *Form View for Substance* tab. It should look like in the Figure 2.32.

The last step of the tutorial is to add a new object into the database. For instance, we can use some ingredients from the catalogue of a company like Givaudan (https://www.givaudan.com/). Most of the ingredients of the amber odor are already in the database, except Ambermax. The trademark of the product was deposited in 2008 (serial number 79063059).

The substance is first added, then the bibliographic reference for Givaudan. The join table is updated last, because it is composed of foreign keys: primary keys must be created first for foreign keys to be valid.

Note, that the fingerprint pattern is complicated to fill in. It would be a great addition to the form to add a software part to compute them from the chemical structure and the organoleptic description.

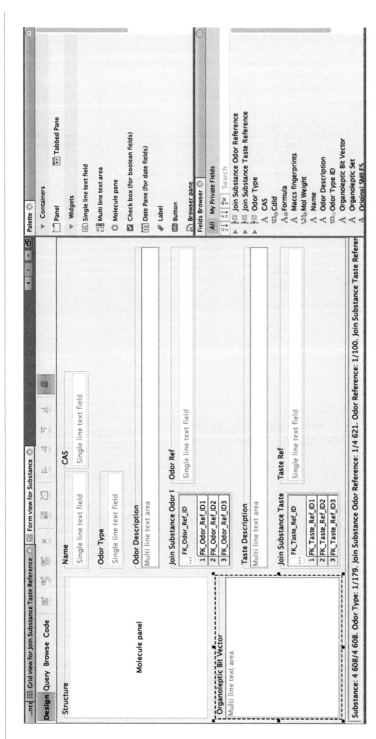

Figure 2.27  Design of a form to explore the database. Buttons at the top left of the interface are used to switch between the different modes of the interface: Design to configure the appearance, Query for interrogating the database, Browse to explore the database answers and Code to program actions on the form using a computer language.

(*Continued*)

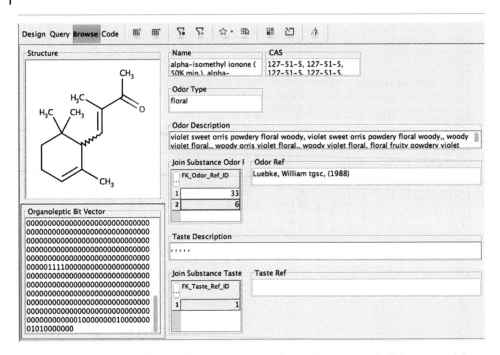

Figure 2.28  The entry 151 illustrate the many-to-many relationship between the Substance and the Odor Reference tables.

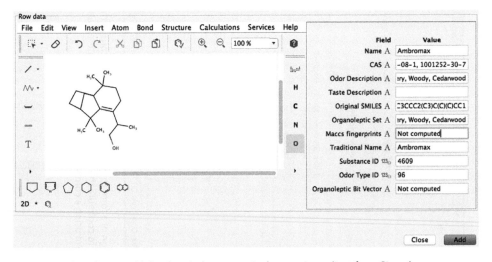

Figure 2.29  Interface to add the chemical structure Ambermax ingredient from Givaudan.

| Field | Value |
|---|---|
| Odor Ref ID 123₀ | 102 |
| Odor Ref A | Givaudan, 2016 |

Figure 2.30  Interface to add the Givaudan 2016 catalogue of ingredient to the database.

| Field | Value |
|---|---|
| FK_Substance_ID 123₀ | 4609 |
| FK_Odor_Ref_ID 123₀ | 102 |

Figure 2.31 Adding a join entry between the substance Ambermax and the entry Givaudan.

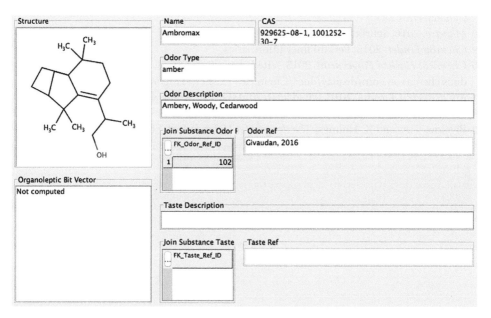

Figure 2.32 Ambermax added to the database. Computing fingerprints could be an interesting addition to the interface.

## Conclusion

The tutorial has described the building and management of a sophisticated database using the Chemaxon tool InstantJChem. Illustrated concepts are not bounded to this particular tool. The tutorial focused mainly on the import of data and creation of tables, building simple many-to-one relationships between tables, create sophisticated many-to-many relationships, creating forms, and adding new data.

## References

1 *InstantJChem*. 2016, Chemaxon.
2 Codd, E.F., *A relational model of data for large shared data banks*. Communications of the ACM, 1970, **13**(6),377–387.
3 *Derby*. 2015, Apache foundation.
4 *MySQL*. 2015, Oracle.
5 Rich, B., *Oracle Database*. 2015, Oracle.

6 Pansanel, J., E.-G. Schmidt, and C. Steinbeck, *ChemiSQL*. 2008: SourceForge. p. The ChemiSQL project intends to federate several open source chemical cartridge projects (Mychem, Orchem, and Pgchem). The project provides documentation, examples, and an unified graphical user interface for using these chemical cartridge.

7 *Molecular Operating Environment (MOE)*. 2016, Chemical Computing Group Inc.: Chemical Computing Group Inc., 1010 Sherbooke St. West, Suite #910, Montreal, QC, Canada, H3A 2R7.

8 *Maestro*. 2016, Schrödinger, LLC: New York, NY.

9 *ChemBioFinder*. 2015, PerkinElmer Informatics.

10 *Chemical Hashed Fingerprint*. 2015, [cited 2016 20/03/2016]; Available from: https://docs.chemaxon.com/display/docs/Chemical+Hashed+Fingerprint.

11 *My Academic License*. 2016, [cited 2016 20/03/2016]; Available from: https://www.chemaxon.com/my-chemaxon/my-academic-license/.

12 Bigelow, C.A. and K. Holmes, *The design of a Unicode font*. Electronic Publishing, 1993, **6**(3), 289–305.

3

# Handling of Markush Structures

*Timur Madzhidov, Ramil Nugmanov, and Alexandre Varnek*

*Goal*: use Markush enumeration for generating sets of molecules with common substructures.

*Software*: ChemAxon MarvinSketch, ChemAxon JChem (cxcalc plugin). Version 15.8.3 was used for tutorial creation.

## Theoretical Background

*Markush structure* (also called *generic structure* or simply *Markush*)[1] is essentially the structure involving R-groups, where a part of the molecule is defined by a series of alternatives. They were developed to use in patent claims, but also widely used in chemistry texts to represent series of compounds with common core structure. In chemoinfomatics Markush structures are widely used for molecules enumeration and combinatorial library representation.

Usually Markush structures are depicted using R groups, in which the side chain can be a radical, for example, "cyclohexyl," or general type, for example, "alkyl." However, other types of variabilities are often used: variable position of substituent, stereochemistry, and repetition of a given group.

*Tutorials in Chemoinformatics*, First Edition. Edited by Alexandre Varnek.
© 2017 John Wiley & Sons Ltd. Published 2017 by John Wiley & Sons Ltd.
Companion website: www.wiley.com/go/varnek/chemoinformatics

This general depiction of the molecules may be useful in claiming intellectual property rights over a focused chemical space zone. Markush structures were named after Dr Eugene A. Markush who pioneered to use them in 1924 to circumvent USPTO's "rule against."[2]

In chemoinformatics, Markush structure is a useful way of representation of compounds families allowing dense storage of structures from combinatorial libraries and its fast enumeration.

## Step-by-Step Instructions

| Instructions | Comments |
| --- | --- |
| ***Drawing simple Markush structure and its library size estimation*** <br> *Drawing core structure (pyridoxine).* | |
| | This molecule will be a core of Markush. |
| • In the starting interface of MarvinSketch (Figure 3.1), click on the button *single bond* (left panel) and then click on the *drawing area*. | |
| • On the bottom panel of starting interface, select the benzene ring, then click on drawn atom of ethane with any atom of ring (Figure 3.2). | |
| • On the right panel click on *periodic table and more*, then in opened window (Figure 3.3) select nitrogen atom, then close window and click on ortho-atom of benzene ring. | |
| • Proceed similarly until you get the pyridoxine molecule. | |
| *Addition of variable substituents* | *Drawing simple substitution groups and generating structures. This step will be repeated below several times.* |
| • Update Markush core by drawing bond to phenolic oxygen and replace drawn atom by R1 group from *Periodic table/Advanced tab* window. | Specification of variable substituent in core molecule. |
| • On free space draw a few new molecules (here we used ethane, propane, ethanol, and methyl ethyl ether). | *Tips and tricks:* to add R1 group to instead of given atom type "r" + "1" on keyboard while mouse pointer positioned over atom. |
| • Right click on terminal carbon atoms of every molecule and select *R-group attachment*. | Addition of substituents: ethyl-, propyl-, hydroxymethyl-, ethoxymethyl-. |
| • Select all drawn molecules (substituents) by *rectangle selector* (top panel) and click to R1 on *Periodic table/Advanced tab* window (or type "r" + "1" on keyboard). | Result shown on Figure 3.4. |

File Edit View Inser Aton Bonc Structur Calculatio Service Help

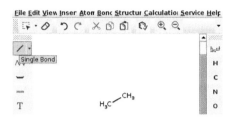

Figure 3.1  Interface of MarvinSketch.

File Edit View Inser Aton Bonc Structur Calculatio Service Help

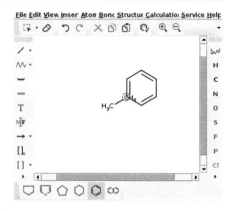

Figure 3.2  Drawing of toluene.

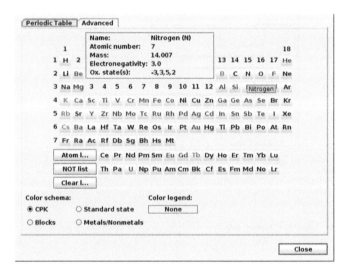

Figure 3.3  "Periodic table and more" window interface.

File Edit View Insert Atom Bond Structure Calculations Services Help

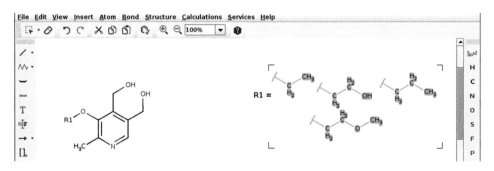

Figure 3.4  Drawing substitution groups.

*(Continued)*

| Instructions | Comments |
|---|---|

*Library size estimation and structures enumeration*

- Unselect all structures. Open Markush enumeration window (**Structure/Markush enumeration**), select "Markush library size" and then click OK.
- Select "Sequential enumeration," click OK in options window. You can save structures using **File/Save All**.

*Library size estimation* is useful to understand how much structures correspond to a given Markush structure.

"**Markush library size**" option returns the estimated number of possible structures. Returned library size is equal to number of combinations of added substituents (4 in our case).

"**Sequential enumeration**" option returns all possible structures for a given Markush structure. Result shown in Figure 3.5.

"**Random enumeration**" option returns given number of randomly generated structures.

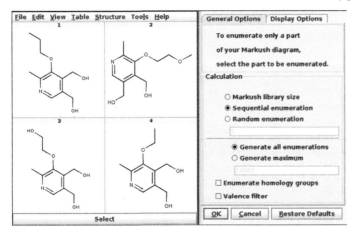

Figure 3.5 Generate structures.

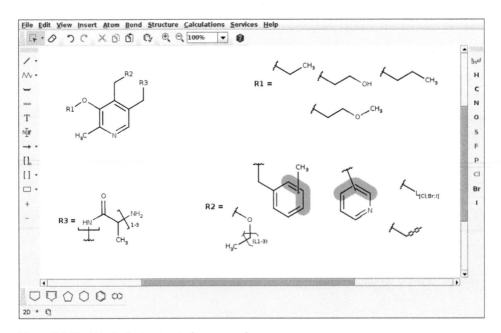

Figure 3.6 Final Markush structure to be prepared.

| Instructions | Comments |
|---|---|

**More complicated types of Markush structures**

Final result of this part of work is shown in Figure 3.6.

*Group position variation addition*

*Group position variation* useful for variation of ortho-, para-, meta- groups of rings or alkyl isomers, and so on.

- Draw toluene molecule and add R-group attachment point on aliphatic carbon (using **R-group attachment** tool) to form benzyl group. Then select (use rectangle or lasso selection tool from top panel) 2,3,4 atoms of benzene ring, right click on any selected atom or bond and select menu **Add/Position variation bond.**

Result:

- Draw pyridine. Select 2,3,4 atoms of pyridine ring and add position variation bond using menu **Add/Position variation bond**, then right click on atom of appeared bond near the ring and select **R-group attachment**, bond with attachment point appear, then remove unnecessary methyl group.

Result:

*Atom type variation addition*

*Atom type variation* is needed when several atom types can exist in a certain position.

- Draw methyl R-group (draw methane and add attachment point), then click **Periodic table** (**Periodic table** tab), choose **Atom list**, select elements by clicking on periodic table (select F, Cl, Br, I), then click on carbon atom of methyl R-group.

Result:

*Chain length variation addition*

*Chain length variation* is simple way for variation of homological groups (e.g., ethoxy-, propoxy-, butoxy-)

- Draw ethoxy R-group, then right click on methylene atom and select **link node/L1-3**

*Note: link node on atom with r-group attachment doesn't work.*

Result:

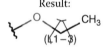

*Bond variation addition*

*Note: you can select also **single or aromatic**, **double or aromatic** and **any** bond types.*

- Draw ethyl R-group, then select **single or double bond** in bond menu on the left panel and click on ethyl bond.

Result:

*Oligomers addition*

*Note: combinations of different variation types are possible.*

- Draw alanine amide and add R-group attachment to amide nitrogen.
- Select repeated unit (amide group and α- carbon atom with R-group), then right click on any selected atom or bond and choose **Group**.
- In **Group** window select type **Repeating unit with repetition ranges**, set **Repetition ranges** to 1-3 and click OK.

*Tips and tricks:* for generating peptide group with various amino acids (as shown on Figure 3.7):

- draw in free space all needed substructures and group it as R4,
- replace methyl group in alanine with R4.

*(Continued)*

| Instructions | Comments |
|---|---|

### *Final Markush drawing and enumeration*

- Select peptide group and assign them to R3 group (click **Periodic Table / Advanced / R3**).

  *R-group assigning*

- Select all unselected groups and assign them to R2 group (click **Periodic Table / Advanced / R2**).

- Replace alcohol oxygens with R2 and R3 group.

  Result is shown in Figure 3.6.

- Count number of molecules corresponding to Markush structure (**Structure / Markush enumeration / Markush library size**).

  180 structures will correspond to the drawn Markush structure

- Add one more structure to R1. Estimate number of structures.

- Save Markush structure (**File/Save as,** check that *mrv* fomat is selected). This file will be used for structure enumeration in command line, see below).

  225 will correspond to Markush structure. Even such a small change in Markush structure significantly increases number of possible structures. One can play around this Markush structure and see that virtually small complication of its structure can lead to drastic increase of library size.

- Generate molecules (Figure 3.5).

- While "Markush Enumeration" window is opened, one can save set of molecules (if needed): **File/Save All** (to save all molecules), **File/Save Selection** (to save selected molecules).

Graphical user interface is not performant enough for generating large datasets. Command line interface could be more useful for generation of large molecular datasets.

### *Command line interface*

Open cmd.exe on MS Windows or bash/any other ol linux/mac, go to folder ~ **ChemAxon/ JChem/bin**

**cxcalc** generates dot-separated set of molecules in SMILES format by default.

Option "**-f sdf**" changes output file format to MDL SDF. But option "**-o path/to/result.file**" in this case may not work (it's probably a bug of used version) and resulted file is written to the terminal. ">"redirects the stream to result.sdf file.

- copy *mrv* file with created Markush structure in this folder

- to get enumerated structures in SMILES format type

  *cxcalc -o result.smi markush.mrv enumerations*

- to get structures in SDF format type

  *cxcalc enumerations -f sdf markush. mrv > result.sdf*

*Note:* check PATH environment if cxcalc not found and add path to JChem/bin directory or replace **cxcalc** word in command line by full path to corresponding cxcalc script in JChem/bin directory.

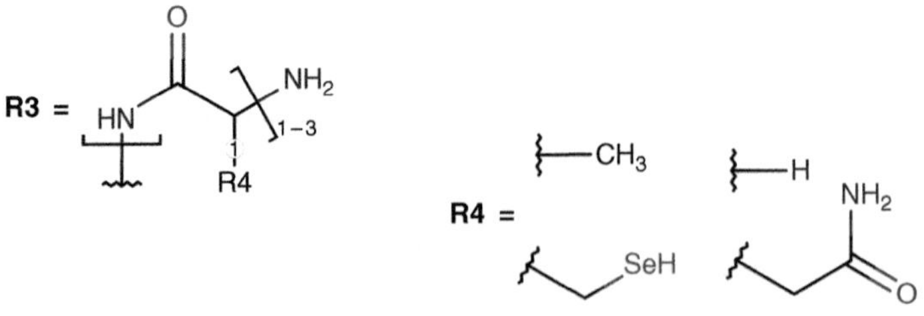

Figure 3.7 Aminoacid variation in peptide group.

## Conclusion

Markush enumeration is a simple and effective way for generating datasets for virtual screening or patent filing and search. However, the degrees of freedom supported by the approach (nature & position of substituents, repetition, stereochemistry) must be limited to the strictly relevant minimal number of options, or else the number of possible compounds matching the markush will become intractably large, due to combinatorial explosion.

## References

1 Barnard, J.M. Representation of Molecular Structures – Overview./Handbook of Chemoinformatics. ed. by J. Gasteiger//Wiley-VCH. – 2003.
2 Pat. US1506316 A. Pyrazolone dye and process of making the same/E.A. Markush//1924.

4

# Processing of SMILES, InChI, and Hashed Fingerprints

*João Montargil Aires de Sousa*

*Goal*: Illustrate the generation of SMILES strings and InChI identifiers, and demonstrate their use as keys for database access. Illustrate the generation of chemical hashed fingerprints and the influence of fingerprint parameters on the discrimination capability.

*Software*: ChemAxon Jchem package version 6.3.0, 2014, ChemAxon (http://www.chemaxon.com)

*Data*: A dataset of 10,000 molecular structures were retrieved from the ZINC database (http://zinc.docking.org, 2015) and adapted. The file is named 10000.sdf, and is in folder SMILES. A file 10stru.smi is also available in the same folder.

## Theoretical Background

The representation of molecular structures in electronic formats is a requirement for the storage of chemical information, visualization of molecular models, administration and access to databases, establishment of structure-property relationships, or the calculation of observable properties. For database systems, molecular representations must be used that are non-ambiguous and unique. In a non-ambiguous format, one representation corresponds to only one molecular structure. Uniqueness requires that one molecule has only one possible representation.

Chemical line notations represent molecular structural formulas by sequences of ASCII characters. Line notations can be very compact, can be used as keys for database access, and provide an easy way to communicate molecular structures, for example, embedded in a text message or entered in a text field of a web site form. This tutorial covers two useful linear notations, which can be applied as unique non-ambiguous molecular representations: SMILES strings and InChI identifiers.

Differently from linear notations, hashed fingerprints aim at encoding molecular features so that similarity between pairs of molecules can be rapidly assessed. They are fixed-length sequences of binary values (*bits*) that encode the presence of sequences of atoms and bonds in a structure. They are unique (one molecule always yield one and only one fingerprint), but they are ambiguous (two different molecules may yield the

*Tutorials in Chemoinformatics*, First Edition. Edited by Alexandre Varnek.
© 2017 John Wiley & Sons Ltd. Published 2017 by John Wiley & Sons Ltd.
Companion website: www.wiley.com/go/varnek/chemoinformatics

same fingerprint). Hashed fingerprints are generated so that the similarity between two molecules is related to the number of bits with the same value in the two fingerprints. They are mainly used for the analysis of similarity and diversity within large databases of molecules, and for similarity queries. Although hashed fingerprints cannot confirm the presence of a substructure in a molecule, they can confirm the absence of a substructure.

## Algorithms

The **SMILES notation** (*Simplified Molecular Input Line Entry System*)[1] provides a compact intuitive representation of molecules by strings of ASCII characters. For example, n-propanol can be represented by "CCCO." One can readily suspect that atoms are represented by their symbols and bonded atoms follow each other in the sequence. The main rules to generate a SMILES string are briefly presented in a simplified way:

1) Atoms are represented by their atomic symbols enclosed in square brackets, []. Elements in the "organic subset" (B, C, N, O, P, S, F, Cl, Br, and I) may be written without brackets.
2) Adjacent atoms are assumed to be connected to each other. Single, double, triple, and aromatic bonds are represented by the symbols –, =, #, and :, respectively (single and aromatic bonds may be omitted, as well as implicit hydrogen atoms). For example, $C = CC$ represents propene.
3) Aromaticity can be specified with lower-case atomic symbols.
4) Branches are specified by enclosing them in parentheses. For example, acetone can be represented as CC(=O)C.
5) Cyclic structures are represented by breaking one bond in each ring. The bonds are numbered in any order, designating ring opening (or ring closure) bonds by a digit immediately following the atomic symbol at each ring closure. For example, the string c1ccccc1 represents benzene.
6) Configuration around double bonds is specified by the characters / and \.
7) The configuration around tetrahedral centers is specified by @ or @@ written as an atomic property following the atomic symbol of the chiral atom inside the square brackets, and it is based on the order in which neighbors occur in the SMILES string. Looking from the first neighbor of the chiral atom to the chiral atom, the symbol "@" indicates that the three other neighbors appear anticlockwise in the order that they are listed. "@@" indicates that the neighbors are written clockwise.

The **InChI** identifier [2] was developed by IUPAC to be a representation standard yielding a unique textual label for any chemical substance. InChIs comprise different layers and sub-layers of information separated by slashes (/). Each InChI strings starts with the InChI version number followed by the main layer. This main layer contains sub-layers for chemical formula, atom connections, and hydrogen atoms. The identity of each atom and its covalently bonded partners provide all of the information necessary for the main layer. The main layer may be followed by additional layers, for example, for charge, isotopic composition, tautomerism, and stereochemistry.[3]

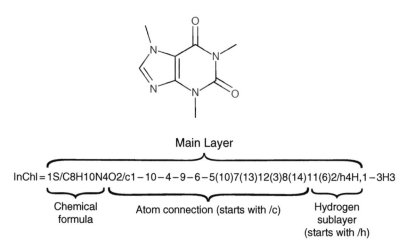

Main Layer

InChI = 1S/C8H10N4O2/c1−10−4−9−6−5(10)7(13)12(3)8(14)11(6)2/h4H,1−3H3

Chemical formula     Atom connection (starts with /c)     Hydrogen sublayer (starts with /h)

Figure 4.1 The InChI identifier of caffeine.

All layers and sub-layers (except for the chemical formula sub-layer of the main layer) start with "/?" where ? is a lower-case letter to indicate the type of information held in that layer.[3]

The generation of the identifier starts with a normalization step (in which information is selected and separated into layers), followed by canonicalization (in which atoms are labeled not depending on how the structure was entered) and serialization (in which the string is generated based on the canonic labels).[3]

The InChI for caffeine is shown in the Figure 4.1.

A chemical **hashed fingerprint**[1] is a bit string (a sequence of $0$ and $1$ digits) encoding the presence of sub-structures. Each sequence of atoms and bonds (pattern) in a molecule activates a number of bits. The number of bits activated by each structure ($b$) is a parameter of the fingerprint. An algorithm determines which bits are activated by which sequences of atoms/bonds. The maximum size ($n$) of sequences considered is another parameter of the fingerprint, as well as the size of the fingerprint ($f$, the total number of bits). The generation of a fingerprint involves the following steps:

- Initialization of all bits in the fingerprint with the 0 value.
- Enumeration of all patterns representing each group of atoms and bonds connected by paths up to $n$ bonds long ($n$ is specified by the user). For example, for alanine (SMILES: CC(N)C(O) = O) the 0-bond patterns are C, N, and O, while the 1-bond patterns are CC, CN, CO, and C = O.
- Each pattern serves as a seed to a pseudo-random number generator (it is "hashed"), the output of which are the position of a set of bits in the fingerprint. These bits then take the value 1.

The same pattern activates the same set of bits in any molecule, and the algorithm can handle any pattern. There is no pre-defined set of substructures or patterns to be encoded. But there may be collisions, that is, the same bit may be activated by different patterns, and fingerprints cannot be reversed, that is, the structure cannot be induced from the fingerprint. Hydrogen atoms are omitted and stereochemistry is generally not considered.

Before generating fingerprints for a set of molecules, the user must define the size of the fingerprint, the maximum size of patterns to be enumerated, and the number of bits activated by each patter. These parameters influence the required memory to store the fingerprints and their capability of discriminating between different molecules.

Measures of similarity between molecules can be defined in terms of similarities between fingerprints. The Euclidean and Tanimoto coefficients are commonly used for this purpose.

## Step-by-Step Instructions

1) Write, without the assistance of any chemistry software, SMILES strings for the following molecules:

a)

b)

c)

d)

2) Copy the SMILES strings of the previous question to the MarvinSketch program and check if they are correct. (Hint: you may paste a SMILES string directly onto the working area of MarvinSketch).

3) Draw a molecular structure in MarvinSketch and obtain the SMILES string (Edit → Copy As Smiles, and paste into some text editor; or Edit → Source, View → SMILES).

4) Draw molecular structures for the following SMILES representations:
   a) CCCCBr
   b) CC(CO)CCCN
   c) C#CCc1ccccc1
   d) CCC(C)(F)OC(=S)OC

5) Copy the SMILES strings of the last question to MarvinSketch and check if you drew them correctly.

6) Now you will analyze a data set of 10,000 molecules, provided in the standard MDL SDFile format, to verify a) if paracetamol is included, and b) if there are duplicated molecules. You will make use of SMILES strings and InChI identifiers. Open the ChemAxon Standardizer program, browse and select the 10000.sdf file.

7) Configure the task by selecting "Remove Explicit Hydrogens" and "Mesomerize," specify an output file named 10000.smiles, and Run.

8) Draw the structure of paracetamol in MarvinSketch, save as MDL SDfile, and repeat exactly the procedures 6 and 7. Open the output file with some text editor and copy the SMILES string of paracetamol.

9) Open the 10000.smiles file of step 7 with a text editor and search for the SMILES string of paracetamol. You should find it in line 1234.

10) Copy the 10,000 SMILES strings into a worksheet. You will verify whether there are duplicated structures. Add a column with an ID number for each compound (e.g., from 1 to 10,000). You can use the VLOOKUP function of the worksheet to search for the first SMILES string in the array of 10,000 strings, and then drag the formula. Alternatively, you can sort the lines by the SMILES string (so that if there are duplicated strings they will be placed in contiguous cells), implement a formula to check if one cell is equal to the cell below, and drag the formula.

11) You should identify two pairs of duplicated structures: in lines 2998 and 3998, and in lines 8156 and 8651.

12) Repeat step 7, but without selecting "Mesomerize" in Standardizer, and repeat step 10 with the new version of the 10,000 SMILES strings. Now you will not detect the duplicated structures. In fact, string 2998 would have the aromatic system specified with lower-case characters, while string 3998 would use alternating single and double bonds in the aromatic system. In the second pair, one structure would have a double $S = O$ bond, while there is separation of charges in the other.

13) Repeat step 7, but without selecting "Mesomerize" in Standardizer, and specify an output file named 10000.inchi. In this case you will generate InChI identifiers for all the molecules. Search for duplicated structures in a worksheet (but now using the InChI identifiers). You will find again the two pairs of duplicated structures. The InChI algorithm normalized the structures so that the duplicates could be found.

14) Download file 10stru.smi. It contains 10 different structures in the SMILES format. You may visualize them, for example, with ChemAxon MarvinView (Figure 4.2).

15) The ChemAxon GenerateMD program can calculate binary hashed fingerprints. Open the command line terminal to use that program and go to the directory where the 10stru.smi file was saved.

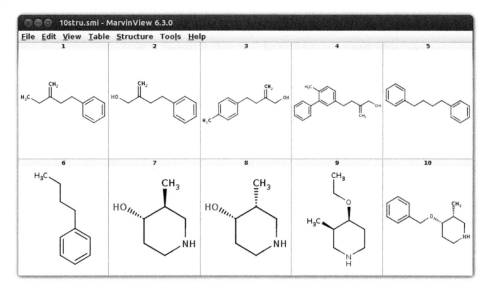

**Figure 4.2** Visualization of the 10 molecular structures in file 10stru.smi.

16) Enter the command

> *generatemd c 10stru.smi -k cf -f 64 -n 3 -o 10stru_a.fp -2*

where "f" is the option for the fingerprint size (in this example 64 bits), "n" is the option for the maximum size of patterns (in this example 3), the output file is specified after the "o" option, and "-2" indicates a binary format (0/1 string).

17) Open the output file (10stru_a.fp) with a text editor or a worksheet and check if all the 10 fingerprints are different. Three pairs of molecules obtained the same fingerprint: 2/3, 5/6, and 7/8. Molecules 7 and 8 are stereoisomers, thus they will never be distinguished by fingerprints because stereochemistry is not considered. For the other two pairs, all the sequences with size 3 occurring in one molecule also occur in the other, so the maximum size of the patterns is not enough to distinguish between them.

18) Generate new fingerprints with maximum size of patterns = 5 bonds for higher discrimination ability. Enter the command:

> *generatemd c 10stru.smi -k cf -f 64 -n 5 -o 10stru_b.fp -2*

Inspect the new output file and observe that molecules 2 and 3 got different fingerprints. But molecules 5 and 6 still have the same fingerprint, although there are patterns of size 5 occurring in molecule 5 but not on molecule 6. This is due to collisions, and can be solved by increasing the size of the fingerprint.

19) Specify fingerprint size = 256 to increase the discriminating power of the fingerprints:

> *generatemd c 10stru.smi -k cf -f 256 -n 5 -o 10stru_c.fp -2*

Open the output file and see that the new parameters enabled the discrimination of all molecules except the pair of stereoisomers.

20) Now try to increase the number of bits activated by each pattern (option "b") from 2 (default) to 5, instead of increasing the fingerprint size. You should use the following command

> *generatemd c 10stru.smi -k cf -f 64 -n 6 -b 5 -o 10stru_d.fp -2*

You can see that increasing the number of activated bits ("1") keeping fingerprint size the same, reduced the discrimination capability—the three pairs of molecules (2/3, 5/6, and 7/8) could not be distinguished again.

## Conclusion

The tutorial demonstrates how to calculate SMILES strings and InChI identifiers from SDFiles, and how to use them to store representations of molecular structures, search for a molecule in a collection, and detect duplicated molecules. The importance of normalization and canonicalization for unique representations is highlighted. The generation of chemical hashed fingerprints for the assessment of molecular similarity is explained, and the impact of calculation parameters on the discrimination ability of fingerprints is illustrated.

## References

1 Daylight Theory Manual v. 4.9, Daylight Chemical Information Systems, Inc., http://www.daylight.com/dayhtml/doc/theory
2 The IUPAC International Chemical Identifier (InChI), http://www.iupac.org/inchi
3 InChI Trust, http://www.inchi-trust.org

Part 2

Library Design

5

# Design of Diverse and Focused Compound Libraries

*Antonio de la Vega de Leon, Eugen Lounkine, Martin Vogt, and Jürgen Bajorath*

*Goal*: Create a standardized compound library from publicly available data sources and derive diverse subsets and focused libraries from it.

*Software/Code*: Python using packages NumPy, RDKit, scikit-learn, and matplotlib, along with custom-made scripts to illustrate each section of the tutorial. Further information about the installation of the packages and the scripts provided can be found in the Introduction section.

*Data*: Will be created at the beginning of the tutorial.

## Introduction

This chapter focuses on the creation and analysis of libraries of chemical compounds from public data sources. Preprocessing steps necessary for the consistent handling of molecule data, especially if data from different sources are selected, are detailed along with the calculation of molecular descriptors and fingerprints.

It is shown how chemical libraries can be generated using different strategies. Two complementary approaches are detailed. First, strategies for selecting subsets from large compound libraries using partitioning and diverse subset selection are shown. Second, the construction of focused compound libraries based on a set of chemical fragments, which are combined randomly to generate novel compounds, is described. These fragments can be obtained through retrosynthetic analysis of an original compound data set.

The generation of compound libraries can be seen as a strategy to sample chemical space in a certain way. Diverse libraries aim to cover a larger section of chemical space evenly while focused libraries seek to provide good coverage of a narrowly defined region of chemical space. Given limited resources for experimental evaluation of compounds goals are to find a compromise between size of the generated library and coverage of the underlying chemical space that is to be explored.

*Tutorials in Chemoinformatics*, First Edition. Edited by Alexandre Varnek.
© 2017 John Wiley & Sons Ltd. Published 2017 by John Wiley & Sons Ltd.
Companion website: www.wiley.com/go/varnek/chemoinformatics

## Data Acquisition

There are several publicly available compound databases that contain potency annotations. Two of the biggest are ChEMBL[1] and PubChem.[2] ChEMBL is a repository of bioactive compounds obtained from the literature. PubChem meanwhile has a large number of assay results that provide both active and inactive compound records. A set of compounds tested against CDC-like kinase 4 was obtained from PubChem (AID1771). This data record contains 356 active and 747 inactive molecules. In order to retrieve their records:

1) Go to PubChem (https://pubchem.ncbi.nlm.nih.gov/).
2) Under the BioAssay tab enter "AID1771" to search for the desired assay.
3) If the AID given was valid, PubChem will show different information related to the assay such as target, protocol, and comments given by the authors.
4) To obtain the active molecules, either select the "Structures SDF" download option from the "Tested substances (ACTIVE)" item under the "Download" menu or select the "Active" link for "Tested Substances" to first see a list of all active substances. Here, use the button "Structure Download" to take you to the download page. Save the active molecules as SDF with compression type "None". (Note, that designs and layouts of web pages and interfaces are prone to change, and the steps required for downloading might change.)
5) Repeat the instructions to obtain the inactive molecules by selecting the "inactive" link under "Tested Substances".

## Implementation

For the implementation of algorithms and functions in this tutorial Python in combination with the chemoinformatics toolkit RDKit[3] is used. The toolkit RDKit provides the capabilities to handle and manipulate molecular structures in Python. A comprehensive introduction, as well as installation instructions, can be found in the online documentation from the RDKit homepage (http://rdkit.org/docs/index.html). Additionally NumPy,[4] scikit-learn,[5] and matplotlib[6] will be used for numerical analysis and plotting. These modules can be installed with the pip command from a command line interface:

```
pip install numpy
pip install matplotlib
pip install scikit-learn
```

To create the examples shown throughout this chapter Python 2.7.6 was used in combination with RDKit (release 2015.03.1), NumPy 1.9.1, scikit-learn 0.15.2, and maptlotlib 1.4.3. Several Python scripts are provided for download in order to perform many of the steps explained throughout this tutorial and form the basis of the snippets of code detailed in the next sections. These files include:

- `libraryCreation.py`: reads active and inactive compound data sets, standardizes their structures and outputs a combined file.
- `generateDescriptors.py`: reads a compound data set, generates a set of 14 molecular descriptors and saves them in a tabulator-separated file.
- `showCorrMatrix.py`: reads a descriptor file and shows a correlation matrix between all descriptors.
- `generateDiverseSubset.py`: reads a descriptor file, performs a principal component analysis (PCA) and partitioning using the descriptor values and generates a diverse subset of compounds for each partition cluster.

- `generateScaffoldLibrary.py`: reads a compound data set, fragments all molecules into cores and R-groups, and creates molecules based on random core and R-group combinations.
- `generateBRICSlibrary.py`: reads a compound data set, fragments all molecules based on retrosynthetic rules, and creates molecules based on random combinations of molecular fragments.

## Compound Library Creation

**Script** `libraryCreation.py`
The SDF files for active and inactive compounds can be combined to obtain a single compound library to work on. First each SDF file is read and a list of molecule objects created. RDKit has a simple function called `SDMolSupplier` that takes a file name and allows access to each molecule in the SDF file using a typical Python for-loop. If the molecule could not be created (if the structure was nonsensical for example), it will return a `None` object rather than a `Mol` object. Each compound is marked with a special tag `ActiveTag` to separate actives from inactives in the final library. This can be done by setting specific properties of the molecule. Each Mol object allows the access and modification of its properties through the `GetProp` and `SetProp` functions.

```
from rdkit import Chem
suppl = Chem.SDMolSupplier(activeFile)
actives = []
for mol in suppl:
    if mol is None:
        continue
    mol.SetProp('ActiveTag', "active")
    actives.append(mol)
```

The same procedure is repeated with the inactive file. Afterward, two lists have been created: `actives` and `inactives` containing molecule objects that will be written to the same SD file. A molecule writer can be created with `SDWriter`. By default, all molecular properties will be written but this behavior can be changed with the `SetProps` function, which takes a list of property tags to write to the SD file. Both original SD files contain many fields from the PubChem database that will not be useful for our purposes, therefore in the combined library only the structure, the compound ID (CID), and a label designating a molecule as either "active" or "inactive" will be retained. SD files are able to store coordinate information for the atoms of a molecule. These can either be three-dimensional coordinates of a conformation or two-dimensional coordinates describing the layout, which some programs will use to depict the structure. Therefore, before a molecule is written two-dimensional layout coordinates can be generated using the function `Compute2DCoords`, a feature that is especially useful when importing molecules from SMILES[7] descriptions, which contain no layout information. If the process is successful, a file with all molecules and only two fields will be created.

```
from rdkit.Chem import AllChem
writer = Chem.SDWriter(bothFile)
writer.SetProps(['CID','ActiveTag'])
for mol in actives:
    AllChem.Compute2DCoords(mol)
```

```
          writer.write(mol)
for mol in inactives:
    AllChem.Compute2DCoords(mol)
    writer.write(mol)
writer.close()
```

Because each compound data source might use different protocols when creating the structure of a molecule, it is recommended to standardize molecules (also known as "washing") after the library is created. This is especially important when creating a mixed library from different data sources or when libraries from different sources are compared. The standardization protocol adopted here consists of the following steps:

1) removal of some metal atoms/ions
2) removal of small fragments
3) adjustment of formal charges
4) rebalancing hydrogen counts

In some cases, metal ions and their organic counterions are represented as covalent bonds in a molecule's description, instead of being represented as disconnected charged molecules. Because of this, they would not be removed when salts and other disconnected small components are deleted from the structure description as part of standardization. A function can be defined to delete these incorrect covalent bonds. These bonds are defined as a single bond between a metal and another atom of organic compounds where the metal atom has a bond order of one and both atoms have no formal charge. Each type of atom is defined based on its atomic number.

```
metals = [3, 11, 19, 37, 55]
organics = [6, 7, 8, 9, 15, 16, 17, 34, 35, 53]
```

RDKit employs the concept of immutablity, which means that the structure of a molecule object of class Mol cannot be changed after it has been created. To modify a molecular structure, a RWMol is needed and can be created from a previously existing molecule object.[3] RWMol has several functions to create and delete atoms and bonds, such as, RemoveBond or AddAtom. All bonds in the molecule are checked to see if their atoms are contained in the lists of metals and organics. If the necessary conditions are fulfilled, the bond is stored in a list of bonds to be deleted.

```
mol = Chem.RWMol(mol)
bondsToDel = []
for bond in mol.GetBonds():
    if bond.GetBondType() ! = Chem.BondType.SINGLE:
        continue
    a1 = bond.GetBeginAtom()
    a2 = bond.GetEndAtom()
    if a1.GetFormalCharge() ! = 0 or a2.GetFormalCharge() ! = 0:
        continue
    if a1.GetAtomicNum() in metals:
        if a2.GetAtomicNum() not in organics:
            continue
        if a1.GetDegree() ! = 1:
            continue
        bondsToDel.append(bond)
```

```
    elif a2.GetAtomicNum() in metals:
        if a1.GetAtomicNum() not in organics:
            continue
        if a2.GetDegree()!=1:
            continue
        bondsToDel.append(bond)
```

When all bonds have been checked, each bond that was saved must be deleted. The function RemoveBond removes the bond indicated by the indices of its two atoms. Once the molecule has been modified, it can be converted back into a typical Mol object to prevent further unintentional modification of the structure.

```
for bond in bondsToDel:
    mol.RemoveBond(bond.GetBeginAtom().GetIdx(), bond.GetEnd
Atom().GetIdx())
return Chem.Mol(mol)
```

Once metal atoms have been disconnected, it is standard procedure to remove salts and other small disconnected components retaining only the largest fragment. In RDKit this can be achieved using the SaltRemover function. This function reads a pre-specified list of salts and removes any component that matches any of the given structures. By default it will keep the largest component, even if this component was also found in the salt list.

```
from rdkit.Chem import SaltRemover
remover=SaltRemover.SaltRemover()
mol=remover.StripMol(mol)
```

The next step is to remove charges resulting from (de-)protonation and the removal of salts in the previous steps.[4] This step can be done using SMARTS[8] patterns. SMARTS are an extension of the SMILES notation to define molecular patterns. A set of commonly charged groups (such as amines or thiols) is described through SMARTS patterns and substituted with corresponding uncharged groups. Some examples are shown below, obtained from the RDKit support information.

```
('[n+;H]','n'),                # Imidazoles
('[N+;!H0]','N'),              # Amines
('[S-;X1]','S'),               # Thiols
('[$([N-;X2]S(=O)=O)]','N'),   # Sulfonamides
```

For each group, the first SMARTS defines the charged atom of that group. The second element is the uncharged atom that should be replaced in the group. For each molecule, a substructure search will be performed for each pattern using the HasSubstructMatch function from the molecule object. If the pattern is found, it is replaced by the second element with the ReplaceSubstructs function.

```
for reactant, product in reactions:
    while mol.HasSubstructMatch(reactant):
        rms=AllChem.ReplaceSubstructs(mol, reactant, product)
        mol=rms[0]
```

Once charges have been adjusted hydrogens might need to be added or removed to correctly reflect the formal charge and valence requirements of the individual atoms. With the function SanitizeMol valences are checked, and if not fulfilled hydrogens are adjusted. Stereochemistry, aromaticity, and hybridization can also be independently assigned with `AssignStereochemistry`, `SetAromaticity`, and `SetHybridization` respectively. Finally, partial charges are computed using the `ComputeGasteigerCharges` function. These steps complete the normalization procedure and ensure that descriptor and fingerprint calculations will be consistent for compounds retrieved from distant sources.

```
Chem.SanitizeMol(mol)
Chem.AssignStereochemistry(mol)
Chem.SetAromaticity(mol)
Chem.SetHybridization(mol)
AllChem.ComputeGasteigerCharges(mol)
```

## Compound Library Analysis

### Descriptor Calculation

**Script:** `generateDescriptors.py`

Based on the compound structure, various descriptors can be calculated. Descriptors are mathematical formulae and algorithms that numerically quantify certain characteristics of a molecule. Descriptors can be categorized as either 1D, 2D, or 3D. Simple descriptors like the molecular weight or atom count descriptors do not need complete structural information and can be calculated based on the molecular formula alone and are categorized 1D. Most descriptors require knowledge of the structural graph and are categorized as 2D. These include simple topological descriptors like ring or bond counts but also physicochemical properties like molar refractivity or the octanol/water partition coefficient which are estimated using models based on the graph structure. 3D descriptors require information of the conformation of a molecule that either has been experimentally determined or been calculated. The most frequently used descriptors are 1D and 2D, as they do not require conformational information.

RDKit can calculate many descriptors (a complete list is given in the RDKit Documentation). Each descriptor is associated with its own function inside the `Descriptors` module and takes a molecule object as a parameter and returns the descriptor value. In the script `generateDescriptors.py` a list of exemplary descriptors to calculate is saved in `DescriptorLabels` and in `DescriptorDict` the function that calculates each descriptor is associated with the descriptor label:

```
from rdkit.Chem import Descriptors
DescriptorLabels = ["MW", "LogP", # etc.
                   ]
DescriptorDict = {"MW": Descriptors.MolWt,
                  "LogP": Descriptors.MolLogP,
                  # etc.
                  }
```

The list of descriptor labels is used to calculate the descriptors for a set of molecules:

```
import numpy as np
data = []
for mol in suppl:
cid = mol.GetProp("CID")
    lab = mol.GetProp("ActiveTag")
    tmp = [cid, lab]
    for desc in DescriptorLabels:
        func = DescriptorDict[desc]
        tmp.append(func(mol))
    data.append(tmp)
data = np.asarray(data)
```

The above snippet calculates the set of descriptors and creates a matrix of descriptor values; more specifically, it creates a "NumPy" (numeric Python) `array`. This matrix also contains the compound ID and the activity tag. A NumPy `array` can be saved directly to a text file through the `savetxt` function. The format in which the values are saved is given, along with the name of the output file, the separator between columns and the header to add at the beginning of the file.

```
header = ["CID", "ActiveTag"] + DescriptorLabels
outputFile = sys.argv[2]
np.savetxt(outputFile, data, fmt = "%s", comments = "",
   header = "\t".join(header), delimiter = "\t")
```

The last snippet creates a file in which the values of the descriptors are linked to the molecule ID and its activity label. Any spreadsheet program or analysis toolkit can now be used to analyze the values. In the following, Python will be used for consistency.

### Normalization of Descriptor Values

Descriptors can have large differences between their numerical values. However, when the descriptors are used for chemical space generation, the relative positions of molecules are often more interesting than the original values. When the values of one descriptor are much greater than the others, calculations such as distances will be heavily biased toward descriptors with large numerical values. In order to provide a balanced representation of chemical space, the original values are scaled. One way of scaling is to calculate "z-scores," where each descriptor value is expressed as a unitless value giving the distance from the mean as number of standard deviations:

$$z = \frac{x - \mu}{\sigma}$$

NumPy can be used for the standardization. Each `array` object has several useful functions, like `mean` and `std` that directly calculate the mean and standard deviation, respectively, of the values of the array. NumPy can also perform element-wise mathematical operations over arrays without having to use explicit loop constructs.[1] In this

---

1 These concepts resemble concepts from Matlab or R and allow the efficient manipulation of large amounts of numerical data in interpreted languages.

manner, the expression `(vals-vals.mean())/vals.std()` gets every element in `vals` and obtains the z-score without having to explicitly loop through each value. Because of the way z-scores are calculated, all z-scores have a mean of 0 and a variance of 1 transforming all different descriptors to similar numerical ranges.

```
n = data.shape[1]
for i in range(2,n):
    vals = data[:,i].astype("float")
    m = vals.mean()
    s = vals.std()
    data[:,i] = (vals-m)/s
outputFile = sys.argv[3]
np.savetxt(outputFile, data, fmt = "%s", comments = "",
    header = "\t".join(header), delimiter = "\t")
```

**Script:** `showCorrMatrix.py`, `showScatterPlot.py`

The first control calculation that is usually done is to compute the pairwise correlations between descriptors. If two descriptors are correlated, they provided similar information. Therefore the information they carry is redundant and will bias the calculations performed in the chemical space by overemphasizing that information. To obtain an overall view of the correlation between the calculated descriptors a correlation matrix is calculated. A simple way to do this is to take all pairs of descriptor values and calculate the correlation coefficient with `corrcoef`. This function returns the correlation matrix between multiple observations.

The correlation matrix can be plotted using the function `imshow` in Matplotlib. First the figure must be defined with `figure`. The imshow function takes a matrix of numbers and assigns each number a specific color, using a colormap (cmap). Then the plotting area is divided into the same number of rows and columns as the matrix, and each cell is filled with the corresponding color. To make the coloring scheme clearer, the value of each cell is printed with the function `text` from Matplotlib and the complete color spectrum and the numerical values they correspond to are displayed with the `colorbar` function. Once the figure has been created, it can be displayed with show or saved with `savefig`.

```
import matplotlib.pyplot as plt
fig = plt.figure(figsize = (10,10))
plt.imshow(100*corrM, cmap = bl_wh_bl, vmin = -100,
interpolation = "None")
for i in range(n):
    for j in range(n):
        c = int(100*corrM[i, j])
        if c > 50 or c < -50:
```

```
        plt.text(i, j, c, va='center', ha='center',
color="white")
        else:
            plt.text(i, j, c, va='center', ha='center')
plt.colorbar(shrink=0.5, ticks=np.arange(-100,110, 50))
plt.show()
```

Figure 5.1 shows the resulting graphical depiction of the correlation matrix of seven descriptors for the sample data set. The different shading of the cells allows the visual identification of strongly and weakly correlated descriptors.

A more detailed visualization of the distribution of two (or three) descriptor value distributions can be obtained with the aid of 2D (3D) scatterplots. These visualizations offer a very intuitive overview of a limited subset of the original descriptor data. From the correlation matrix, uncorrelated and correlated pairs of descriptors can be selected and scatterplots can be used to visualize the different joint distributions. Scatterplots can be created in matplotlib with the function scatter, which allows the additional specification of the size or the color of each data point (by giving a list of values) individually or for all points (by giving only one value). Furthermore, Matplotlib allows a figure to hold more than one plot, by using the function add_subplots of the Figure object. Below, the parameters 121 and 122 represent the first and second plot of a grid with 1 row and 2 columns.

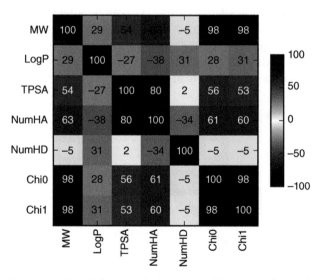

Figure 5.1 **Correlation matrix.** Uncorrelated descriptors have values close to zero. Values near 1 or −1 represent maximal correlation. Dark cells represent highly correlated descriptor pairs, while light cells indicate little or no correlation.

```
colors = []
for i in range(data.shape[0]):
    if data[i,1] == "active":
        colors.append(blue)
    elif data[i,1] == "inactive":
        colors.append(red)
fig = plt.figure(figsize = (11,5))
ax = fig.add_subplot(121)
plt.scatter(mw, fCsp3, color = cols, s = 2)
ax = fig.add_subplot(122)
plt.scatter(mw, chi, color = cols, s = 2)
```

The resulting scatter plots between two different pairs of descriptors are shown in Figure 5.2. The scatter plot of two mostly uncorrelated descriptors is shown on the left while a scatter plot of two highly correlated descriptors is shown on the right. As mentioned before, the value ranges of the descriptors after standardization are very similar although their original values where very different.

Correlation is immediately visible in the right scatterplot. Almost all data points are placed in a diagonal line. In contrast, the data points in the left scatterplot are much better distributed away from the diagonal, occupying most of the space in the plot. In this example, the information from molecular weight and the Chi-0 descriptors is redundant, but not that from molecular weight and the fraction of $sp^3$ carbons.

### Decorrelation and Dimension Reduction

**Script:** generateDiverseSubset.py

We have explained in the previous section why correlated descriptors might create skewed chemical space representations, which limit their utility. Principal component analysis (PCA) aims to create an alternative chemical space representation using a decorrelated basis for representation of the molecules as characterized by their descriptor values. Each principal component is a linear combination of the original descriptors,

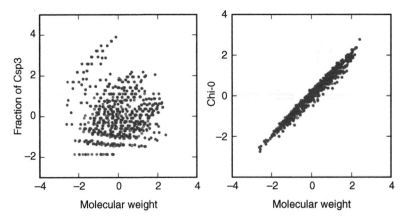

Figure 5.2 **Scatterplots.** Scatterplots for two different pairs of descriptors are shown. Active molecules are colored blue and inactive molecules are colored red.

all principal components are orthogonal to each other, and the coordinate values of the different principal components are uncorrelated to each other.

Graphically, the principal components are a set of orthogonal axes onto which the data points are projected. Principal components are ordered based on the amount of original data variance they model. The more variance they conserve, the better they separate the data points when they are projected onto the component. The first principal component (PC1) is the composite variable that best separates the molecules. Any other linear combination of descriptors will be inferior at this task.

If the original data has correlated descriptors, it is possible that the first two principal components account for a large portion of the original variance. In this case, a scatter-plot based on these two components will give a good visual indication of how the data points are distributed in the chemical space.

In our example, PC1 and PC2 account for 55% and 20% of the original variance, respectively. In total, a projection of the data points into a PC1-PC2 chemical space would keep three quarters of the variance while allowing visual inspection through 2D scatterplots. Principal components can be obtained with scikit-learn, which implements the PCA function. When initialized, the number of desired components can be given. Then the PCA model is fit with the descriptor values using the `fit` function of the `PCA` object. Once fit, arbitrary data can be projected onto the principal components with the `transform` function. In this example, the original data will be transformed by projecting them into a two-dimensional space spanned by the first two principal components, PC1 and PC2, resulting in a scatter plot.

```
from sklearn.decomposition import PCA
pca = PCA(n_components = 2)
pca.fit(descVals)
pcaVals = pca.transform(descVals)
fig = plt.figure(figsize = (3,3))
plt.scatter(pcaVals[:,0], pcaVals[:,1], color = cols, s = 2)
```

A scatter plot with respect to principal components is known as a score plot. Figure 5.3 shows the score plot for the first two principle components. By definition, the two components are uncorrelated, which is also visually apparent, as expected. Interestingly, active and inactive compounds seem to distribute themselves somewhat differently in this chemical space. This might point to differences in descriptor values between active and inactive molecules and allow the identification of regions of chemical space where active compounds are enriched.

## Partitioning and Diverse Subset Calculation

**Script:** generateDiverseSubset.py

### Partitioning

Partitioning takes a chemical space and divides it into subsets or clusters. This is done by dividing each descriptor into several categories (or bins) based on a series of threshold values. These thresholds can be chosen to be either uniformly distributed over the

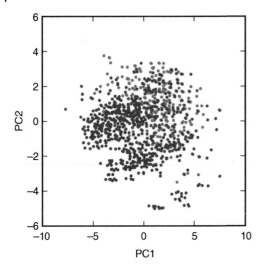

Figure 5.3 **Score plot.** A score plot of the sample data set using the first two principal components is shown. The molecules are colored as described in the legend of Figure 5.2.

descriptor value range, or by creating bins that contain a similar amount of molecules. For each descriptor, every data point is categorized according to the bin it belongs to. Compounds that fall in the same bin for all descriptors belong to the same cluster. Therefore, clusters can be understood as "hypercubes"[2] in high-dimensional chemical space. They can also be obtained and visualized by partitioning the principal component-based space. A part of the partitioning procedure is shown below along with the final result. To represent the partitioning each data point is given an index based on which cluster it belongs to. Then the 2D PC plot is modified by using a different color scheme, a colormap that distributes the different numerical values along the range of possible values of hue.

```
clus = []
for i in range(pcaVals.shape[0]):
if -8 <= pcaVals[i,0] < -4 and 0 <= pcaVals[i,1] < 4:
        clus.append(1)
elif -8 <= pcaVals[i,0] < -4 and -4 <= pcaVals[i,1] < 0:
        clus.append(2)
    #continue for all clusters
fig = plt.figure(figsize = (3,3))
cm = plt.get_cmap('hsv')
plt.scatter(pcaVals[:,0], pcaVals[:,1],
            color = cm(np.array(clus)*25-25), vmin = 1,
vmax = 10, s = 2)
```

Figure 5.4 shows the partitioning of the sample data set based on a two-dimensional principal component space. The gray lines that divide the PC-based space

---

2 Or, more correctly, "hypercuboids."

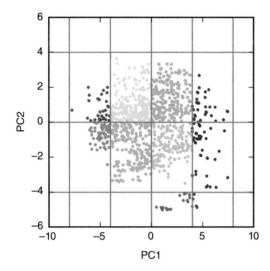

**Figure 5.4  Partitioning.** The partitioning of the two-dimensional space given by the first two principal components into equal-sized bins is shown.

represent the threshold values. In total, molecules are divided into 10 clusters, each colored in a different hue. Unpopulated bins point toward unexplored regions of chemical space.

### Diverse Subset Selection

Diverse subset selection is a useful technique when the original compound library is too big and desired algorithms cannot be applied because of running time constraints or when selecting compounds for experimental testing given restraints on the resources available. In this case, it is beneficial to obtain a subset of the whole library that behaves in a similar fashion to the original library, while reducing its size, that is, a set of compounds that well represents the chemical space as covered by the original set. The distribution of the subset compounds in the chemical space should be, on the whole, similar to the distribution of the whole data set. Most diverse subset algorithms require the computation of all pairwise similarity (or distance) values. On large data sets this can be very time consuming. Partitioning can help to reduce the complexity by computing pairwise similarities only within each cluster. This can provide a significant speed up to computation.

In order to create a diverse subset, a distance metric must be defined so that two data points can be compared and judged to be more or less similar. The function `norm` in the linear algebra module of NumPy is used to calculate Euclidean distances between two numerical arrays. RDKit implements two diversity based subset algorithms, the `MaxMinPicker` based on the MaxMin algorithm and the `HierarchicalClusterPicker` based on a clustering procedure. Once a "picker" is initialized, each cluster created by the partitioning is given to the picker and 10% of the compounds are returned that maximize diversity. The snippet below creates a new file with the diverse molecules that have been selected from each cluster.

```
from rdkit.SimDivFilters.rdSimDivPickers import MaxMinPicker
def distFunc(i,j):
    desc1 = subdata[i,2:].astype("float")
    desc2 = subdata[j,2:].astype("float")
    return np.linalg.norm(desc1-desc2)

picker = MaxMinPicker()
diverse = []
for i in np.unique(clus):
    idx = np.where(clus==i)[0]
    subdata = data[idx,:]
    pickIndices = picker.LazyPick(distFunc,
        subdata.shape[0], subdata.shape[0]/10)
    diverse.extend(subdata[pickIndices,:])
diverse = np.array(diverse)
np.savetxt(outputFile, diverse, fmt = "%s", comments = "",
    header = header, delimiter = "\t")
```

## Combinatorial Libraries

### Combinatorial Enumeration of Compounds

**Script:** generateScaffoldLibrary.py

   Large compound libraries (such as those for virtual screening) need not be simply large collections of previously described molecules. Combinatorial enumeration of compounds can produce a large number of compounds from a restricted set of molecular fragments. In this case the chemical space is not defined by a collection of molecules but by a set of fragments. A simple approach might be to define a set of scaffolds with specific attachment points and a set of R-groups. For example, three scaffolds each containing four attachment points and five R-groups together define $3 \cdot 4 \cdot 5 = 60$ molecules. Thus, eight molecular structures and a set of rules to combine them can generate 60 virtual compounds on the fly.

   It is possible to obtain a set of molecular fragments by fragmenting the molecules of a compound library. All compounds in the library are fragmented and the resulting fragments are used by recombining them. As illustrated in Figure 5.5, one possible way to fragment molecules is to separate them into their scaffolds and sidechains. RDKit

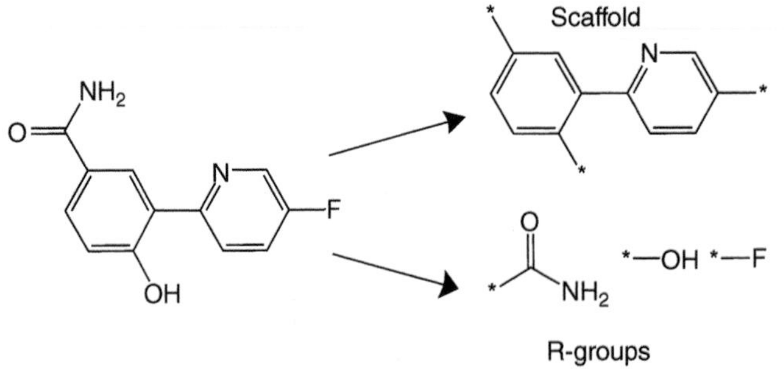

Figure 5.5 **Compound fragmentation.** Shown is the decomposition of a sample molecule into a scaffold and several R-groups.

allows the decomposition of a molecule into its scaffold with the `MurckoScaffold` module. This module contains the `GetScaffoldForMol` function, which returns the Bemis-Murcko scaffold as a molecule object. To get the scaffold and the R-groups with dummy atoms representing their attachment points, the functions `ReplaceCore` and `ReplaceSidechains` can be used, which will replace the core and the side chains with dummy atoms represented by "[*]" in SMILES. In the molecule objects these dummy atoms are represented by atoms with an atomic number of zero.

```
from rdkit.Chem.Scaffolds import MurckoScaffold as BMS
for mol in suppl:
    bms = BMS.GetScaffoldForMol(mol)
    rgroups = Chem.ReplaceCore(mol, bms)
    core = Chem.ReplaceSidechains(mol, bms)
```

Once a set of different scaffolds and R-groups are obtained, they can be combined systematically or in a random fashion to create novel molecules structurally related to the original ones. In RDKit this combination of fragments can be described using SMIRKS strings.[9] SMIRKS are a SMILES-like representation designed to represent chemical reactions. It contains two parts separated by ">>", the first describes the reactants and the second the then products. Atom mapping between reactants and products is achieved by indexing atoms with numbers following ":". The SMIRKS [*:0][#0].[#0][*:1]>>[*:0][*:1] recognizes two dummy atoms in two disconnected components that are connected by a single bond to the rest of the fragment and creates a single bond between the neighbors of the dummy atoms while removing the dummy atoms from the products. SMIRKS can be read by RDKit using the function `ReactionFromSmarts`. Once created, the reaction can be applied to a list of molecule objects with the `RunReactants` function. If more than one possible set of reactants is found, all possible products from all possible reactant combinations will be generated systematically. In the example below, a scaffold is combined with random R-groups until all its dummy atoms have been substituted. This step is repeated for the scaffold of each molecule in the data set. Thus, in this exemplary application, for a given compound library a new combinatorial library is generated with the same number of compounds having the same scaffolds but where the side chains have been randomized.

```
rxn = AllChem.ReactionFromSmarts(
    "[*:0][#0].[#0][*:1]>>[*:0][*:1]")
for coreSmi in cores:
    while "[*]" in coreSmi:
        i = random.randint(0, len(frags)-1)
        ps = rxn.RunReactants(
                [Chem.MolFromSmiles(coreSmi),
                 Chem.MolFromSmiles(frags[i])])
        i = random.randint(0, len(ps)-1)
        coreSmi = Chem.MolToSmiles(ps[i][0])
    mol = Chem.MolFromSmiles(coreSmi)
```

Retrosynthetic Approaches to Library Design

**Script:** `generateBRICSlibrary.py`

Computationally combining molecular fragments can lead to virtual molecules that cannot be synthesized. To address this issue, specific compound fragmentation methods have been developed that produce chemically meaningful fragments and combination rules. One such methodology is BRICS,[10] a set of fragmentation rules based on common chemical reactions. That way, synthesizing a molecule from the fragments should be feasible. RDKit offers an implementation of BRICS-based fragmentation and molecule building in its BRICS module. First the molecule is decomposed into fragments using the BRICSDecompose function. The result is given as a list of fragment SMILES. These fragments are added to a set to remove duplicate structures and transformed into molecule objects with the MolFromSmiles function.

```
from rdkit.Chem import BRICS
frags = set()
for mol in suppl:
    res = BRICS.BRICSDecompose(mol)
    for frag in res:
        frags.add(frag)
fragMols = [Chem.MolFromSmiles(frag) for frag in frags]
```

Once a set of molecule fragments is obtained they can be used to build new molecules with the help of BRICSBuild. This function requires a list of molecule objects and this is the reason for the previous conversion from SMILES to Mol objects. Furthermore, a maximal depth parameter can be set. This represents the maximum number of fragments that are allowed to be strung together. As many fragments will have two or more attachment points, they might be matched together in increasingly large and complex molecules. The larger the depths and the larger the number and the size of virtual product molecules are, the longer the runtime of the algorithm will be. It is important to note that the BRICSBuild function is implemented as a generator so that not all possible combinations are necessarily generated and the process can therefore be stopped during computation with the break keyword. This can be useful if a specific number of molecules is desired, rather than all possible products.

```
writer = Chem.SDWriter(molFile)
for mol in BRICS.BRICSBuild(fragMols, maxDepth=mDepth):
    AllChem.Compute2DCoords(mol)
    writer.write(mol)
```

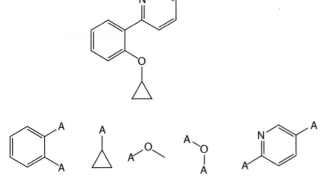

Figure 5.6 **BRICS based fragmentation.** The sample molecule shown can be fragmented into five different fragments using retrosynthetic fragmentation rules.

Figure 5.6 shows an example where a molecule was fragmented to yield five fragments. Fragmentation occurred around the ether groups and between aromatic carbons bonded through an aliphatic exocyclic single bond. When these five fragments were combined with default settings (max depth of 3), 134 molecules were constructed.

## References

1 Gaulton, A.; Bellis, L. J.; Bento, A. P.; Chambers, J.; Davies, M.; Hersey, A.; Light, Y.; McGlinchey, S.; Michalovich, D.; Al-Lazikani, B.; Overington, J. P. ChEMBL: a Large-Scale Bioactivity Database for Drug Discovery. *Nucleic Acids Research* **2012**, *40*, D1100–1107.

2 Wang, Y.; Xiao, J.; Suzek, T. O.; Zhang, J.; Wang, J.; Zhou, Z.; Han, L.; Karapetyan, K.; Dracheva, S.; Shoemaker, B. A.; Bolton, E.; Gindulyte, A.; Bryant, S. H. PubChem's BioAssay Database. *Nucleic acids research* **2012**, *40*, D400–D412.

3 RDKit: Open-source Cheminformatics., http://www.rdkit.org.

4 Van der Walt, S.; Colbert, S. C.; Varoquaux, G. The NumPy Array: A Structure for Efficient Numerical Computation. *Computing in Science & Engineering* **2011**, *13*, 22–30.

5 Pedregosa, F. et al. Scikit-learn: Machine Learning in Python. *Journal of Machine Learning Research* **2011**, *12*, 2825–2830.

6 Hunter, J. D. Matplotlib: A 2D Graphics Environment. *Computing in Science & Engineering* **2007**, *9*, 90–95.

7 Weininger, D. SMILES, a Chemical Language and Information System. 1. Introduction to Methodology and Encoding Rules. *Journal of Chemical Information and Computer Sciences* **1988**, *28*, 31–36.

8 Daylight Chemical Information Systems, Inc. SMARTS - A Language for Describing Molecular Patterns. Aliso Viejo, CA, 2008.

9 Daylight Chemical Information Systems, Inc. SMIRKS - A Reaction Transform Language. Aliso Viejo, CA, 2008.

10 Degen, J.; Wegscheid-Gerlach, C.; Zaliani, A.; Rarey, M. On the art of compiling and using 'drug-like' chemical fragment spaces. *ChemMedChem* **2008**, *3*, 1503–1507.

Part 3

Data Analysis and Visualization

# 6

## Hierarchical Clustering in R

*Martin Vogt and Jürgen Bajorath*

*Goal*: Illustrate the usage of clustering methods in R as an example of unsupervised learning.

*Software/Code*: R 3.2.2, RStudio. The script `calcTc.R` contains functions for determining Tanimoto coefficients of fingerprints.

*Data*: Three compound data sets consisting of 40 5-HT serotonin receptor ligands (file `5HT.sdf`), 35 tyrosine kinase inhibitors (file `TK.sdf`), and 48 HIV protease inhibitors (file `HIV.sdf`) originally collected from the literature.[1] For the compound data MACCS fingerprints[2] and 20 numerical descriptor values have been calculated using MOE[3](files `data.fp-maccs.dat` and `data.desc.dat`).

## Theoretical Background

The goal of cluster analysis of compound data sets is to generate an organization of compounds into different clusters (also called groups or communities) so that compounds within a cluster are, with respect to pre-defined characteristics or descriptors, more similar to each other than to compounds in other clusters. As prerequisite of cluster analysis, distances between (or similarities of) compounds need to be quantified on the basis of a given representation. How data is clustered depends on the following three aspects:

1) The representation of the data.
2) A metric to assess distances between data points using a specific representation.
3) The algorithm used for clustering.

Two of the most popular representations of molecules for numerical analysis are numerical descriptors (with continuous value ranges) and binary fingerprint representations. Descriptors frequently encode physicochemical properties of molecules like the octanol/water partition coefficient, molar refractivity, or molecular weight but also capture topological properties characterizing the shape of a molecule. Compounds can be categorized using literally thousands of different descriptors[4] many of which can be calculated using open-source or commercial software packages like the Molecular

*Tutorials in Chemoinformatics*, First Edition. Edited by Alexandre Varnek.
© 2017 John Wiley & Sons Ltd. Published 2017 by John Wiley & Sons Ltd.
Companion website: www.wiley.com/go/varnek/chemoinformatics

Operating Environment (MOE).[3] Fingerprints, on the other hand, typically encode either the presence or absence of structural features of a molecule as binary values. Many different fingerprint designs exist,[5] ranging from fingerprints encoding a fixed small set of features like MACCS[2] and fingerprints encoding pharmacophore properties[6] to connectivity fingerprints that systematically enumerate certain structural features atom environments.[7]

In descriptor spaces, distances are usually measured by metric functions like the Euclidean distance using raw or normalized data. For binary fingerprints, the by far most popular method for assessing molecular similarity is the Tanimoto coefficient (Tc).[8] The Tc is defined as the ratio of the number of common features of two molecules to the total number of features in either molecule and falls in the range 0 to 1. The Tc can be converted into a distance metric by taking the complement $1 - Tc$. This distance metric is known as the Soergel distance.

## Algorithms

A large number of clustering methods exist that differ, for one, in the notion of what constitutes a "good" cluster and, for the other, in the efficiency with which the clusters are determined.[9] This tutorial focuses on hierarchical clustering, which is considered to be a relatively slow method with a run time of $O(n^3)$, that is, cubic with respect to the number of objects to cluster, and only suited for data sets of moderate size. As the name indicates hierarchical clustering methods construct a hierarchy of clusters. This is done either following a "bottom up" approach by starting with single object clusters and successively merging clusters considered to be proximal, or in a "top down" approach starting with all objects in a single cluster and proceeding by successively splitting a cluster into more clusters of smaller size. This latter approach is known as divisive clustering and the former as agglomerative clustering, which will be explained in detail below.

Centroid-based methods form a conceptually different subgroup of clustering methods. Here, each cluster is represented by a centroid and objects are assigned to a cluster based on their closeness to the centroid. The most prominent of these methods is $k$-means clustering, which aims to identify $k$ centroids in such a way that the squared distances of objects to the centroid become minimal. The optimization problem itself is NP-hard, however, the problem lends itself to a straightforward iterative heuristic algorithm:

1) Initially, distribute $k$ centroids at random.
2) (Re-)assign each object to one of the $k$ centroids based on its minimum distance.
3) Update each centroid to the average of the objects assigned to it.
4) Repeat.

The algorithm requires an embedding of the data points in a Euclidean space as is given by (normalized) continuous descriptor vectors. It is not suited for fingerprint representations whose distance relationships are characterized by pairwise similarities and not by an embedding in an $n$-dimensional space.

Agglomerative clustering can be described by the following steps:

1) Start by assigning each object to its own cluster.
2) Select the two clusters that are "closest" together.

3) Merge the two clusters.
4) Repeat until only one cluster is left.

The second step requires further specification. In order to specify closeness the concept of distance between objects needs to be extended to distance between clusters. This can be done in a number of different ways, each implementing different concepts of the notion of a cluster.

Three popular choices to assign distances between two clusters A and B are:

1) Choose the distance between the two objects in each cluster closest to each other, that is, $d_{SL}(A,B) = \min\{d(a,b) \mid a \in A, b \in B\}$. This approach is known as single-linkage clustering.
2) Choose the distance between the two objects in each cluster farthest away from each other, that is, $d_{CL}(A,B) = \max\{d(a,b) \mid a \in A, b \in B\}$. This approach is known as complete linkage.
3) Choose the average distance between objects of the two clusters, that is, $d_{AL}(A,B) = \dfrac{1}{|A||B|} \sum_{a \in A} \sum_{b \in B} d(a,b)$. This approach is known as average linkage clustering.

Which of the definitions should be considered preferred and will yield a "good" clustering will depend on the data and how it is distributed as well as the notion of what a good clustering should look like, depending on goals and applications.

Especially in chemoinformatics another approach known as Ward's criterion for joining clusters has been popular for agglomerative clustering. According to Ward's criterion one considers the total variance of all objects within each cluster. That is, for clusters $A_1, ..., A_m$ consider $\sum_{i=1}^{m} \text{var}(A_i)$. Each time two clusters are merged the total variance will increase. Ward's criterion states that two clusters should be joined that result in the smallest increase of the total variance.

## Instructions

This tutorial has been prepared using R 3.2.2. R is a programming language and interactive software for performing statistical analysis, data mining, machine learning, and visualization and can be obtained from http://www.r-project.org. Originally, R uses a command line interface and the tutorial can be completed using just this interface, but a number of GUI interfaces and IDEs exist for R that make interacting with R more user-friendly and can be used for the purpose of this tutorial. For instance, RStudio is a popular open-source cross-platform IDE that can be obtained from http://www.rstudio.com.

For three sets of compounds MACCS fingerprints and 24 numerical descriptors have been determined and saved in two files, `data.fp-maccs.dat` and `data.desc.dat`, respectively. In these files, data is organized in tab-separated columns. The first column contains a character indicating the activity of the compound (H for 5HT receptor ligands, V for HIV protease inhibitors, and T for tyrosine kinase inhibitors). In `data.fp-maccs.dat` the second column contains the fingerprint as a list of features (integers) separated by commas and in `data.desc.dat` each of the descriptor values is given in a separate column.

In the following, lines prefixed by > and + should be typed into the R command line (without the prefix). Lines without the prefix indicate expected output or contents of source files.

## Hierarchical Clustering Using Fingerprints

- After starting R make sure your working directory is set to the directory containing the data and script file `calcTc.R` either by launching R from the correct directory, using the `setwd` command from within R, or by selecting the directory from the menu if using an IDE.

```
> data.fp <- read.table('data.fp-maccs.dat',sep='\t',
+              head=TRUE,colClasses=c('factor','character'))
```

- `read.table` can be used to read in a variety of data text files by specifying header lines, column separators and types of columns. For the fingerprint file, the first column is of type factor, which defines a categorical variable. In our case it will either be H, V, or T. R is not able to parse fingerprints directly and the function reads them as character strings for now. The data is read into a structure called *data frame*. Data frames are a central concept in R for storing tabulated data.
- *Note:* Single or double quotes can be used to delimit character strings. Identifiers consist of alphanumeric characters (letters and digits) and the two special characters '_' and '.'. In contrast to Java and other languages the dot "." has no special meaning. `data.fp` is just a simple variable name and it is conventional to separate words that make up a variable name by dots.
- *Note:* Variable assignment is done using the '=', '<-', or '->' operators. '=' and '<-' can be used synonymously in variable assignment. Many functions in R have a lot of parameters, many of which are optional. For clarity, it is common to use named parameters where a parameter is explicitly assigned to an argument using '='.

```
> summary(data.fp)
class       MACCS
  H:40    Length:123
  T:35    Class :character
  V:48    Mode  :character
```

- The summary of the data reports the total number of compounds of each category.

```
> fp <- data.fp$MACCS
> fp.classes <- data.fp$class
```

- The fingerprints and classes are separated into two different vectors.

```
> source('calcTc.R')
```

- The R code contained in `calcTc.R` is read in and parsed. In order to perform cluster analysis using MACCS the distances between each pair of fingerprints are needed. As a distance measure, the Soergel distance should be used, that is, $1 - Tc$. `calcTc.R` contains a user-defined function for determining the Tc of two fingerprints given as strings:

```
calc. TC <- function(fp1, fp2){
    # convert string fingerprint to vector of bit set on
    fp1 <- unlist(strsplit(fp1, ","))
    fp2 <- unlist(strsplit(fp2, ","))
    Na <- length(fp1)
    Nb <- length(fp2)
    Nab <- length(which(!is.na(match(fp1, fp2))))
    tc <- if (Nab==0) 0 else Nab/(Na+Nb-Nab)
    tc
}
```

- The hierarchical clustering functions of R require that all pairwise distances between objects are calculated. This can be accomplished by creating a matrix of all pairwise Tc coefficients and then generating a distance matrix. The following function of calcTc.R calculates all the pairwise Tc values of a vector of fingerprint strings and stores the results in a matrix:

```
calc.TC.matrix <- function(fp.vector){
    tc.matrix <-matrix(0,length(fp.vector),length(fp.vector))
    for (i in 1:length(fp.vector)){
        for(j in i:length(fp.vector)){
            tc.matrix[i,j]=calc.TC(fp.vector[i],fp.vector[j])
        }
    }
    diag.tc.matrix <- diag(tc.matrix)
    tc.matrix <- tc.matrix + t(tc.matrix)
    diag(tc.matrix) <- diag.tc.matrix
    tc.matrix
}
```

- calc.TC.matrix returns a matrix of all pairwise Tc coefficients.

```
> fp1 <- "0,1,2,3,4,5"
> fp2 <- "4,5,6,7,8,9"
> calc.TC(fp1,fp2)
[1] 0.2
> fp3 <- "1,2,3,4,5,6,7,8"
> fp4 <- "3,4,5,6"
> calc.TC.matrix(c(fp1,fp2,fp3,fp4))
          [,1]         [,2]         [,3]         [,4]
[1,] 1.0000000    0.2000000    0.5555556    0.4285714
[2,] 0.2000000    1.0000000    0.5555556    0.4285714
[3,] 0.5555556    0.5555556    1.0000000    0.5000000
[4,] 0.4285714    0.4285714    0.5000000    1.0000000
```

- Exemplary function calls.

```
> tc.matrix <- calc.TC.matrix(fp)
> dist.fp <- as.dist(1-tc.matrix)
```

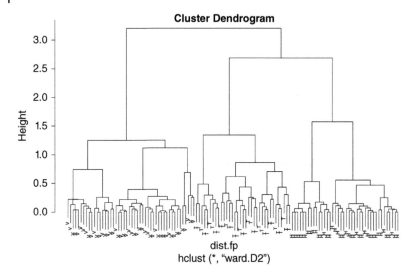

Figure 6.1 Hierarchical clustering dendogram. MACCS fingerprints.

- All pairwise Tcs of the fingerprint data are determined and converted to a distance matrix.

```
> cluster.fp <- hclust(dist.fp,method = 'ward.D2')
```

- Hierarchical clustering is performed.

```
> plot(cluster.fp,labels=fp.classes,cex=0.7)
```

- The dendrogram of the clustering is plotted (Figure 6.1).
- The height where clusters are merged refers to the "distance" between the clusters. Because hierarchical clustering per se does not yield a fixed number of clusters the clustering tree has to be "cut" at a certain height to yield a specific separation into clusters. The visualization can help to identify appropriate levels at which to separate the clusters. Large vertical stretches without a merge indicate that the distance between clusters to be merged is high and suggest that cutting a tree at that level will yield clusters that are well separated. From the plot one can see that 3 is actually a good number of clusters and these can be generated using the cutree function:

```
# divide data into 3 clusters
> memberships = cutree(cluster.fp,k=3)

# check if all clusters are correct
> fp.classes[memberships==1]
  [1] H H H H H H H H H H H H H H H H H H H H H H H H H H H H H H
 [31] H H H H H H H H H H
Levels: H T V
```

```
> fp.classes[memberships==2]
 [1] V V V V V V V V V V V V V V V V V V V V V V V V V V V V V V
[31] V V V V V V V V V V V V V V V V V V
Levels: H T V
> fp.classes[memberships==3]
 [1] T T T T T T T T T T T T T T T T T T T T T T T T T T T T T T
[31] T T T T T
Levels: H T V
```

- Because clustering is an unsupervised learning method it is not able to assign classes to the clusters but from the output we see that cluster 1 corresponds to 5HT, 2 to HIV, and 3 to TK. The clustering can be repeated using different methods like single or complete linkage and the results can be compared.

## Hierarchical Clustering Using Descriptors

```
# get descriptor data
> data.desc <- read.table(file="data.desc.dat",
     header=TRUE,sep="\t")
> desc.classes <- data.desc$class
# remove class labels before clustering
> desc <- data.desc[,-1]
```

- Retrieve the descriptor data from the file and remove the class label from the descriptor data.

```
> dist.desc <- dist(desc,method="euclidean")
> cluster.desc <-hclust(dist.desc,method = 'ward.D2')
> plot(cluster.desc,labels=desc.classes)
> memberships.desc = cutree(cluster.desc,k=3)
> desc.classes[memberships.desc==1]
 [1] H H H H H H H H H H H H H H H H H H H H H H H H H H T T T T T T
[31] T T T T
Levels: H T V
> desc.classes[memberships.desc==2]
 [1] H H H H H H H H H H H H H H H H H V T T T T T T T T T T T T T T
[31] T T T T
Levels: H T V
> desc.classes[memberships.desc==3]
 [1] V V V V V V V V V V V V V V V V V V V V V V V V V V V V V V
[31] V V V V V V V V V V V V V V V V V V T T T T T T T T
Levels: H T V
```

- The Euclidean distance matrix is calculated and hierarchical clustering is performed. The plot command is used to look at the clusters (Figure 6.2).

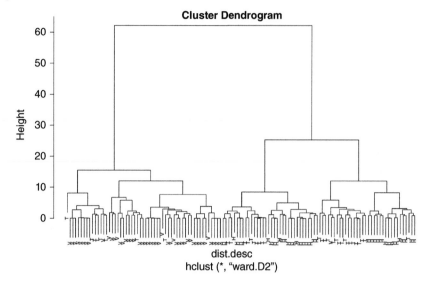

**Figure 6.2** Hierarchical clustering dendogram. Raw descriptor values.

- The plot suggests two or three clusters and we again generate three clusters to compare the quality of clustering with the fingerprint-based clustering.
- As can be seen the clusters contain compounds from different classes. The first two clusters contain mainly a mix of 5HT and TK compounds while the final cluster contains most HIV compounds and some TK compounds.
- In order to quantitatively assess the quality of a clustering with respect to a known correct classification (or another clustering) a number of quantitative measures have been developed based on the following idea. For any pair of compounds having the same (or different) class labels one can determine if they have been clustered together or separated into different clusters. Thus, one can consider all pairs of compounds and determine the ratio of pairs that have correctly been assigned to either the same or to different clusters depending on whether they belong to the same category or not. This quantity is known as the Rand Index[10] yielding a value in the range 0 to 1. A value of 1 is obtained if the clustering corresponds perfectly to a known data classification.
- The R package "flexclust" provides a function for determining the Rand index.

```
> library(flexclust)
> comPart(desc.classes,memberships.desc,c("RI"))
       RI
0.7683593
```

- A problem with using raw descriptor data is that different descriptors will typically generate numerical data on very different scales. For example, molecular weight measured in Dalton will generate values in the hundreds for small molecules whereas the octanol/water partition coefficient measured on a logarithmic scale ($\log P_{oct/wat.}$) will have small single digit units. This means that numerical distances of raw descriptor values are dominated by descriptors having the largest variance, which depends entirely on the unit of measurement.

- To circumvent this problem, the descriptor data can be scaled to create *dimensionless* numbers that do not depend on the unit of measurement. A popular method to accomplish this is to use unit variance scaling. For a descriptor $x$ data is rescaled by subtracting the mean and dividing by the standard deviation:

$$x' = \frac{x - \bar{x}}{\sigma(x)}$$

```
> desc.scaled = scale(desc)
> dist.desc.scaled = dist(desc.scaled,method="euclidean")
> cluster.desc.scaled <-hclust(dist.desc.scaled,
     method = 'ward.D2')
> plot(cluster.desc.scaled,labels=desc.classes)
> memberships.desc.scaled = cutree(cluster.desc.scaled,k=3)
# check if all clusters are correct
> desc.classes[memberships.desc.scaled==1]
 [1] H H H H H H H H H H H H H H H H H H H H H H H H H H H H H H
[31] H H H H H H H H H H V T T T
Levels: H T V
> desc.classes[memberships.desc.scaled==2]
 [1] V V V V V V V V V V V V V V V V V V V V V V V V V V V V V V
[31] V V V V V V V V V V V V V V V V
Levels: H T V
> desc.classes[memberships.desc.scaled==3]
 [1] V T T T T T T T T T T T T T T T T T T T T T T T T T T T T T
[31] T T T
> comPart(desc.classes,memberships.desc.scaled,c("RI"))
      RI
0.9488205
```

- The dendrogram of the clustering (Figure 6.3) indicates either two or four clusters; however, for comparison with previous clusterings, again, three clusters are generated. The clusters are now dominated by compounds of a single class and reflect the different biological activities of the compounds quite well. This is also reflected in the increased Rand Index compared to the clustering of the unscaled descriptors.

## Visualization of the Data Sets

- High-dimensional data like fingerprints or descriptor vectors cannot be directly visualized so that pairwise distance relationships and thus potential community structures become apparent. For visualization, the data has to be embedded into a two-dimensional plane. Ideally, the embedding should preserve the distance relationships as closely as possible. Principal component analysis (PCA) is one of the most popular methods for dimension reduction and visualization of high-dimensional data in a coordinate-based Euclidean space, and can be applied to the descriptor data.

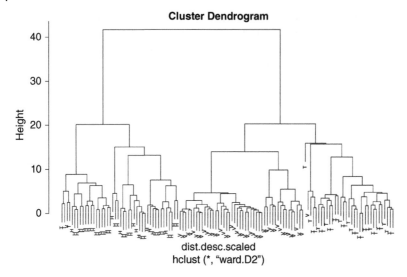

Figure 6.3 Hierarchical clustering dendogram. Scaled descriptor values.

```
> pca <- princomp(desc)
> summary(pca)
Importance of components:
                          Comp.1      Comp.2      Comp.3      Comp.4
Standard deviation     4.5586143 1.13374646 0.61719800 0.50841610
Proportion of Variance 0.8966076 0.05545856 0.01643561 0.01115258
Cumulative Proportion  0.8966076 0.95206614 0.96850174 0.97965432
...
> x <- pca$scores[,1]
> y <- pca$scores[,2]
> plot(x,y,col=c('red','green','blue')[desc.classes])
```

- The summary of the PCA shows that two principal components account for 95% of the total variance of the data. This is an indication that the two-dimensional plot (Figure 6.4) reflects the original distances in the descriptor space quite well. The parameter col=c('red', 'green', 'blue')[desc.classes] will color the points according to their class: 5HT, red; TK, green, HIV, blue. Regions for compounds of different activity can be distinguished, although there is no clear separation between the classes.

```
> pca.scaled <- princomp(desc.scaled)
> summary(pca.scaled)
Importance of components:
                          Comp.1     Comp.2      Comp.3      Comp.4
Standard deviation     3.0023701 1.5631828 1.31555813 1.08609503
Proportion of Variance 0.4544057 0.1231785 0.08724396 0.05946356
Cumulative Proportion  0.4544057 0.5775841 0.66482811 0.72429167
...
> x.scaled <- pca.scaled$scores[,1]
> y.scaled <- pca.scaled$scores[,2]
> plot(x.scaled,y.scaled,col=c('red','green','blue')[desc.classes])
```

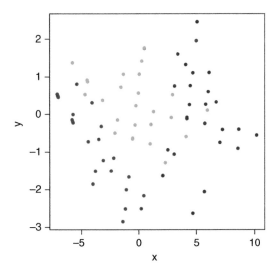

Figure 6.4 Scatterplots of the compound data. Scatterplots of two-dimensional projections of the compound data using PCA of descriptor values. Dots are colored according to their class: 5HT, red; TK, green; HIV, blue.

- Using scaled descriptor values, the summary of the PCA shows that two principal components account for only 58% of the total variance of the data. This indicates that the two-dimensional plot (Figure 6.5) does not reflect the original distances in descriptor space very well.

```
> library(MASS)
> m <- as.matrix(dist.fp)
> e<-array(1,dim(m))
> diag(e) <- 0
> dd <- as.dist(m+e*0.0001)
> mds.fp <- isoMDS(dd)
> plot(mds.fp$points,
    col=c('red','green','blue')[fp.classes])
```

- PCA cannot be applied to fingerprints if distances between fingerprints are derived from a distance matrix and are not given by Euclidean distances of coordinate vectors. If only pairwise distances are known, multidimensional scaling (MDS) is a method that seeks to find an optimal two-dimensional embedding of data points that preserves distances as well as possible. The R library "MASS" provides an implementation of MDS.

- Fingerprints of very similar compounds might frequently be identical yielding a distance of 0 between them. However, the implementation of MDS in R does not allow pairs with 0 distance. To avoid this problem a very small value is added to the distance between each pair of objects. Figure 6.6 shows the result of MDS using a distance matrix based on the Tc of MACCS fingerprints. The three classes are clearly separated and occupy disjoint regions of the chemical space as represented by MACCS fingerprints.

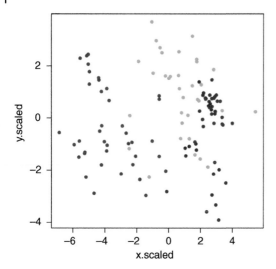

**Figure 6.5** Scatterplots of the compound data. Scatterplots of two-dimensional projections of the compound data using PCA of scaled descriptor values. Dots are colored according to their class: 5HT, red; TK, green; HIV, blue.

## Alternative Clustering Methods

R provides functions for a variety of different clustering methods. Many non-hierarchical methods like *k*-means require a parameter that pre-defines the number of clusters. In addition to *k*-means clustering, the R library "cluster" provides a number of clustering algorithms. Partitioning Around Medoids (PAM)[11] is a variant of *k*-means algorithm that can be applied to distance matrices. We use the following code to compare the performances using the Rand index. Note that *k*-means and PAM are non-deterministic and results can vary from call to call.

```
# Perform k-means on raw descriptor values
> kmeans.desc <- kmeans(desc,3)
> comPart(desc.classes,kmeans.desc$cluster,c("RI"))
       RI
0.7655604

# Perform k-means on scaled descriptor values
> kmeans.desc.scaled <- kmeans(desc.scaled,3)
> comPart(desc.classes,kmeans.desc.scaled$cluster,c("RI"))
       RI
0.873384
> library(cluster)

# Perform PAM on raw descriptors
> pam.desc <-pam(desc,k=3)
> comPart(desc.classes,pam.desc$cluster,c("RI"))
       RI
0.7627616
```

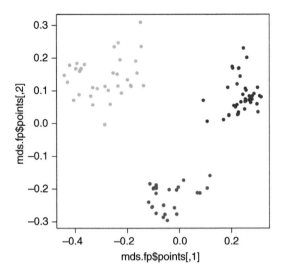

**Figure 6.6** Scatterplots of the compound data. Scatterplots of two-dimensional projections of the compound data using multidimensional scaling of MACCS fingerprints. Dots are colored according to their class: 5HT, red; TK, green; HIV, blue.

```
# Perform PAM on scaled descriptors
> pam.desc.scaled <-pam(desc.scaled,k=3)
> comPart(desc.classes,pam.desc.scaled$cluster,c("RI"))
       RI
0.9285619

# Perform PAM on fingerprint distance matrix.
> pam.fp <-pam(dist.fp,k=3,diss=TRUE)
> comPart(fp.classes,pam.fp$cluster,c("RI"))
       RI
0.9770758
```

- The methods perform slightly worse than hierarchical clustering but are computationally more efficient for larger data sets.

## Conclusion

R can be used for clustering and cluster visualization of compound data sets using descriptors or fingerprint representations. Fingerprints cannot be handled by standard R and custom code is required to generate distance matrices. The data sets used in this tutorial consist of compounds active against different targets that are structurally diverse and can be easily distinguished using MACCS fingerprint-based clustering. The small set of 20 descriptors is not capable of accurately separating the data sets. A larger more comprehensive set of descriptors can be expected to perform better. The embedding of compounds in descriptor spaces has to take into account that coordinates represent different physicochemical or topological properties and often have incomparable units. Scaling is important to derive unit-less values and hence render distances independent of units of measurement.

## References

1 Xue, L.; Godden, J.; Gao, H.; Bajorath, J. Identification of a preferred set of molecular descriptors for compound classification based on principal component analysis. *J Chem Inform Comput Sci* 1999, *39*, 699–704.

2 MACCS Structural keys; Accelrys Inc.: San Diego, CA, USA.

3 Molecular Operating Environment (MOE). http://www.chemcomp.com, Chemical Computing Group: Montreal, Canada. (accessed: 16.09.2015)

4 Todeschini, R.; Consonni, V. *Handbook of Molecular Descriptors*. Wiley-WCH: Weinheim, 2000.

5 Cereto-Massagué A. et al. Molecular fingerprint similarity search in virtual screening. *Methods* 2014, *71*, 58–63.

6 McGregor, M.J.; Muskal, S.M. Pharmacophore fingerprinting. 1. Application to QSAR and focused library design. *J Chem Inform Comput Sci* 1999, *39*, 569–574.

7 Rogers, D.; Hahn, M. Extended-connectivity fingerprints. *Journal of Chemical Information and Modeling* 2010, *50*, 742–754.

8 Willett, P. Chemical similarity searching. *J Chem Inform Comput Sci* 1998, *38*, 983–996

9 Jain, A.K.; Murty, M.N.; Flynn, P.J. Data clustering: A review. *ACM Computing Surveys* 1999, *31*, 264–323.

10 Rand, W.M. (1971). Objective criteria for the evaluation of clustering methods. *Journal of the American Statistical Association* 1971, *66*, 846–850.

11 Kaufman, L.; Rousseeuw, P.J. Clustering by means of medoids. In *Statistical Data Analysis Based on the $L_1$–Norm and Related Methods*, edited by Y. Dodge. North-Holland 1987, 405–416.

7

# Data Visualization and Analysis Using Kohonen Self-Organizing Maps

*João Montargil Aires de Sousa*

*Goal*: Illustrate the application of Kohonen self-organizing maps (SOMs), and the related counterpropagation neural networks, to the visualization and classification of multivariate data using a set of molecules assigned to five types of biological activity. *Software*:

1) CDK Descriptor Calculator (v1.4.6, CDK v1.5.10), available at http://www.rguha. net/code/java/cdkdesc.html
2) JATOON (v2.2), available at http://joao.airesdesousa.com/jatoon
3) Weka (v3.7.3), available at http://www.cs.waikato.ac.nz/ml/weka

*Data*: The dataset of Jorissen and Gilson[1] consisting of 250 compounds and their known protein target was retrieved from cheminformatics.org. The targets are CDK2 (cyclin-dependent kinase 2), COX2 (cyclooxygenase 2), FXa (coagulation factor Xa), PDE5 (phosphodiesterase 5), and A1A (alpha-1A adrenoceptor). The SMILES strings, protein targets, and 176 molecular descriptors calculated with the CDK Descriptor Calculator were stored in a file named jor.xls, in folder SOM. Examples of files generated during the tutorial are also provided in the same folder.

## Theoretical Background

Kohonen SOMs (or Kohonen neural networks)[2] learn by unsupervised training, distributing objects through a grid of so-called neurons, on the basis of the objects' features (descriptors, or attributes). This is an unsupervised method that projects multidimensional objects into a 2D surface (a map), that is, it reduces multidimensional information to two-dimensions, maintaining the topology of the information. A SOM can reveal similarities between objects, mapped into the same or neighbor neurons. Objects may be provided for training labeled with classes. In that case, if the descriptors globally contain relevant information enabling class differentiation, then clustering by class will emerge, and such a trained SOM can be possibly used for classifying new objects. The opposite situation is scattering and mixing of classes on the surface, and that indicates no relationship between the profile of attributes and the classes of objects.

*Tutorials in Chemoinformatics*, First Edition. Edited by Alexandre Varnek.
© 2017 John Wiley & Sons Ltd. Published 2017 by John Wiley & Sons Ltd.
Companion website: www.wiley.com/go/varnek/chemoinformatics

Typical applications of Kohonen SOMs in chemoinformatics have been the classification of molecules according to biological activity, selection of data sets encompassing large diversity, identification of redundant features (with transposed features' matrices), and generation of molecular descriptors by mapping of molecular components (e.g., bonds and points of the molecular surface).

Kohonen SOMs do not provide a numerical output, but instead a position on a map, possibly associated to a class. In order to enable an output, Counterpropagation Neural Networks (CPNN)[3] link a Kohonen SOM to a second layer of neurons (output layer) that acts as a look-up table and stores output (numerical) data provided in the training set. The CPNN method is considered a semi-supervised technique.

The SOM algorithm (explained below) is typically based on Euclidean distances between neurons and objects (e.g., molecules) represented by descriptors. Therefore, the numerical range of a descriptor can influence its impact on the mapping. Normalization of the data is thus required previous to training.

## Algorithms

Each neuron of a Kohonen SOM contains as many elements (weights) as the number of input features for the objects to be mapped. The topology of a SOM is toroidal, that is, the left side is continuing the right side and the bottom neurons are considered to be adjacent to those at the top. Before self-organization (training) starts, the weights take random values. Learning consists of adjusting the weights during the training phase, and this is a competitive process: every time an object from the training set is presented to the network, all the neurons compete, but only one neuron wins—the neuron with the most similar weights compared to the descriptors of the object. This is the central or winning neuron (see Figure 7.1). Similarity can be defined in terms of the Euclidean distance. It is said that the winning neuron was activated by the object, and it gets the weights adjusted to become even more similar to the features of the presented

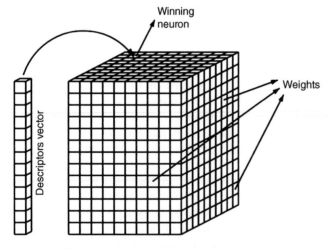

Figure 7.1 The architecture of a Kohonen neural network: each column in the grid represents a neuron, and each box in such a column represents a weight (a number). Each neuron has as many weights as the input descriptors for the objects to be mapped.

object. Not only are the weights of the winning neuron adjusted, but also those of the neurons in its neighborhood. The extent of adjustment depends on the topological distance to the winning neuron—the closer a neuron is to the central neuron the larger is the adjustment of its weights. The objects of the training set are iteratively fed to the map, the weights corrected, and the training is stopped when a pre-defined number of cycles (*epochs*) is attained.[2,4]

Once the network has been trained, the entire training set is sent again through the network. Each neuron can then be classified according to the majority class of the objects that were mapped onto it, and empty neurons can be classified according to the objects that activated their neighbors.

During the training phase, activation of winning neurons and correction of weights are exclusively based on the descriptors—no information is used concerning classes or outputs. This is what makes learning *unsupervised*. Only after the training, when the whole training set is finally mapped, is the information concerning the classes used, in order to assign classes to neurons.

When new objects are to be classified by a trained SOM, its descriptors are presented to the network and the winning neuron is determined. If that neuron has an associated class, the object can be classified into that class.

A counterpropagation neural network has a Kohonen layer on the top of an output layer. To every neuron in the Kohonen layer corresponds one neuron in the output layer. During the training, the winning neurons are chosen by the Kohonen layer (exactly as in a Kohonen SOM).

But then, not only are the weights of the Kohonen layer adapted but also the weights of the output layer are as well in order to become closer to the output values of the presented object. The output layer may include one or more levels depending on the number of output properties in the data. After the training, the CPNN can produce an output for an object—the winning neuron is activated and the corresponding weights in the output layer are taken as the prediction (Figure 7.2).[3,4]

In the JATOON software employed in this tutorial, the following specific algorithm is followed.[4,5] The weights at level $i$ are initialized with random values between $a$-$sd$ and $a + sd$, where $a$ and $sd$ are respectively the average and the standard deviation of the $i$th variable of the objects in the training set. During the training, weights are corrected according to

$$w_{i,new} = w_{i,old} + (obj_i - w_{i,old})(0.4995(epoch_{max} - epoch) / (epoch_{max} - 1) + 0.0005)$$
$$(1 - d / (dL + 1)),$$

where $w_{i,new}$ is the $i$th weight of the neuron after the correction, $w_{i,old}$ is the $i$th weight of the neuron before the correction, $obj_i$ is the $i$th variable of the object, $d$ is the distance from the winning neuron to the neuron being corrected ($d \le dL$), and $dL$ is defined by $dL = ILS - epoch.ILS / epoch_{max}$, where $ILS$ is the initial learning span, $epoch_{max}$ is the epoch at which the training will be stopped, and $epoch$ is the current epoch. The $ILS$ and $epoch_{max}$ are user-defined parameters.

## Instructions

1) In the first exercise of this tutorial you will build a **Kohonen neural network (or self-organizing map, SOM)** that automatically classifies molecular structures according to five types of biological activity. First of all, the SOM will provide a way

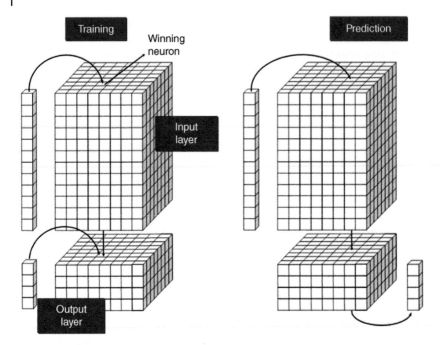

**Figure 7.2** Architecture of a counterpropagation neural network, and its operation in the training and prediction modes.

to visualize the data set in terms of similarities between molecules and to investigate relationships between profiles of descriptors and types of biological activity. Download file jor.xls containing 176 molecular descriptors calculated by CDK for 250 molecules, and their biological activity.

2) Before submitting the data to a SOM, each descriptor must be normalized, so that its impact on the results does not depend on its numerical range. You can use a very simple normalization scheme with the Weka "Normalize" filter to linearly scale each descriptor into (0,1). Open the jor.xls file in a spreadsheet, delete the first three columns and save as CSV file (e.g., jor_desc.csv, this file is also provided for your assistance). Open the file with Weka in the "Preprocess" tab, click on the "Choose" button under "Filter", and select "filters" → "unsupervised" → "attribute" → "Normalize," and "Close." Click on the "Apply" button on the right (Figure 7.3).

3) Save the normalized descriptors (button "Save") as CSV (e.g., jor_norm.csv, this file is also provided for your assistance) and open the file in a spreadsheet. Copy the ID and SET columns of the initial jor.xls file into the spreadsheet. In order to be used by the JATOON software next, you need to replace the labels corresponding to the activity class by uppercase letters (A1A → A, CDK2 → B, COX2 → C, FXa → D, PDE5 → E). Sort the lines according to the "SET" column, to separate training and test set molecules.

4) Copy the training set molecules into a text file (e.g., jor_tr.dat). You should copy all the descriptors and the activity columns, but not the first line with the headers. Do the same for the test set (e.g., into jor_te.dat file). Both files are provided for your assistance.

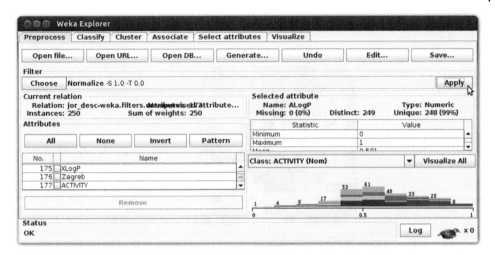

Figure 7.3 Normalization of the descriptors with the Weka program. This is a critical step previously to train Kohonen maps.

5) Open the JATOON.jar file with a Java program (e.g., the Oracle Java Runtime). Now open the training set file (Menu "File" → "Read data"). The program will automatically suggest a size for the network (15x15), the initial learning span (7) and 300 epochs. Click on "Train Kohonen NN" to train the SOM.

6) After the training, the map is displayed with each neuron colored according to the class of the (majority of) molecules that activated it. The program generates the map with the activations obtained with the whole training set at the end of the training. Black neurons correspond to undecided neurons, that is, neurons with tied winner classes. You should observe quite a reasonable separation of classes, particularly for classes A (A1A), C (COX2), and D (FXa). Note that the surface is toroidal, which means that left neurons are neighbors of right neurons, and top neurons are neighbors of bottom neurons (Figure 7.4).

7) You may train the network again, or experiment with different parameters. Even with the same parameters, you will not obtain exactly the same map because of random decisions during the training (initial values of the weights and order of presentation of objects).

8) The observed trend for clustering by class shows that in this data set a) molecules of the same class have in general similarities in descriptors' profiles (so they are mapped into close neurons) and b) molecules of different classes have globally different descriptors' profiles (so they are mapped into separated regions). In such a case, the SOM can be used to classify new molecules.

9) Click on "Map Objects" and select the test set (jor_te.dat) to map the molecules of the test set into the trained SOM. When the map is displayed each molecule is labeled according to the last column in the file. You can easily verify if they were correctly classified by using labels corresponding to the true classes. Probably ca. 4 classifications are wrong/undecided among the 27 molecules of the test set. The errors will mostly involve molecules of classes B and E, which were indeed the least clustered. But you should observe that in most cases the wrongly classified molecules were mapped quite close to the true region.

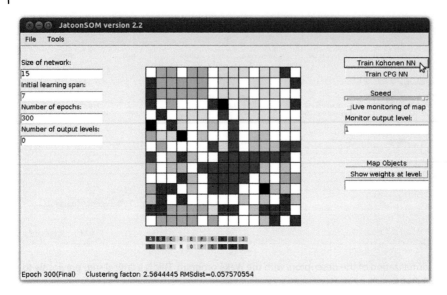

Figure 7.4 The JATOON main panel after training the Kohonen network.

10) There may be different reasons for a "wrong" classification. It may happen that the descriptors and the SOM algorithm could not learn a specific important feature responsible for the biological activity of a molecule (you should remember that the SOM uses non-supervised learning and no selection/weighting of descriptors). But a "wrong" classification may also be a hint that the experimental data is wrong, or that the molecule may also exhibit a second type of biological activity which had not been investigated. The SOM algorithm implies that the classification of a molecule is based on the classes of training set molecules with similar descriptors. You may inspect which molecules have more similar descriptors (and fall in the same or close neurons) by mapping all the molecules together (incorporate the molecule ID in the label, place all the molecules into one file, and map).

11) In order to save predictions, go to the "Tools" menu and choose "Predict." Then select the file with the descriptors and labels, and the file where predictions are to be saved. You can then process predictions as you wish, for example, in a spreadsheet.

In the next exercise you will approach the same problem with a **counterpropagation neural network**.

12) As there are five types of biological activity in this data set, the activity may be encoded with five numbers, instead of a character. We will use "1 0 0 0 0" for class A (A1A), "0 1 0 0 0" for class B (CDK2), and so on. Change your files accordingly in a spreadsheet. Note that JATOON requires that there is no header line, and the last column must be a label used for display. So copy the molecule ID merged with the class character into the last column. Save new training and test set files. Example files jor_CPG_tr.dat and jor_CPG_te.dat are provided for your assistance. A CPNN will be trained with five output weights, each one corresponding to a class. The network will be trained so that output weights of a neuron will provide numerical indications of its belonging to each class.

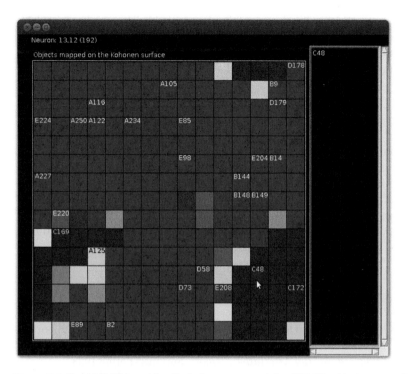

**Figure 7.5** The JATOON panel for displaying output weights (CPNN), with the map colored according to weights in the third output level, and test set objects projected on its surface.

13) Open the training set file in JATOON as before, and change the "Number of output levels" to 5. Click on "Train CPG NN." After the training, the program maps all the objects and the result is displayed with each neuron colored according to the average of the first output property for the objects that fell onto it.

14) Visualize the output weights for a class by entering the level corresponding to it below the button "Show weights at level." Click on that button. Because the network was trained with 176 descriptors, the output weights for the first class are at level 177. Do the same for the other classes (levels 178-181) and observe that the different classes have high output weights in specific regions of the map. That was expected since we had seen good clustering of classes with Kohonen maps.

15) Map the objects of the test set on the map, with neurons colored according to an output level. For example, for the third output class (class C) fill in with "3" below "Monitor output level", click on "Map Objects" and select the test set. "Correct" class C classifications are obtained for objects of class C falling into blue neurons (these correspond to neurons with output weights near 1) (Figure 7.5).

16) Open the "Tools" menu and choose "Predict." Select the test set and a new file to save the numerical predictions. Numerical predictions enable various possibilities for interpretation. For example, one may interpret the maximum of the five values as a prediction of the corresponding class, or may assume no prediction when the maximum is below some threshold.

17) Compare the predictions with those obtained in step 11. The results should be essentially the same, because the Kohonen layer (the input layer) was adapted in the same way. Discrepancies may be due to random fluctuations. More stable results may be obtained by training several networks and combining their results.

## Conclusion

The tutorial demonstrates how to train Kohonen self-organizing maps and counter-propagation neural networks, and illustrates how to use them for classification tasks and data visualization. Unsupervised learning enables the identification of global similarities between objects' profiles, and the exploration of relationships between those profiles and classes of the objects. The possibility of training CPNN with numerical output data associated to the network topology provides numerous ways of encoding problems and interpreting results, both in classification and regression tasks.

## References

1 Jorissen, R.N.; Gilson, M.K. Virtual Screening of Molecular Databases Using a Support Vector Machine. *J. Chem. Inf. Model.* 2005, *45*(3), 549–561.
2 Kohonen, T. *Self-Organization and Associative Memory*. Springer: Berlin, 1988.
3 Hecht-Nielsen, R. Counterpropagation networks. *Appl. Opt.* 1987, *26*(23), 4979–4984.
4 Gasteiger, J.; Zupan, J. Neural Networks in Chemistry. *Angew. Chem. Int. Ed. Engl.* 1993, *32*, 503–527.
5 Aires-de-Sousa, J. JATOON: Java Tools for Neural Networks. *Chemometr. Intell. Lab.* 2002, *61*(1–2), 167–173.

Part 4

Obtaining and Validation QSAR/QSPR Models

8

Descriptors Generation Using the CDK Toolkit
and Web Services

*João Montargil Aires de Sousa*

*Goal:* Illustrate the generation of molecular descriptors with freely available software, and the quick visual inspection of correlations between descriptors and biological endpoints.
*Software:*

1) CDK Descriptor Calculator (v1.4.6, CDK v1.5.10), available at http://www.rguha. net/code/java/cdkdesc.html
2) E-DRAGON web service, available at http://www.vcclab.org/
3) OCHEM web service, available at http://ochem.eu

*Data:* A dataset of 235 molecular structures with results of fish toxicity (LC50 values) were retrieved from the EPA Fathead Minnow Acute Toxicity Database (EPAFHM, Feb 2008) restricted to compounds acting by baseline narcosis.[1] The file is named descriptors_fish_LC50.xls, and is in folder DESCRIPTORS. An example file generated during the tutorial is also provided in the same folder. A dataset of 31 steroids for which corticosteroid binding globulin (CBG) receptor affinity is available[2] was retrieved from http://www2.chemie.uni-erlangen.de/services/steroids. The file is named descriptors_steroids.sdf, and is in folder DESCRIPTORS.

## Theoretical Background

Molecular descriptors are attributes of a molecule, representations derived from its structural information, typically numbers that codify features of the molecular structure.[3,4] Simple molecular descriptors are, for example, the molecular weight and the number of atoms. A wide variety of molecular descriptors have been proposed and are implemented in available software. They range from the very simple to highly complex mathematical transformations of the molecular structural information. Molecular descriptors are designed to capture specific aspects of the molecular structure that are related to observable properties. They are mostly used to establish quantitative structure-property and structure-activity relationships (QSPR and QSAR), in which statistical and machine learning algorithms fit models to predict observable properties (e.g., biological activity) from molecular descriptors. Results of experiments can also be used

*Tutorials in Chemoinformatics*, First Edition. Edited by Alexandre Varnek.
© 2017 John Wiley & Sons Ltd. Published 2017 by John Wiley & Sons Ltd.
Companion website: www.wiley.com/go/varnek/chemoinformatics

as molecular descriptors, such as partition coefficients, or spectroscopic properties, and they can be relevant to predict other less accessible endpoints.

Molecular descriptors are classified according to the type of information they encode. Constitutional descriptors encompass global molecular properties such as the molecular weight, number of atoms, number of bonds, number of rotatable bonds, or the sum of atomic volumes. Fragment descriptors count the occurrence of functional groups and sub-structures previously defined. Topological descriptors consider the molecular structure as a mathematical graph—and graph theory can then be applied to generate graph invariants that are used as descriptors.

Geometrical descriptors that encode 3D features of molecules are called "3D descriptors" and require information on 3D atomic coordinates. Descriptors of this type can distinguish between different conformations and diastereoisomers.

## Algorithms

Three examples of molecular descriptors are presented with their algorithms.

**The Wiener index**[5] is a topological descriptor defined as the sum of all distances between pairs of (non-hydrogen) atoms in the molecule. The distance is defined in terms of number of bonds in the shortest path between the two atoms. It can be calculated by summing all elements of the distance matrix and dividing by 2. The Wiener index provides an indication of ramification, and is a very simple approach to the van der Waals surface.

**2D autocorrelation vectors**[3,4] are defined by

$$a(d) = \sum_{i=1}^{N}\sum_{j=1}^{N} \delta(d_{i,j} - d) p_j p_i$$

$$\delta = \begin{cases} 1 \forall d_{i,j} = d \\ 0 \forall d_{i,j} \neq d \end{cases}$$

in which $N$ is the number of atoms in the molecule, $d_{i,j}$ is the topological distance between atoms $i$ and $j$ (number of the bonds in the shortest path), and $p$ is an atomic property of atoms $i$ and $j$ (e.g., partial charge, or polarizability). One number is calculated for each value of $d$. Topological autocorrelation vectors are invariant with respect to translation, rotation, and conformation of the molecule.

An example of a 3D descriptor is the Radial Distribution Function code (**RDF descriptors**).[6] It is defined by

$$g(r) = \sum_{i=1}^{N-1}\sum_{j=i+1}^{N} p_i p_j e^{-B(r-r_{i,j})^2}$$

in which $N$ is the number of atoms in the molecule, $p_i$ is an atomic property of atom $i$, $r_{i,j}$ is the 3D distance between atoms $i$ and $j$, and $B$ is an adjustable parameter. The graphical representation of $g(r)$ vs $r$ can be easily interpreted—each pair of atoms contributes to a region of the plot, the region is centered on the value of their interatomic distance, the contribution is proportional to the product of their atomic properties, and the width of the region is related to the $B$ parameter. Sampling the function at a specified number of $r$ values yields the same number of descriptors.

With any of these three descriptors, no alignment of molecules is required, and a fixed number of descriptors can be defined for all molecules of a data set independently of the size of each molecule.

## Step-by-Step Instructions

1) Download file descriptors_fish_LC50.xls. It contains 235 structures in SMILES format and the corresponding LC50 values for the fathead minnow (the higher the LC50 value the lower the toxicity).
2) Extract only the SMILES into a text file and save it.
3) Download the CDK Descriptor Calculator and start it.
4) Choose the Input File ("Browse" → select the file with the SMILES strings), and a new file (Output File) to store the descriptors.
5) In the "Options" menu, you can select the format of the output file in "Output Method." For example, "Comma delimited" will produce columns separated by commas.
6) For this exercise, select constitutional and topological descriptors. Click on "Go"—Figure 8.1.
7) Open the output file in a spreadsheet. Verify that numbers are correctly recognized as numbers, for example, the scientific notation format. Each row corresponds to a molecule, and each column to a molecular descriptor. The first row contains abbreviated names of the descriptors. You may obtain more details about the descriptors calculated by CDK at http://sourceforge.net/projects/cdk/
8) Remove the first column and add a last column with the toxicity values available in the initial file. Save as comma-separated file. An example is provided for your assistance (descriptors_fish_LC50.csv).
9) Open the Weka software, select "Explorer," go to the "Preprocess" tab and click on "Open file." Select the file with the descriptors. Ensure that non-available values are correctly recognized (in CDK they are represented by "NA").
10) Select the "Visualize" tab (at the top) and inspect correlations between descriptors and toxicity (LC50). You can use this tool to visually identify correlations between the biological activity and a specific descriptor, or inter-correlations between descriptors. For example, observe how the XLogP descriptor is related to the toxicity: within this dataset, a small value of the descriptor is clearly an indication of low toxicity— Figure 8.2. You can also observe how the Wiener index (WPOL) is more correlated with the number of bonds (nB) than with the number of atoms (nAtom).

In the second exercise, the **web service E-DRAGON** is used to calculate RDF descriptors.
11) Go to http://www.vcclab.org, choose E-DRAGON and "start the program." Click on 'upload data' and choose the file descriptors_steroids.sdf with 31 molecular structures of steroids in the MDL SDFile format. Upload the file. (Note that the file includes 3D atomic coordinates, as required to calculate 3D descriptors.) Choose "MDL sdf" in the menu for the selection of the input file format. In the list of molecular descriptors, select "RDF descriptors." Click on "submit your task."

**Figure 8.1** Calculation of molecular descriptors with the CDK Descriptor Calculator.

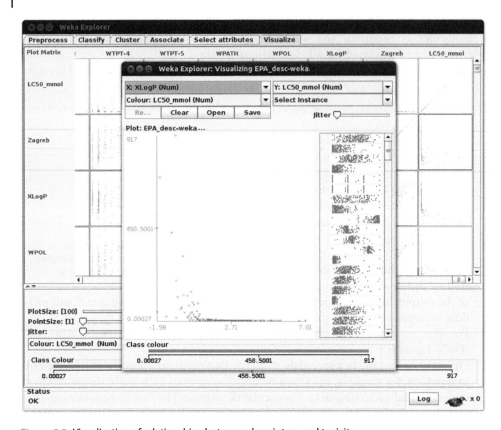

Figure 8.2  Visualization of relationships between descriptors and toxicity.

12) After the calculation is finished, go to the top menu "Task Manager" and choose "Results as text." Click on "inspect dragon log" to check if the tasks proceeded without problems. Click on "open results.txt in a browser." Copy the results to a spreadsheet in such a way that there are as many rows as the compounds (plus the first row with the headers) and as many columns as the descriptors. Save the file. RDF descriptors are calculated by E-Dragon with several atomic properties: mass (m), van der Waals volume (v), Sanderson electronegativity (e), polarizability (p), and unweighted (u). You may want later to use a machine learning algorithm to cluster the molecules and to inspect if similar molecules are clustered together on the basis of RDF descriptors.

You may obtain more details about the E-DRAGON descriptors at http://www. talete.mi.it/products/dragon_molecular_descriptors.htm

Another example of a web service for the calculation of molecular descriptors is the **Online Chemical Modeling Environment (OCHEM)** at http://ochem.eu. The third exercise demonstrates how to use it.

13) Go to http://ochem.eu, choose the "Models" menu and "Calculate descriptors." Log in and upload the file saved in "2." with 235 structures in the SMILES format. Click "Next."

14) Several procedures for preprocessing the molecules are available. For this data set they are not required. Click "Next".

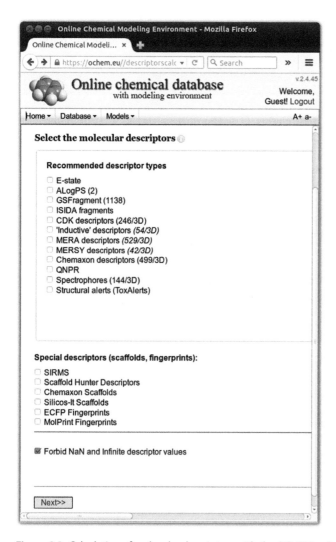

Figure 8.3 Calculation of molecular descriptors with the OCHEM web service.

15) Several descriptors can be selected for calculation—Figure 8.3. Try, for example, E-state descriptors of "Atom indices" type. Click "Next."
16) After the calculation is finished, click on "Proceed to the descriptors page" and choose the contents and format of the output file. Click on the button of your choice and download the file with the descriptors.

## Conclusion

The tutorial demonstrates how to calculate a wide range of molecular descriptors starting from data sets of molecules in the SMILES or MDL SDFile formats strings, using three different software tools: CDK Descriptor Calculator, E-DRAGON, and OCHEM.

The quick visualization of correlations between descriptors and dependent variables, as well as inter-correlations between descriptors is illustrated with the Weka software.

## References

**1** C. L. Russom, S. P. Bradbury, S. J. Broderius, D. E. Hammermeister, R. A. Drummond, *Environmental Toxicology and Chemistry* **1997**, *16*(5), 948–967. Data available at http://www.epa.gov/ncct/dsstox

**2** A.C. Good, S. So, W. G. Richards, *J. Med. Chem.* **1993**, *36*, 433–438.

**3** R. Todeschini, V. Consonni, R. Mannhold (Series Editor), H. Kubinyi (Series Editor), G. Folkers (Series Editor), *Molecular Descriptors for Chemoinformatics*, 2nd Edition, Wiley-VCH: Weinheim, 2009, ISBN: 978-3-527-31852-0.

**4** J. Gasteiger, T. Engel, (Eds.) *Chemoinformatics: A Textbook*, Wiley-VCH: Weinheim, 2003.

**5** H. Wiener, *J. Am. Chem. Soc.* **1947**, *69*(1), 17–20.

**6** M. C. Hemmer, V. Steinhauer, J. Gasteiger, *Vibrational spectroscopy* **1999**, *19*(1), 151–164.

# 9

## QSPR Models on Fragment Descriptors

*Vitaly Solov'ev and Alexandre Varnek*

The tutorials illustrate quantitative structure-property relationship (QSPR) modeling by the ISIDA_QSPR program,[1–4] realizing Multiple Linear Regression (MLR) analysis on the base of ISIDA Substructure Molecular Fragment (SMF) descriptors. ISIDA SMF descriptors are counts of the occurrence of subgraphs (fragments) in a molecule, where each descriptor element is associated to one of the detected possible fragments, complying with the user-proposed fragmentation scheme (fragment type, size, etc.). The program builds MLR models combining forward[3] and backward[4] stepwise variable selection techniques.

The ISIDA_QSPR program is a graphical interface piloting this workflow and supporting graphical analysis of the results linked to the compound structures. It runs under the Windows operating system (ISIDA_QSPR.exe). It is strongly recommended to use of a non-system disk for the ISIDA_QSPR directory.

The following exercises are considered in the tutorial:

1) *Individual MLR model* - multiple linear regression on a single SMF descriptor set with descriptor selection, property predictions on a test set.
2) *Fragment analysis of the individual MLR model*: fragment contributions in modeling property, a pairwise correlation matrix for fragment contributions and the similarity of molecules according to SMF.
3) *External n-fold cross-validation.*
4) *Consensus modeling* based on the ensemble of Multiple Linear Regression models involving various types of SMF descriptors.
5) *Property predictions and virtual screening.*

The tutorial includes step-by-step instructions indicated by the vertical blue line on the left side.

## Abbreviations

AD          Applicability domain
CM          Consensus model
EdChemS     The sketcher of the MOL files

*Tutorials in Chemoinformatics*, First Edition. Edited by Alexandre Varnek.
© 2017 John Wiley & Sons Ltd. Published 2017 by John Wiley & Sons Ltd.
Companion website: www.wiley.com/go/varnek/chemoinformatics

| EdiSDF | SDF manager of the ISIDA_QSPR program |
|---|---|
| *F* | The Fischer's criterion |
| *FIT* | The Kubinyi fitness criterion |
| FVS | Forward variable selection |
| HIV | The human immunodeficiency virus |
| *HRF* | The Hamilton R-factor percentage |
| ISIDA | In SIlico design and data analysis |
| LMO | Leave-many-out |
| LOO | Leave-one-out |
| *MAE* | Mean absolute error |
| MLR | Multiple linear regression |
| *n* | The number of data points |
| n-fold CV | n-Fold cross-validation |
| *Q* | Leave-one-out cross–validation correlation coefficient |
| QSPR | Quantitative structure–property relationships |
| *R* | The Pearson's correlation coefficient |
| $R_{det}^{2}$ | Squared coefficient of determination |
| *RMSE* | Root-mean squared error |
| *s* | Standard deviation |
| SDF | Structure data file |
| SMF | Substructural molecular fragments |
| SVD | Singular Value Decomposition |
| *TC* | The Tanimoto similarity coefficients |
| TIBO | Tetrahydroimidazobenzodiazepinone derivatives |
| $Y_{calc}$ | The fitted property |
| $Y_{exp}$ | The modeling property |

## DATA

The following Structure-Data Files (SDF) are used in this tutorial: **AHIV-TIBO.SDF** and **TEST_AHIV-TIBO.SDF.** The first file contains experimental values of anti-HIV activity log $(1/IC_{50})$ of 57 tetrahydroimidazobenzodiazepinone (TIBO) derivatives,[5] where $IC_{50}$ is the concentration (mol/L) of the TIBO compound inhibiting 50% of the HIV-1 reverse transcriptase activity. The second file **TEST_AHIV-TIBO.SDF** contains five TIBO derivatives for which experimental anti-HIV activities are available and seven virtual TIBO derivatives generated by the CombiLIB / EdChemS tool.[6, 7] Experimental values of anti-HIV activity are represented by the log_1_C_exp. The input files must be located in the ISIDA_QSPR program directory.

**Exercise 9.1**   Individual MLR model: modeling set up and output analysis.
*Goal:* Building an individual MLR model based on SMF descriptors. The user needs only input files in SDF format to perform the modeling on the training set and predictions on the test set.

Click the **Single Model** button of the ISIDA_QSPR program (Figure 9.1) to open the **Single Model Calculations** dialog box (Figure 9.2). The dialog box includes the **Data** panel for data input set up, the **Descriptors** panel for selection of the SMF descriptor type, the **Model** panel for modeling set up, and the **Validation** internal and external model validation set up (Figure 9.2).

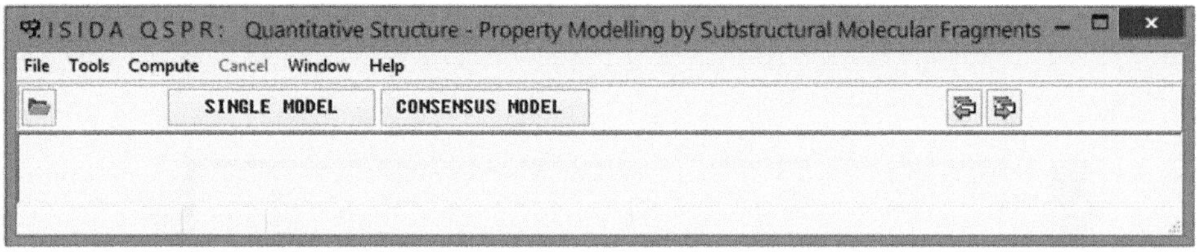

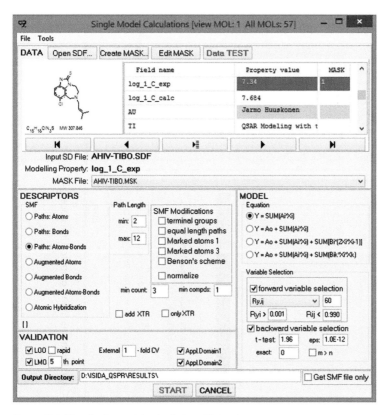

Figure 9.1 The ISIDA_QSPR Desktop.

Figure 9.2 The single model calculations Dialog box.

## ISIDA_QSPR input

Input data for ISIDA_QSPR should be prepared in Structure-Data File format,[8] where the modeled property is represented by a data field. The property data field should be specified for all records in SDF, although the values of the property for the test compounds may be absent. Molecular structures may be represented as 2D or 3D structures. As a rule, hydrogen atoms of the structures are not specified, although molecules with explicit hydrogen atoms are supported.[9, 10] The input file must be located in the ISIDA_QSPR program directory.

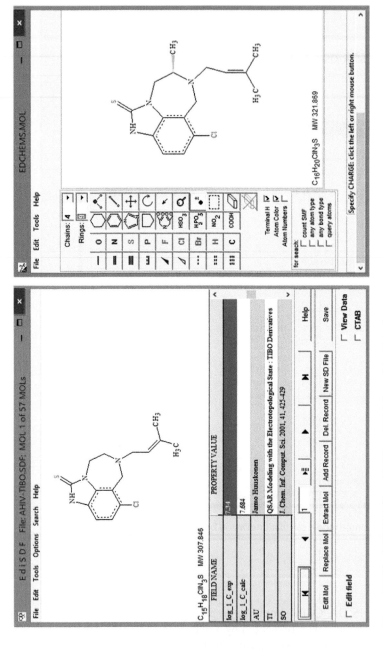

Figure 9.3  The EdiSDF (left) and EdChemS (right) graphical interface.

From the **Data** panel of the single model building Dialog box (Figure 9.2), click on the **Open SDF** button and proceed to opening AHIV-TIBO.SDF file. The **Input SD File: AHIV-TIBO.SDF** label appears in the **Data** panel (Figure 9.2). The table in the **Data** panel includes the information stored in **Field name** and **Property value** items (Figure 9.2) of the AHIV-TIBO.SDF file. Click on the **log_1_C_exp** cell of the **Field name** column to select the modelling property ($Y_{exp}$). The **Modelling Property: log_1_C_exp** label appears in the **Data** panel (Figure 9.2).

The SD File can be edited using EdiSDF tool. Click **Tools → SDF Editor** of the ISIDA_QSPR main menu (Figure 9.1) to open the EdiSDF manager (Figure 9.3). This versatile tool allows one to add new entries to the SD File, to add new data fields or edit existing ones, or to edit the structures (Figure 9.3). The current SDFs do not need any fixing, but we encourage the reader to try to use this self-explanatory tool to complete or edit corrupted SD files.

### Data Split Into Training and Test Sets

ISIDA_QSPR can split the initial data set into two subsets: training and test sets for model building and validation, respectively. The program uses a MASK file (*.msk) to indicate the train / test status for each compound. By default, ISIDA_QSPR produces mask files in which every N-th compound is kept out for testing, systematically starting from the M-th compound in the list ($M \leq N$).

From the **Data** panel of the single model calculations Dialog box (Figure 9.2), click on the **Create MASK** button. The Create Mask Dialog box appears (Figure 9.4). Click on the Create MASK with TEST SET radio button; enter 5 in the **each** edit box and 5 in the **starting from** edit box. In the case shown on Figure 9.4, each fifth compound will be used for the test set. Click on the **START** button to save or overwrite the mask file AHIV-TIBO.MSK in the ISIDA_QSPR directory. The **MASK File: AHIV-TIBO.MSK** label appears in the **Data** panel (Figure 9.2). Click on the **Data TEST** button to verify the input data. The **Information** dialog box appears with message: "Input data files are in internal agreement." Click on the **OK** button to close the **Information** dialog box.

### Substructure Molecular Fragment (SMF) Descriptors

The ISIDA_QSPR program includes a module for descriptor generation. ISIDA SMF, or simply SMF, descriptors[2, 5, 11–16] are counts of subgraphs (fragments) in a molecular graph. Each descriptor is associated with one of the fragments generated within the user-defined fragmentation scheme (fragment type, size, etc.). The program can handle

Figure 9.4 The Create Mask Dialog box.

two main types of fragments (Figure 9.5): topological paths (I) and atom-centered fragments (atoms with nearest connected neighbors) (II). Either of these schemes supports indication of the atom and bond types (AB), the atom types only (A), the bond types only (B). The atom type can have different attributes: atom symbol only, hybridization state, Benson's notation,[17] and special mark[10, 15] (Figure 9.5 and 9.6).

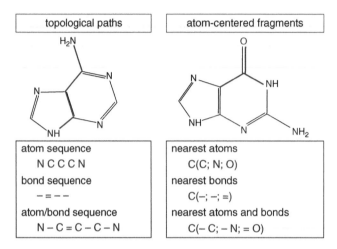

| topological paths | atom-centered fragments |
|---|---|
| atom sequence<br>N C C C N | nearest atoms<br>C(C; N; O) |
| bond sequence<br>– = – – | nearest bonds<br>C(–; –; =) |
| atom/bond sequence<br>N – C = C – C – N | nearest atoms and bonds<br>C(– C; – N; = O) |

Figure 9.5 Two main classes of ISIDA SMF fragments: topological paths (I) and atom-centered fragments (II).

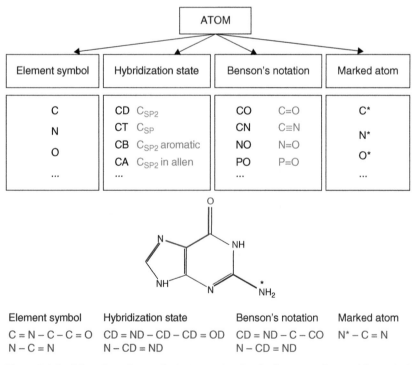

| ATOM | | | |
|---|---|---|---|
| Element symbol | Hybridization state | Benson's notation | Marked atom |
| C<br>N<br>O<br>... | CD  $C_{SP2}$<br>CT  $C_{SP}$<br>CB  $C_{SP2}$ aromatic<br>CA  $C_{SP2}$ in allen<br>... | CO  C=O<br>CN  C≡N<br>NO  N=O<br>PO  P=O<br>... | C*<br>N*<br>O*<br>... |

| Element symbol | Hybridization state | Benson's notation | Marked atom |
|---|---|---|---|
| C = N – C – C = O<br>N – C = N | CD = ND – CD – CD = OD<br>N – CD = ND | CD = ND – C – CO<br>N – CD = ND | N* – C = N |

Figure 9.6 (top) Atomic attributes of substructural molecular fragments. (bottom) Example demonstrating different atomic labels used for the atom/bond paths containing five (in red) or three (in blue) atoms.

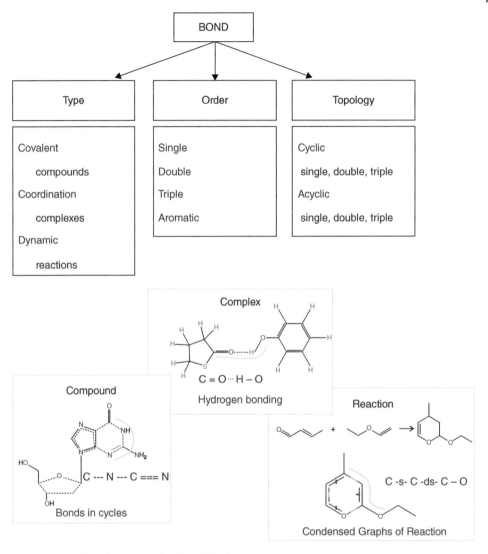

Figure 9.7 The bond attributes for ISIDA SMF descriptors.

The bond attributes support special typing (covalent for σ-bonds, coordinating for noncovalent bonds, and dynamic for reactions), order (single, double, triple, aromatic) and topology (cyclic or acyclic bond) (Figure 9.7). For topological paths, their optimality (shortest or all paths), length (minimal and maximal, by defaults ranged between 2 and 15 atoms), and explicitness (all atoms are indicated or terminal atoms are indicated only) can be selected.

From the **Descriptors** panel of the single model calculations Dialog box (Figure 9.2), click the **Paths: Atoms-Bonds** radio button, enter 2 in the **min** edit box and 12 in the **max** edit box for minimal and maximal path lengths, respectively. Please, verify that the *SMF Modifications* group boxes of the **Descriptors** panel are not checked. Use by default 3 in the **min count** edit box and 1 in the **min compds** edit box (Figure 9.2). For this exercise, choose shortest topological paths with explicit consideration of atom and bond types. Select the minimal ($m_{min}=2$) and maximal ($m_{min}=12$) numbers of atoms in the paths. Notice that the program will also generate all intermediate paths with $m$ atoms: $m_{min} \leq m \leq m_{max}$. Please, verify that the **Get SMF file only** check box is not selected in the right bottom corner of the dialog box (Figure 9.2). This option is used only for the purpose to generate the SMF descriptors file and to store it in the working directory.

### Regression Equations

Multiple linear regression analysis is applied to build relationships between the variables $x_i$ (SMF descriptors) and a dependent variable $y$ (modeled property). Four types of equations are considered:

$$y = \sum_i a_i x_i + \Gamma \tag{9.1}$$

$$y = a_0 + \sum_i a_i x_i + \Gamma \tag{9.2}$$

$$y = a_0 + \sum_i a_i x_i + \sum_i b_i \left(2x_i^2 - 1\right) + \Gamma \tag{9.3}$$

$$y = a_0 + \sum_i a_i x_i + \sum_{i,k} b_{ik} x_i x_k + \Gamma \tag{9.4}$$

Here, $a_i$ and $b_i$ ($b_{ik}$) are fragment contributions, $x_i$ is the count of the $i$-th type fragment. The free term $a_o$ is fragment independent. An extra term $\Gamma = \Sigma c_m D_m$ can be used to describe any specific feature of the compound using external descriptors $D_m$; by default $\Gamma = 0$. The parameters $a_i$ and $b_i$ are determined using the singular value decomposition (SVD) method.[18]

From the **Model** panel (Figure 9.2), click on the **Y=SUM(Ai*Xi)** radio button to select linear fitting equation without the free term.

### Forward and Backward Stepwise Variable Selection

Combined forward and backward stepwise techniques have been used to select the most pertinent variables from initial pool of the generated SMF descriptors.[3, 4] Initially, the forward variable selection (FVS) algorithm is applied to pre-select the user-defined number $m_p < n$ of the most relevant variables, where $n$ is the size of training set. The FVS employs the known equations for the correlation coefficients between the response variable $y$ and one, two, and three variables[19] in combination with the FSMLR algorithm.[20] Accordingly, three sub-algorithms (FVS-1, FVS-2, and FVS-3) have been used. At step $p$, the FVS procedure defines a new response variable $y^{(p)} = y^{(p-1)} - y_{calc}$, where $y_{calc} = c_0 + c_i x_i$ (FVS-1), $y_{calc} = c_0 + c_i x_i + c_j x_j$ (FVS-2) or $y_{calc} = c_0 + c_i x_i + c_j x_j + c_k x_k$ (FVS-3), $p = 1, 2, 3,...$ and $y^{(0)} = y_{exp}$. Thus at every step, one ($x_i$), two ($x_i, x_j$) or three variables ($x_i, x_j$ and $x_k$) are selected ensuring maximal correlation coefficients ($R_{y,i}, R_{y,ij}$ or $R_{y,ijk}$ correspondingly) between the variable(s) and $y^{(p)}$. The steps are repeated until the number of selected variables $m_p$ reaches a user-defined value.

Optionally, variables $x_m$ with small correlation coefficient with $y^{(p)}$ ($|R_{y,m}| < R^0_{y,m}$), those highly correlated with other variables $x_i$ ($|R_{i,m}| > R^0_{i,m}$) or "rare" fragments (i.e., found in less than $q$ molecules, here $q < 3$) can be eliminated. Here $R^0_{y,m}$ and $R^0_{i,m}$ are the user-defined thresholds. Then backward stepwise variable selection algorithm[3] eliminates the variables with low $t_i = a_i / \Delta a_i$ values for the models (1) and (2), where $\Delta a_i$ is a standard deviation for the coefficient $a_i$ at the $i$-th variable in the model. First, the program selects the variable with the minimal $t_{min} < t_0$, then it builds a new model excluding this variable. This procedure is repeated until $t \geq t_0$ for all selected variables. Here $t_0$ is the tabulated value of Student's criterion. By default, $t_0$ equals 1.96.

> From the **Model** panel (Figure 9.2), check the **forward variable selection** and **backward variable selection** check boxes in the **Variable Selection** panel. Select the $R_{y,ij}$ item from the dropdown list of the FVS algorithm combo box. On the right of the combo box, enter 60 in the edit box for the number of pre-selected variables as the percentage of the training set size. Enter 0.001 in the $R_{yi}$ edit box and 0.99 in the $R_{ij}$ edit box for the correlation coefficient thresholds. Enter 1.96 in the **t-test** edit box, 1E-12 in the **eps** edit box and 0 in the **exact** edit box. Make sure that the **m > n** check box is not selected (Figure 9.2).

## Parameters of Internal Model Validation

One can distinguish internal and external validation. The former corresponds to the procedure—leave-one-out (LOO) or leave-many-out (LMO) cross-validation—performed after completing variables selection on the entire set. External validation in $n$-fold cross-validation or on a selected set is always performed on the data never used at any step of the modelbuilding.

> From the **Validation** panel (Figure 9.2), check the **LOO** check box to calculate the leave-one-out (LOO) cross-validation correlation coefficient ($Q^2$). Make sure that the **rapid** check box is not selected. Check the **LMO** check box for the calculation of the leave-many-out (LMO) cross-validation correlation coefficient. Enter 5 in the **i-th point** edit box for the LMO calculations: each fifth data point is discarded followed by the modelbuilding on the remaining training data and to use discarded objects for the model validation. Enter 1 in the **External n-fold CV** edit box to perform the modeling without external n-fold cross-validation (n-fold CV).

## Applicability Domain (AD) of the Model

The applicability domain (AD) of the model defines an area of chemical space where the model is presumably accurate. Three types of AD definitions can be used either simultaneously or individually: fragment control, bounding box,[21] and "quorum control."[22] Bounding box approach considers AD as a multidimensional descriptor space confined by minimal and maximal occurrences of the descriptors involved in an individual model (AD1). Fragment control consists in discarding predictions for the compounds containing descriptors not occurring in the initial SMF pool generated for the training set (AD2). "Quorum control" is a threshold for the number of models accepted by AD1 and AD2. If this number is lower than a user defined threshold, the consensus prediction is ignored.

From the **Validation** panel (Figure 9.2), check the **Appl. Domain1** and **Appl. Domain2** check boxes for the fragment control and bounding box of model applicability domain, respectively.

## Storage and Retrieval Modeling Results

The output files are saved in a user-selected directory by clicking **Output Directory** button (Figure 9.2). The Open dialog box appears, where click **Open** button to select the directory, for instance, C:\ISIDA_QSPR\RESULTS. The output files can always be opened by clicking File → Open → *.out file in the ISIDA_QSPR main menu (Figure 9.1). Typically, the *.out file includes the name of the input SD file name as substring and begins with the date and the time of the performed calculations.

## Analysis of Modeling Results

To build the model, click **Start** button in **Single Model Calculations** dialog box (Figure 9.2). The program creates nine output files: four plain text and five files with the graphical representation of results, see their description below.

The *.TXT file (here, <*date_time*>_*AHIV-TIBO.TXT*) contains the following information concerning the QSPR model:

a) initial list of the SMF descriptors,
b) set up parameters,
c) groups of concatenated fragments always occurring in the same combination in each compound of the training set,
d) statistical parameters of the multiple linear regression (MLR) including Pearson's correlation coefficient $R$, Fischer's criterion $F$, root mean squared error $RMSE$, mean absolute error $MAE$, the leave-one-out cross–validation correlation coefficient $Q^2$,
e) SMF descriptors involved in the MLR equation, regression equation coefficients (SMF contributions) $a_i$ and their random errors $\Delta a_i$ for the 95% confidence interval,
f) a pairwise correlation matrix for SMF contributions,
g) singular values $s_i$ obtained in SVD calculations(see section 1.4),
h) Table of experimental ($Y_{exp}$) and fitted ($Y_{calc}$) property, residuals $Y_{exp}$ - $Y_{calc}$ for the training set.

The *.SMF file (<date_time>_AHIV-TIBO.SMF) contains full set of generated SMF descriptors and their counts in the training set molecules:

```
Full Set of Fragments.
     1.                                        C*C
     2.                                        C-N
     3.                                        C-C
     4.                                        C=S
     5.                                        C-Cl
...
```

```
MATRIX: Compound (Line) × Fragment Count (Column).
         1    2    3    4    5    6    7    8    9    10   11   12 ...
    1    6    8    5    1    1    1    7    4    6    1    2    4 ...
    2    6    8    5    1    1    1    7    4    6    1    2    4 ...
    3    6    9    7    0    0    1    7    4    6    1    4    6 ...
    4    6    8    6    0    0    1    7    4    6    1    2    6 ...
    6    6    8    5    1    1    1    7    4    6    1    2    5 ...
    7    6    8    7    1    1    0    7    4    6    1    2    5 ...
    8    6    8    7    0    0    0    7    4    6    1    2    5 ...
    9    6    8    7    0    0    1    7    4    6    1    2    5 ...
   11    6    8    7    0    0    1    7    4    6    1    2    5 ...
   12    6    8    6    0    0    1    7    4    6    1    2    5 ...
...
```

The *.MF file (<date_time>_AHIV-TIBO.MF) includes a list of SMF selected for the model and related descriptor's values for the training set molecules

```
Set of Fragments for the Model.
    12.                                              C-C-N
    32.                                          C*C*C-N-C
    36.                                          C-C-N-C-N
    54.                                      Cl-C*C*C*C-N
    83.                                C*C*C-C-N-C-C=C-C
...
```

```
MATRIX: Compound (Line) × Fragment Count (Column).
          12   32   36   54   83   88   99   144   167
     1    4    4    1    1    4    1    0    0     0
     2    4    4    1    1    4    1    0    0     0
     3    6    4    1    0    4    1    2    2     0
     4    6    4    1    0    0    1    1    0     0
     6    5    4    2    1    0    1    0    0     0
     7    5    4    2    1    0    0    0    0     0
     8    5    4    2    0    0    0    1    0     2
     9    5    4    2    0    0    1    1    0     2
    11    5    4    2    0    0    1    1    0     1
    12    5    4    1    0    4    1    1    0     0
...
```

The *.DOC file (<date_time>_AHIV-TIBO_Pred.DOC) contains the Table of predicted property (right column) for the compounds of the test set defined by the MASK file. The *Datum* column contains experimental data. If a compound is identified as being outside the AD of the model, the predicted value for this compound is excluded:

```
TABLE P1. Test set: Predicted property log_1_C_exp for the
compounds from the AHIV-TIBO.SDF file.

cmp. no.                Datum              IAB2-120
        5               4.49                 2.94
       10               4.48
```

|    |      |      |
|----|------|------|
| 15 | 5.61 | 5.20 |
| 20 | 5.65 | 5.91 |
| 25 | 5.18 | 5.72 |

...

The *.RAC plot file (here, <date_time>_AHIV-TIBO.RAC) provides with the analysis of residuals ($Y_{exp}$ - $Y_{calc}$) as a function of fitted property ($Y_{calc}$) for the training set (Figure 9.8). The red dotted line corresponds to zero deviation. To visualize the residual value and corresponding molecular structure, move the mouse pointer on the data point (small circle) and click. The structure appears on the yellow background; the internal number of the data point and the $Y_{exp}$, $Y_{calc}$, and ($Y_{exp}$ - $Y_{calc}$) values emerge in the status bar at the bottom of the program window (Figure 9.8).

The *.PLT plot file (here, <date_time>_AHIV-TIBO.PLT) displays the correlation between $Y_{exp}$ and $Y_{calc}$ as well as corresponding linear equation, including the number of data points ($n$), correlation coefficient ($R$), Fischer's criterion ($F$), standard deviation ($s$), squared coefficient of determination ($R_{det}^2$), root-mean squared error ($RMSE$), and mean absolute error ($MAE$) (Figure 9.9a).

Remaining two plots (files <date_time>_AHIV-TIBO.LOO and <date_time>_AHIV-TIBO.LMO) represent the relationships between $Y_{exp}$ and $Y_{pred}$ as well as corresponding linear equations with their statistical parameters for the leave-one-out (Figure 9.9b) and leave-many-out cross-validations.

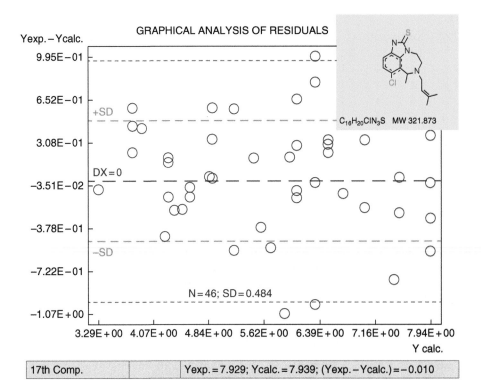

| 17th Comp. | Yexp. = 7.929; Ycalc. = 7.939; (Yexp. − Ycalc.) = −0.010 |
|---|---|

Figure 9.8 The graphical window of residuals' analysis.

(a)

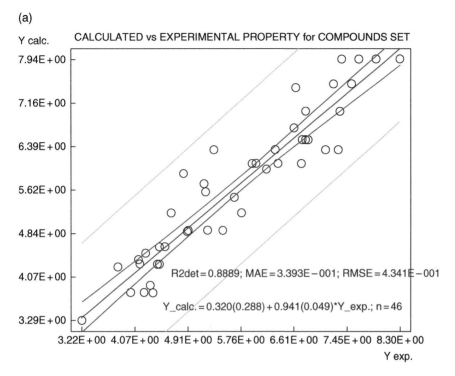

Y calc.

CALCULATED vs EXPERIMENTAL PROPERTY for COMPOUNDS SET

R2det = 0.8889; MAE = 3.393E − 001; RMSE = 4.341E − 001

Y_calc. = 0.320(0.288) + 0.941(0.049)*Y_exp.; n = 46

Y exp.

(b)

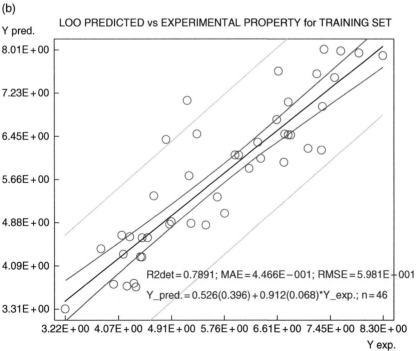

Y pred.

LOO PREDICTED vs EXPERIMENTAL PROPERTY for TRAINING SET

R2det = 0.7891; MAE = 4.466E − 001; RMSE = 5.981E − 001

Y_pred. = 0.526(0.396) + 0.912(0.068)*Y_exp.; n = 46

Y exp.

Figure 9.9 Modeling of anti-HIV activity ($Y = \log(1/IC_{50})$) of tetrahydroimidazobenzo-diazepinone (TIBO) derivatives. The three plots show linear correlation for calculated versus experimental activities obtained for the training set at the fitting (a) and LOO (b) stages as well as for external test set (c).

(c)

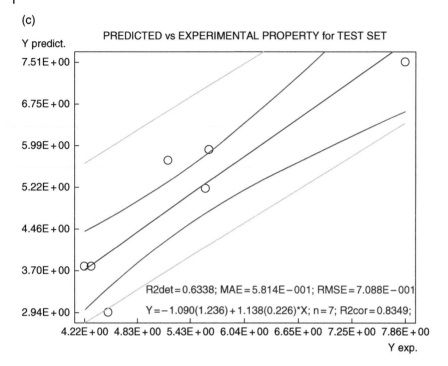

Figure 9.9 (Continued)

The fifth plot displays the results of linear regression analysis, including the plot $Y_{pred}$ versus $Y_{exp}$ for the test set defined by the MASK file. Objects identified as being outside AD of the model are excluded (Figure 9.9c).

Molecular structure corresponding to selected data point on the graphs can be visualized by mouse clicking. The structure appears on the yellow background; the internal number of the data point, the $Y_{exp}$ and $Y_{calc}$ ($Y_{pred}$) values emerge in the status bar at the bottom of the program window.

### Root-Mean Squared Error (RMSE) Estimation

Root-mean squared error $RMSE = \left[ 1/n \sum_{i=1}^{n} (Y_{exp,i} - Y_i)^2 \right]^{1/2}$ characterizes the ability of the model to reproduce quantitatively the experimental data, where $Y_i$ is the fitted $Y_{calc,i}$ or predicted $Y_{pred,i}$ value of the property for the $i$-th data point. Typically, $RMSE$ values increase in the order $RMSE$ (fitting) < $RMSE$ (LOO) < $RMSE$ (external test set), as demonstrated in Figure 9.9. This can be explained by the fact that information about the training set compounds is used at the fitting and partially (at the variables selection step) at LOO or LMO stages whereas the test compounds are never seen at any step of the model building.

**Exercise 9.2**  Analysis of the fragment contributions for individual MLR model
One may expect that the presence of some particular structural motifs increases or decreases the compound potency. In this exercise, we demonstrate how fragment contributions can be analyzed with the help of the MolFrag module (Figure 9.10) which was opened after the calculations of the individual MLR model (Exercise 9.1). This tool provides the user with:

a) the list of fragment descriptors and their contribution, minimal and maximal occurrence in the training set compounds (**Model Parameters** tab),
b) an assessment of pairwise molecular similarity based on fragment descriptors involved in the model (**Similarity** tab),
c) a pairwise correlation of these descriptors (**Correlations** tab),
d) the list of fragment descriptors involved in the model and their contributions for each molecule in the training set (**SMF table** tab).

Some details of these functionalities are given below

The **Model Parameters** tabsheet displays two tables (Figure 9.10). The upper one shows the list of SMF descriptors (molecular fragments) generated for the training set. For each descriptor it reports: identification number (*id*), name (the denomination of the associated fragment), contribution (*contrib.*) and its standard deviation (*SD*), the

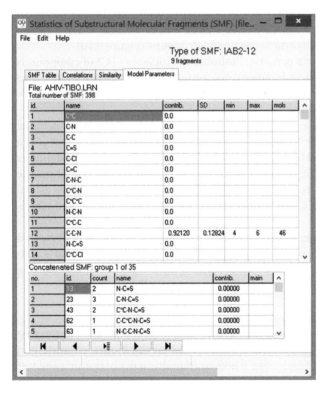

Figure 9.10  MolFrag graphical interface.

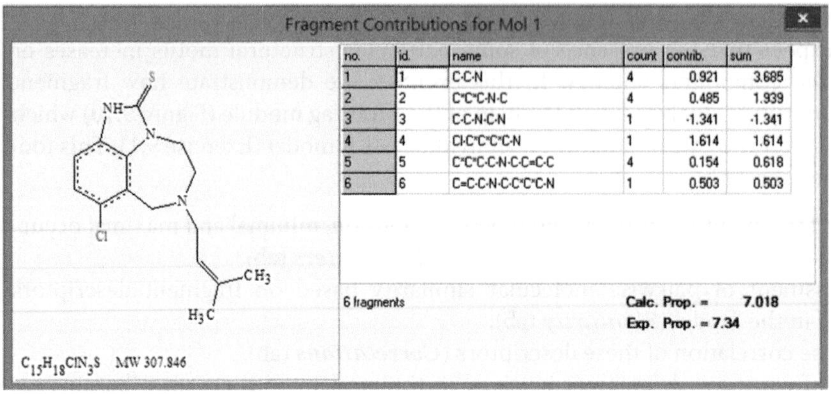


| no. | id. | name | count | contrib. | sum |
|---|---|---|---|---|---|
| 1 | 1 | C-C-N | 4 | 0.921 | 3.685 |
| 2 | 2 | C*C*C-N-C | 4 | 0.485 | 1.939 |
| 3 | 3 | C-C-N-C-N | 1 | -1.341 | -1.341 |
| 4 | 4 | Cl-C*C*C*C-N | 1 | 1.614 | 1.614 |
| 5 | 5 | C*C*C-C-N-C-C=C-C | 4 | 0.154 | 0.618 |
| 6 | 6 | C=C-C-N-C-C*C*C-N | 1 | 0.503 | 0.503 |

Figure 9.11 Selected training set compound, its constituting fragments and their contributions into the modeled property.

minimal (*min*) and maximal (*max*) fragment counts over the training set, and the number of the compounds containing the given fragment (*mols*). The lower table contains the groups of fragments always occurring in the same combination in certain compounds of the training set. The "main" (longest or lexicographically high-order path) fragment in the group is indicated by the " + " sign in the ***main*** column. The navigation buttons at the bottom are used to browse the fragment groups (Figure 9.10).

The ***Similarity*** tabsheet reports paitwise Tanimoto coefficients (TC) of compounds similarity calculated with a help of fragment descriptors involved in the model. The user can enter a *TC* threshold value in the ***TC*** edit box then click on the ***Mark*** button to highlight the Tanimoto coefficients exceeding this threshold.

The ***Correlations*** tabsheet reports a pairwise correlation of the descriptors included in the model. The user can enter a threshold value of the correlation coefficient $|R|$ in the ***R*** edit box then click on the ***Mark*** button to highlight in the ***Correlated fragments*** window all $|R|$ exceeding this threshold. Clicking on any highlighted cell opens the ***Y versus X*** window which reports corresponding linear equation and related statistical parameters squared correlation coefficient (*R2cor*), Fischer's criterion (*F*), and standard deviation (*s*).

The ***SMF Table*** tabsheet summarizes occurrences of molecular fragments involved in the model. Click on a cell corresponding to a particular molecule (e.g., ***mol 1*** cell) opens the ***Fragment Contributions*** window describing its 2D structure, constituting fragments and their contributions into the modeled property (Figure 9.11).

**Exercise 9.3** External n-fold cross-validation.

The external *n*-fold cross-validation procedure is often used[16, 23, 24] as a standard protocol for the estimation of the predictive performance of the model. According to this procedure, an entire dataset is split into *n* non-overlapping pairs of training and test sets. On each fold, a training set covers $(n - 1)/n$ of the data points while related test set covers the remaining $1/n$ of the data points. The model developed on the training set is applied to the corresponding test set. Finally, predictions for all test sets are concatenated and, in such a way, all data points in the entire data set are predicted. Note that the bigger *n*, the larger the training set, meaning that the information available at model training stage—and hence, implicitly, the chance to encounter, at training stage, compounds that are similar to test molecules—is increased. Thus, the bigger *n*, the more "optimistic" cross-validation results become. The most aggressive cross-validation, at $n = 2$, challenges a model trained on half of the original set to predict the other half, and therefore may be too pessimistic—unless very large data sets are used. At the other extreme of the spectrum, LOO cross-validation (which is nothing else but N-fold cross-validation, with N = number of compounds in the entire set) is definitely too optimistic, but generates equations that are closest to the one that could be obtained on hand of the entire set.

### Setting the Parameters

Specify the parameters for the modelling as stated above in the Exercise 9.1 before the section "Analysis of Modeling Results". Then click on the **Create MASK** button of the single model calculations Dialog box (Figure 9.2). The Create Mask Dialog box appears (Figure 9.4). Click on *Create MASK, **No test set*** radio button. Click on the **START** button to save or overwrite the mask file **AHIV-TIBO.MSK** in the ISIDA_QSPR directory using the Save mask file Dialog. Click on the **Data TEST** button to verify if the input data are suitable for the modelbuilding (Figure 9.2). If this is a case, the **Information** dialog box displays a message: "Input data files are in internal agreement". Close the **Information** dialog box.

From the **Validation** panel (Figure 9.2), enter **5** in the **External n-fold CV** edit box to perform the modeling with external five-fold cross-validation (five-fold CV).

### Analysis of n-Fold Cross-Validation Results

Click on the **Start** button of the **Single Model Calculations** dialog box (Figure 9.2) to perform the calculations. The program creates six output files: four plain text files (*.TOM, *.DOC, *.AVE and *.ECV) and two files of the graphical presentation of results.

The **\*.TOM** file contains statistical parameters of the individual MLR model built on every fold of five-fold CV calculations:

```
Five-Fold External Cross-Validation Procedure.


...
File of Mol Structures: AHIV-TIBO.SDF;  57 compounds in
training set.
Modeling Property Name: log_1_C_exp
Mask File:              AHIV-TIBO.MSK
Exter.Descriptors File: -

...
```

| no | fragment type | fitting equation | n | k | R2 | F | FIT | s | HRF | Q2 |
|----|-----------|----------|----|----|----------|-------|-------|----------|-------|----------|
| 1 | IAB2-12 | 0 | 45 | 13 | 0.943843 | 44.82 | 2.846 | 3.69E-01 | 5.410 | 0.901341 |
| 2 | IAB2-12 | 0 | 45 | 15 | 0.964574 | 58.35 | 3.389 | 2.81E-01 | 4.109 | 0.905603 |
| 3 | IAB2-12 | 0 | 46 | 10 | 0.913880 | 42.45 | 3.008 | 4.30E-01 | 6.700 | 0.862763 |
| 4 | IAB2-12 | 0 | 46 | 20 | 0.982215 | 75.57 | 3.528 | 2.18E-01 | 2.847 | 0.937505 |
| 5 | IAB2-12 | 0 | 46 | 9 | 0.888879 | 37.00 | 2.691 | 4.84E-01 | 7.448 | 0.789069 |

For each fold, the following statistical parameters of related individual model are given:

- the number of the data point ($n$) in the training set of five-fold CV,
- the number of fitted parameters (fragment contributions) ($k$),
- squared Pearson correlation coefficient ($R^2$),
- the Fischer criterion ($F$),
- the Kubinyi fitness criterion[19] ($FIT$),
- standard deviation ($s$),
- the Hamilton R-factor percentage[25] ($HRF$), and
- squared LOO cross–validation correlation coefficient ($Q^2$).

The **\*.DOC** file reports the fitted property values for every fold of five-fold CV:

...

TABLE L1. Training set: Calculated property log_1_C_exp for the compounds from the AHIV-TIBO.SDF file.

| | | 1 | 2 | 3 | 4 | 5 |
|---|---|---|---|---|---|---|
| cmp. no. | Exp. | IAB2-120 | IAB2-120 | IAB2-120 | IAB2-120 | IAB2-120 |
| 1 | 7.34 | | 7.39 | 6.98 | 7.33 | 7.02 |
| 2 | 6.80 | 7.06 | | 6.98 | 7.03 | 7.02 |
| 3 | 5.20 | 5.49 | 5.26 | | 5.27 | 5.57 |
| 4 | 4.64 | 4.72 | 4.51 | 4.37 | | 5.20 |
| 5 | 4.49 | 4.33 | 4.51 | 4.37 | 4.39 | |

The **\*.AVE** file contains the average fitted property and its standard deviation calculated from the data given in the described above **\*.DOC** file.

In the **\*.ECV** file, predicted in 5-CV property values (right column) are compared with the experimental ones given in the *Datum* column. If a compound is identified as being outside AD of the model, the predicted value for this compound is excluded (e.g., the case of compound 11):

TABLE P1. Test set: Predicted property log_1_C_exp for the compounds from the AHIV-TIBO.SDF file.

| cmp. no. | Datum | IAB2-120 |
|---|---|---|
| 1 | 7.34 | 7.27 |
| 6 | 6.17 | 5.68 |
| 11 | 4.32 | |
| 16 | 7.11 | 6.75 |
| 21 | 4.84 | 6.12 |

The plots display the relationship between observed $Y_{exp}$ and predicted $Y_{pred}$ (or fitted $Y_{calc}$) property as well as corresponding linear equation and its statistical parameters, including data points for all folds cross-validation. Clicking on selected data point visualizes corresponding molecular structure and the $Y_{exp}$ and $Y_{pred}$ ($Y_{calc}$) values.

**Exercise 9.4**  Consensus model: obtaining and validation.

The ISIDA_QSPR program may generate many different linear models, each involving particular set of SMF descriptors and /or variable selection technique. The individual models are recruited into the consensus model according to two criteria: the LOO cross–validation correlation coefficient $Q^2$ should be larger than a user defined threshold $Q^2_{lim}$ and a residual $(R^2 - Q^2)$ between the squared correlation coefficient $(R^2)$ and $Q^2$ should also be larger than $(R^2 - Q^2)_{lim}$ threshold (Figure 9.12). The program then applies this consensus model (CM) to every test compound, that is, predicts the target property as an arithmetic average of the values estimated by selected individual models. A given individual model doesn't contribute into consensus calculations if it produces the outliers according to Tompson's rule[26] or it can't be applied to a given compound due to applicability domain (AD) problem (Figure 9.12). Three types of AD criteria can be used simultaneously or individually: fragment control, bounding box,[21] and "quorum control."[22]

In this tutorial, we use only a few descriptor types and one fitting equation type leading to generation of 144 individual models. It ensures quick calculations and demonstrates ensemble learning and predictions by consensus model.

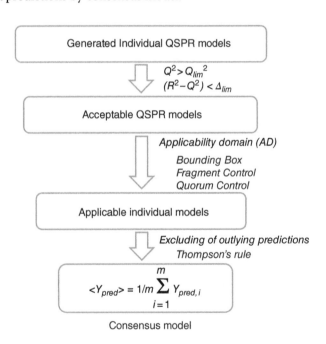

Figure 9.12  Consensus calculations based on ensemble of selected individual models.

**Loading Structure-Data File**

Restart ISIDA_QSPR, upload input SD file and select the modelling property as shown in Exercise 9.1. Click on the **Create MASK** button. The Create Mask Dialog box appears (Figure 9.4). Click on the Create MASK Dialog box, **No test set** radio button. Click on **START** to save or overwrite the mask file **AHIV-TIBO.MSK** in the ISIDA_QSPR directory using the Save mask file Dialog. The **MASK File: AHIV-TIBO.MSK** label appears in the **Data** panel.

Click on **Data TEST** to verify the consistency of input data (Figure 9.2). The **Information** dialog box appears with message: "Input data files are in internal agreement." Close the **Information** dialog box and then the **Single Model Calculations** dialog box by clicking on the **CANCEL** button (Figure 9.2).

### Descriptors and Fitting Equation

Click on the **Consensus Model** button of the ISIDA_QSPR program (Figure 9.1) to open the **Consensus Modelling Calculations** Dialog box (Figure 9.13) used to enter parameters for consensus model. The dialog box includes the **Data** panel for data input, the **Descriptors** panel for selection of the SMF descriptor types, the **Model** panel for options of individual MLR equation types and forward and backward stepwise variable selection techniques, and the **Validation** panel for parameters of internal and external individual model valida-tion (see details about validations in the Exercises 9.1 and 9.3).

From the **Descriptors** panel (Figure 9.13), check the **Paths: Atoms-Bonds** check box and then the **equal length paths** check box, enter 2-3 in the **min** edit box and 7-12 in the **max** edit box for minimal and maximal path lengths. Please verify that there are no addi-tional check marks in the check boxes of the **Descriptors** panel. Use by default 3 in the **min count** edit box and 1 in the **min compds** edit box (Figure 9.13). This set up leads to generation of two descriptor classes: *a*) shortest topological paths with explicit atoms and bonds and *b*) similar shortest paths including paths of equal length. For every class of the

Figure 9.13  The consensus model calculations Dialog box.

sequences, the minimal ($2 \leq n_{min} \leq 3$) and maximal ($7 \leq n_{max} \leq 12$) numbers of constituent atoms ($n$) are defined. The sequences include all intermediate shortest paths with $n$ atoms: $n_{min} \leq n \leq n_{max}$, thus leading to generation of 24 types of fragment descriptors.

From the **Model** panel (Figure 9.13), click on the **Y=SUM(Ai*Xi)** check box for the selection of one linear fitting equation type only.

Variables Selection

From the **Model** panel (Figure 9.13), check the **FVS** (forward variable selection) **methods** and **backward variable selection** check boxes in the **Variable Selection** subpanel. On the right of the FVS methods' combo box, enter 2,3 in the first edit box for the selection of the $R_{y,i}$ and $R_{y,ij}$ variable selection algorithms.[4] In the second edit box, enter 50-70 for the scalable numbers of pre-selected variables presented as the percentage of the training set size ($m$). In this case, $0.5m$, $0.6m$ and $0.7m$ variables will be pre-selected by the $R_{y,i}$ ($R_{y,ij}$) algorithm and sequentially applied to individual model preparations. Enter 0.001 in the $R_{yi}$ edit box and 0.99 in the $R_{ij}$ edit box for the correlation coefficient thresholds. Enter 1.96 in the **t-test** edit box, 1E-12 in the **eps** edit box and 0 in the **exact** edit box. Make sure that the **m > n** check box is not selected (Figure 9.13).

Consensus Model

From the **Model** panel (Figure 9.13), enter 0.7 in the **Q2 lim** edit box for the threshold $Q^2_{lim}$ of minimal LOO cross–validation correlation coefficient ($Q^2$) of acceptable individual models. Enter 0.1 in the **R2 – Q2** edit box for the threshold of maximal residual between the squared correlation coefficient ($R^2$) and $Q^2$ of acceptable individual models.

Model Applicability Domain

From the **Validation** panel (Figure 9.13), user can select three AD approaches: fragment control (AD1), bounding box (AD2), and "quorum control" (AD3). Here, check the **Appl. Domain 1** check box and uncheck the **Appl. Domain 2** and **Appl. Domain 3** check boxes.

n-Fold External Cross-Validation

From the **Validation** panel (Figure 9.13), check the **LOO** and **rapid** check boxes for fast calculation of Q, enter 5 in the **External n-fold CV** edit box for the execution of the external five-fold cross validation.

Saving and Loading of the Consensus Modeling Results

The program saves the output files in user-defined directory by clicking on the **Select Directory** button (Figure 9.13). The Open dialog box appears, where click on **Open** to select a directory, for instance, C:\ISIDA_QSPR\RESULTS.

The output files can always be opened by clicking File → Open → *.out file in the ISIDA_QSPR main menu (Figure 9.1). Typically, the *.out file includes the name of the input SD file name as substring and begins with the date and the time of the performed calculations. Verify that the **Save Models** check box is not selected in the right corner of the dialog box (Figure 9.13).

## Statistical Parameters of the Consensus Model

In order to obtain an ensemble of individual models forming CM, click on **START** in the **Consensus Modelling Calculations** Dialog box (Figure 9.13). The program creates eight output files: six plain text files and two files of the graphical presentation of results, see their description below.

The **\*.TOM** file contains statistical parameters of the individual MLR models for every fold of CV:

```
5-Fold External Cross-Validation Procedure. Table of models.

=>  Subset 1/5

File of Mol Structures: AHIV-TIBO.SDF;  45 compounds in
training set.
Modeling Property Name: log_1_C_exp
Mask File:              AHIV-TIBO.MSK
```

| no | fragment type | fitting equation | n | k | R2 | F | FIT | s | HRF | Q2 |
|----|---------------|------------------|----|----|----------|-------|-------|----------|-------|----------|
| 1 | IAB3-837 | 0 | 45 | 13 | 0.951211 | 51.99 | 3.301 | 3.44E-01 | 5.043 | 0.904357 |
| 2 | IAB2-1036 | 0 | 45 | 13 | 0.943843 | 44.82 | 2.846 | 3.69E-01 | 5.410 | 0.901341 |
| 3 | IAB2-1136 | 0 | 45 | 13 | 0.943843 | 44.82 | 2.846 | 3.69E-01 | 5.410 | 0.901341 |
| 4 | IAB2-1236 | 0 | 45 | 13 | 0.943843 | 44.82 | 2.846 | 3.69E-01 | 5.410 | 0.901341 |
| 5 | IAB2-10a26 | 0 | 45 | 10 | 0.933738 | 54.80 | 3.914 | 3.83E-01 | 5.877 | 0.896698 |

...

For each individual model, the following parameters are reported: the number of the data point ($n$) in the training set at a given fold CV, the number of fitted variables ($k$), squared Pearson correlation coefficient ($R^2$), the Fischer criterion ($F$), the Kubinyi fitness criterion[19] (*FIT*), standard deviation ($s$), the Hamilton R-factor percentage[25] (*HRF*), and squared LOO cross–validation correlation coefficient ($Q^2$). The models are sorted according to $Q^2$ in descending order.

The **\*\_TST\_5fCV.AVE** file reports for the compounds of the test set at the given fold the following parameters: average predicted property values (*Average*) and their standard deviation (*STDEV*) estimated by consensus models and the number of the individual models (*Nm*) used for the *Average* value calculation. If a compound is identified as being outside AD of any individual model, no predictions are reported (e.g., see compound 11):

```
TABLE PA. Test set: Average predicted property log_1_C_exp
```

| cmp. no. | Datum | Average | STDEV | Nm | Dat.- Ave. |
|----------|-------|-------------|------------|-----|------------|
| 1 | 7.34 | 7.23103E+000 | 4.080E-001 | 97 | 1.090E-001 |
| 6 | 6.17 | 6.09287E+000 | 3.144E-001 | 80 | 7.713E-002 |
| 11 | 4.32 | | | 0 | |
| 16 | 7.11 | 6.42392E+000 | 3.050E-001 | 102 | 6.861E-001 |

...

The **.TSP** file contains property values predicted by individual MLR models for every fold of cross-validation:

```
5-Fold External Cross-Validation =>  Subset 1/5
File of Mol Structures: AHIV-TIBO.SDF
Property Name:              log_1_C_exp
Mask File:                  AHIV-TIBO.MSK


TABLE P1. Test set: Predicted property log_1_C_exp

            102 Selected MODELs, Q2 >= 0.7
cmp. no.    Datum     IAB3-8370     IAB2-10360     IAB2-11360    ...
     1      7.34        6.67           7.27           7.27       ...
     6      6.17        5.80           5.68           5.68       ...
    11      4.32                                                 ...
    16      7.11        6.34           6.75           6.75       ...
    21      4.84        5.50           6.12           6.12       ...
...
```

Remaining three text files, similarly to described above *_TST_5fCV.AVE and *.TSP files, contain information about fitted property values for the compounds of training sets at each fold.

The plots display the relationship between observed $Y_{exp}$ and predicted $Y_{pred}$ (or fitted $Y_{calc}$) property as well as corresponding linear equation and its statistical parameters, including data points for all folds cross-validation. Clicking on selected data point visualizes corresponding molecular structure and the $Y_{exp}$ and $Y_{pred}$ ($Y_{calc}$) values.

### Consensus Model Performance as a Function of Individual Models Acceptance Threshold

The consensus model predictive performance depends on recruited individual models which selection, in turn, depends on user-defined threshold $Q^2_{lim}$ for determination coefficient of LOO calculations (see introduction to Exercise 9.4). Treating the output of consensus modeling the user can build a plot of the dependence of determination coefficient $R^2_{det}$ for predictions in $n$-fold CV as a function of $Q^2_{lim}$, which may help to determine an optimal $Q^2_{lim}$ value providing with reasonable $R^2_{det}$.

Click **Tools → R2det vs Q2** of the ISIDA_QSPR main menu (Figure 9.1) to open the **Averaging** tool (Figure 9.14), which enables to explore $n$-fold CV determination coefficient $R^2_{det}$ as a function of $Q^2_{lim}$. In this window check **use Q2 threshold** and **do R2det vs Q2** boxes, in accordance with the **Q2 lim** for Consensus Model, enter 0.7 in the **use Q2 threshold** edit box, enter the increment value 0.05 in the **Step** edit box. In order to load consensus model information, click **File → Open** of the **Averaging** window. In the Open dialog box select the < date_time_ > **AHIV-TIBO_5fCV.TSP** file from the list of the *.TSP files in the directory, where the consensus modelling results are saved (e.g., C:\ISIDA_QSPR\RESULTS directory) and then click **Open**. The Open dialog box appears again. Select the proper < date_time_ > **AHIV-TIBO_5fCV.TOM** file from the list of the *.TOM files in the same directory, click the **Open** button. After the **Averaging** tool performed calculations, click the **Graph** tab (Figure 9.14). The plot displays the relationship between $R^2_{det}$ and $Q^2_{lim}$. One can see that $R^2_{det}$ insignificantly increases with $Q^2_{lim}$: $R^2_{det}=0.881$ at $Q^2_{lim}=0.70$ and $R^2_{det}=0.883$ at $Q^2_{lim}=0.85$. The complementary textual information related to this plot is available in the **Table** tabsheet (Figure 9.14).

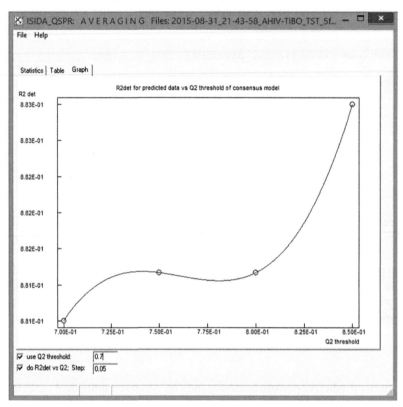

Figure 9.14 The Averaging tool graphic interface.

### Building Consensus Model on the Entire Data Set

This section explains how to build and save the CM on the entire data set. Click the **Consensus Model** button of the ISIDA_QSPR program (Figure 9.1) to open the **Consensus Modelling Calculations** Dialog box (Figure 9.13). Keep all previously used settings (Figure 9.13) except the number of folds and the name of directory for output files. Enter 1 in the **External n-fold CV** edit box to deactivate n-fold CV. Check the **Save Models** check box, click on the **Select Directory** button. In the Open dialog box select a directory (e.g., C:\ISIDA_QSPR\AHIV-TIBO_MODELS). Click **START** to prepare and save a set of individual MLR models.

The program creates and opens five output files: four plain text files and one file of the graphical presentation of results which are similar to the files for training subsets in *n*-fold CV described above in the section "Statistical Parameters of the Consensus Model". The consensus model is described in AHIV-TIBO.TSC and AHIV-TIBO.TOM files containing information about 117 constituting individual models (the *.SPE files).

**Exercise 9.5**   Property predictions and virtual screening using consensus models. This exercise demonstrates the Consensus Predictor program tool,[27] which applies previously obtained consensus models to an external data set. As an input, Consensus Predictor uses chemical structures in SDF format.[8] The input can also include experimental or estimated property values in a data field named as a property aimed to be predicted with the help of stored QSPR model. In this case, this input value will be compared with the predicted one followed by the assessment of prediction performance.

Loading Input Data

Click **Tools → Property Prediction** of the ISIDA_QSPR main menu (Figure 9.1) to open the Consensus Predictor graphical interface (Figure 9.15). Click **LOAD**, then select in the Open dialog box the TEST_AHIV-TIBO.SDF file from the list of the ISIDA_QSPR directory. The < ... >\ISIDA_QSPR\TEST_AHIV-TIBO.SDF string appears under the upper LOAD button indicating that the selected file is downloaded. At the same time, the output file name < ... >\ISIDA_QSPR\TEST_AHIV-TIBO_FMF.TSP appears under the SAVE button.

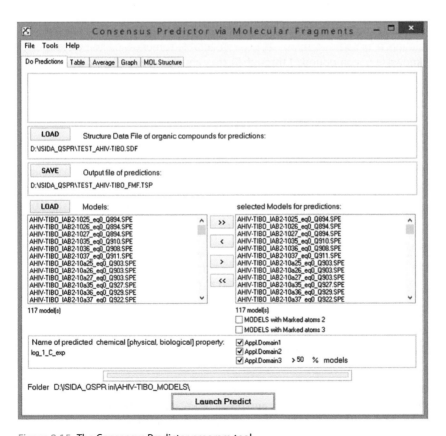

Figure 9.15  The Consensus Predictor program tool.

Click the lower **LOAD** button on the Consensus Predictor window (Figure 9.15). In the Open dialog box open the AHIV-TIBO_MODELS directory containing stored MLR models (see the section "Building Consensus Model on the Entire Data Set"), select and open any *.SPE file from the list. The *.SPE file names appear in the left list box of Consensus Predictor. Click the >> button to select all *.SPE files. The list of the 117 models appears in the right list box. Make sure that the **MODELS with Marked atoms 2** and **MODELS with Marked atoms 3** check boxes are not selected.

The program uses three types of AD definitions to ensure reliable predictions. Check the **Appl. Domain1** (AD1), **Appl. Domain2** (AD2), and **Appl. Domain3** check boxes for the strict AD control. Enter **50** in the edit box of percentage of applicable individual models (i.e., the models for which AD1 and AD2 do not discard the given molecule). If this number is lower than a threshold, the overall CM prediction is ignored.

Click the **Launch Predict** button of Consensus Predictor (Figure 9.15) and agree to overwrite the output *.TSP file if it does exist. Results of the calculations are given in four tabsheets of the Consensus Predictor window. The **Average** tabsheet reports for each molecule the average predicted property value (*Average*), its standard deviation (*STDEV*) estimated by CM and the number of the individual models (*Nm*) used for the *Average* value calculation. If for some compounds, experimental or somehow estimated property values are known (e.g., compounds 1 through 5); they are displayed in the *Datum* column. If a compound is identified as being outside AD, the predicted value for this compound is not given (e.g., for compound 11):

```
TABLE PA. Average predicted property log_1_C_exp
```

| cmp. no. | Datum | Average | STDEV | Nm | Dat.- Ave. |
|---|---|---|---|---|---|
| 1 | 7.92 | 7.96224E+000 | 5.925E-002 | 74 | -4.224E-002 |
| 2 | 7.64 | 7.59814E+000 | 1.160E-001 | 99 | 4.186E-002 |
| 3 | 8.30 | 8.31903E+000 | 6.017E-002 | 74 | -1.903E-002 |
| 4 | 7.86 | 7.50816E+000 | 3.707E-002 | 81 | 3.518E-001 |
| 5 | 7.53 | 7.50816E+000 | 3.707E-002 | 81 | 2.184E-002 |
| 6 | - | 6.67168E+000 | 4.470E-001 | 117 | |
| 7 | - | 7.07149E+000 | 1.500E-001 | 95 | |
| 8 | - | 7.53993E+000 | 1.078E-001 | 87 | |
| 9 | - | 8.98958E+000 | 1.957E-001 | 91 | |
| 10 | - | 7.48592E+000 | 4.201E-001 | 111 | |
| 11 | - | | | 0 | |
| 12 | - | 8.47588E+000 | 2.735E-001 | 98 | |

Click the **Graph** tab. For the compounds 1 through 5 with known activity, a plot displays a linear correlation between observed $Y_{exp}$ and predicted $Y_{pred}$ property and related statistical parameters. Click on selected data point, and then click the **MOL Structure** tab in order to visualize a corresponding molecular structure.

### Analysis of the Fragments Contributions

Click **Tools → Fragment Contributions** of the Consensus Predictor main menu (Figure 9.15) to open the *Fragment Contributions for Molecule* window (Figure 9.16). This window includes 2D structure , its constituting fragments, their occurrence, and contributions in the context of selected individual MLR model. Enter **9** in the **Mol Number** edit box to examine the compound predicted as the most active, select any individual model by the dropdown **Selected Model**.

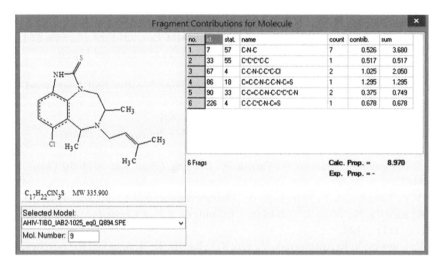

**Figure 9.16** The graphic window of Fragment Contributions for Molecule.

## References

1 Solov'ev, V. P.; Varnek, A. A., *ISIDA (In Silico Design and Data Analysis) program*; version 5.79; 2008–2014. http://infochim.u-strasbg.fr/spip.php?rubrique53; http://vpsolovev.ru/programs/ (accessed 21 August 2014).

2 Solov'ev, V. P.; Varnek, A.; Wipff, G., *J. Chem. Inf. Comp. Sci.* 2000, **40**, (3), 847–858.

3 Solov'ev, V. P.; Varnek, A. A., *Rus. Chem. Bull.* 2004, **53**, (7), 1434–1445.

4 Solov'ev, V. P.; Kireeva, N.; Tsivadze, A. Y.; Varnek, A., *J. Incl. Phenom. Macrocycl. Chem.* 2013, **76**, (1–2), 159–171.

5 Solov'ev, V. P.; Varnek, A., *J. Chem. Inf. Comp. Sci.* 2003, **43**, (5), 1703–1719.

6 Varnek, A.; Solov'ev, V. P., *Comb. Chem. High Throughput Screening* 2005, **8**, (5), 403–416.

7 Solov'ev, V. P.; Varnek, A. A., *EdChemS (Editor of Chemical Structures)*; version 2.6; 2008–2014. http://infochim.u-strasbg.fr/spip.php?rubrique51; http://vpsolovev.ru/programs/ (accessed 21 August 2014).

8 Dalby, A.; Nourse, J. G.; Hounshell, W. D.; Gushurst, A. K. I.; Grier, D. L.; Leland, B. A.; Laufer, J., *J. Chem. Inf. Comput. Sci.* 1992, **32**, (3), 244–255.

9 Solov'ev, V.; Oprisiu, I.; Marcou, G.; Varnek, A., *Ind. Eng. Chem. Res.* 2011, **50**, (24), 14162–14167.

**10** Ruggiu, F.; Solov'ev, V.; Marcou, G.; Horvath, D.; Graton, J.; Le Questel, J.-Y.; Varnek, A., *Mol. Inf.* 2014, *33*, (6–7), 477–487.

**11** Varnek, A. A.; Wipff, G.; Solov'ev, V. P., *Solv. Extr. Ion. Exch.* 2001, *19*, (5), 791–837.

**12** Varnek, A. A.; Wipff, G.; Solov'ev, V. P.; Solotnov, A. F., *J. Chem. Inf. Comput. Sci.* 2002, *42*, (4), 812–829.

**13** Varnek, A.; Fourches, D.; Solov'ev, V. P.; Baulin, V. E.; Turanov, A. N.; Karandashev, V. K.; Fara, D.; Katritzky, A. R., *J. Chem. Inf. Comp. Sci.* 2004, *44*, (4), 1365–1382.

**14** Katritzky, A. R.; Fara, D. C.; Yang, H.; Karelson, M.; Suzuki, T.; Solov'ev, V. P.; Varnek, A., *J. Chem. Inf. Comp. Sci.* 2004, *44*, (2), 529–541.

**15** Varnek, A.; Fourches, D.; Hoonakker, F.; Solov'ev, V. P., *J. Comput. Aid. Mol. Des.* 2005, *19*, (9-10), 693–703.

**16** Solov'ev, V.; Varnek, A.; Tsivadze, A., *J. Comput. Aided Mol. Des.* 2014, *28*, (5), 549–564.

**17** Reid, R. C.; Prausnitz, J. M.; Sherwood, T. K., *The Properties of Gases and Liquids.* McGraw-Hill Book Co: New York, 1977.

**18** Lawson, C. L.; Hanson, R. J., *Solving Least Squares Problems.* Prentice Hall, Englewood Cliffs: New Jersey, 1974.

**19** Kubinyi, H., *Quant. Struct.-Act. Relat.* 1994, *13*, (4), 393–401.

**20** Zhokhova, N. I.; Baskin, I. I.; Palyulin, V. A.; Zefirov, A. N.; Zefirov, N. S., *Doklady Chemistry* 2007, *417*, (2), 282–284.

**21** Solov'ev, V. P.; Oprisiu, I.; Marcou, G.; Varnek, A., *Ind. Eng. Chem. Res.* 2011, *50*, (24), 14162–14167.

**22** Solov'ev, V. P.; Tsivadze, A. Y.; Varnek, A. A., *Macroheterocycles* 2012, *5*, (4-5), 404–410.

**23** Varnek, A.; Kireeva, N.; Tetko, I. V.; Baskin, I. I.; Solov'ev, V. P., *J. Chem. Inf. Model.* 2007, *47*, (3), 1111–1122.

**24** Tetko, I. V.; Solov'ev, V. P.; Antonov, A. V.; Yao, X. J.; Fan, B. T.; Hoonakker, F.; Fourches, D.; Lachiche, N.; Varnek, A., *J. Chem. Inf. Model.* 2006, *46*, (2), 808–819.

**25** Hartley, F. R.; Burgess, C.; Alcock, R. M., *Solution Equilibria.* John Wiley: Chichester, 1980; p 361.

**26** Muller, P. H.; Neumann, P.; Storm, R., *Tafeln der mathematischen Statistik.* VEB Fachbuchverlag: Leipzip, 1979; p 280.

**27** Solov'ev, V. P., *FMF (Forecast by Molecular Fragments)*; version 2.5; 2014. http://vpsolovev.ru/programs/ (accessed 21 August 2014).

10

# Cross-Validation and the Variable Selection Bias

*Igor I. Baskin, Gilles Marcou, Dragos Horvath, and Alexandre Varnek*

*Goal*: To demonstrate the danger of the variable selection bias and the need for external cross-validation for correct assessment of the prediction performance of QSAR models.
*Software*: WEKA
*Data*:

- `alkan-bp-louse100.arff` – dataset with 100 descriptors taking random values
- `alkan-bp-louse1000.arff` – dataset with 1000 descriptors taking random values
- `preselected_descr.arff` – dataset with preselected descriptors
- `knn_descr.arff` – dataset with descriptors selected by the kNN procedure

A database containing the values of the boiling points of 74 alkanes[1] is used. Two sets of descriptors are used: 100 and 1000 descriptors representing random numbers. The *k* Nearest Neighbors (kNN, or IBk in Weka's terminology) and Multi-Linear Regression (MLR) methods will be used for the modeling.

## Theoretical Background

The *n*-fold cross-validation technique[2] is widely used to estimate the performance of QSAR models. In this procedure, the entire dataset is divided into *n* non-overlapping pairs of training and test sets. Each training set covers $(n-1)/n^{th}$ of the dataset while the related test set covers the remaining $1/n^{th}$. Following developments of models with the training set, the predictions for the test set are performed. Thus, predictions are made for all molecules of the initial dataset, since each of them belongs to one of the test sets. This tutorial demonstrates the danger of the variable selection bias and the need for

*Tutorials in Chemoinformatics*, First Edition. Edited by Alexandre Varnek.
© 2017 John Wiley & Sons Ltd. Published 2017 by John Wiley & Sons Ltd.
Companion website: www.wiley.com/go/varnek/chemoinformatics

external cross-validation for correct assessment of the prediction performance of QSAR models based on automatically selected descriptors. In principle, two scenarios are possible:

- selection of descriptors using all molecules from the entire data set followed by $n$-fold cross-validation (*internal CV*);
- selection of descriptors using the molecules from the corresponding training set on each fold of cross-validation (*external CV*).

The models obtained on each fold of the *internal CV* are based on the same set of descriptors, whereas corresponding models for the *external CV* may involve different descriptors. Notice that frequently reported *Leave-One-Out* cross-validation typically represents the *internal CV*. Here, we will show that only *external CV*, in which the information of test compounds is not used for the development of the models, can be used for reasonable assessment of accuracy of predictions (see also [3–5]). The other setups lead to a biased accuracy assessment, which is sometime termed the Freedman paradox.

For this purpose, an experimental setup similar to the original design of Freedman[6] is proposed. A set of random numbers will be used as molecular descriptors to develop a model for the normal boiling point of alkanes. Thus, a procedure leading to "robust" QSAR models, for example, models with high Pearson correlation coefficient $R^2$ based on those descriptors is, definitely not correct one. The tutorial consists of three parts:

- descriptors selection is performed before the model development (internal CV, Figure 10.1A);
- descriptors selection is used to optimize the model (internal CV, Figure 10.1B);
- descriptors are selected independently on each fold of cross-validation (external CV, Figure 10.1C).

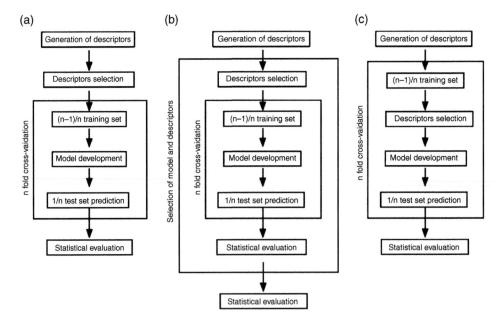

Figure 10.1 Internal (a, b) and External (c) cross-validation procedures.

In order to select descriptors (attributes in Weka's terminology), one should specify a descriptor search algorithm and specify a way to compute the value being optimized in the course of descriptors selection.

The default setting for descriptor search is to use the *BestFirst* algorithm.[7,8] It searches the space of descriptor subsets by greedy hill-climbing augmented with a backtracking facility. The *BestFirst* method may start with the empty set of descriptors and searches forward (default behavior), or starts with the full set of attributes and searches backward, or starts at any point and searches in both directions (by considering all possible single descriptor additions and deletions at a given point). The method is labeled as *weka - > attributeSelection - > BestFirst.*

Another way to assess the utility of a descriptor subset is the wrapper method.[9] In this method, the minimization of the cross-validation error of classification or regression model built on subsets of descriptors is used to guide the descriptors selection. The method is labeled as *weka - > attributeSelection - > WrapperSubsetEval.*

The default method to evaluate the worth of a subset of descriptors consists in considering the individual predictive ability of each one along with the degree of redundancy between the descriptors. Subsets of descriptors that are highly correlated with the property/activity values and having low inter-correlation are preferred.[7] The method is labeled as *weka - > attributeSelection - > CfsSubsetEval.*

It is possible to combine a descriptor selection procedure with a regression or classifiation algorithm as a whole. The classifier method *weka - > classifiers - > meta - > AttributeSelectedClassifier* applies an arbitrary classifier to the data that has been reduced through molecular descriptor selection procedure.[7] Using the modeling setup within a cross-validation loop, the descriptors are selected in each fold on the training set, while the test set is only accessed when the model is achieved.

## Step-by-Step Instructions

1) *Internal Cross-Validation using Pre-selected Descriptors*
In the first part of this tutorial, the naïve strategy (Figure 10.1a) is applied: descriptors selection is performed before the model development. It is shown that this leads to inconsistent results due to the models strongly biased toward the training set.

| Instructions | Comments |
| --- | --- |
| • In the **Preprocess** tab, click on the button **Open File**. In the file selection interface, select the directory Alkan, then the file alkan-bp-louse1000.arff.<br>• In the **Select Attribute** tab, examine the **Attribute Evaluator** and the **Search Method.**<br>• Click on **Start**. | Load the dataset with 1000 descriptors taking random values. For this problem, **BestFirst** (default search method) and **CfsSubsetEval** (default attribute evaluator) combination is as efficient as best variable selection techniques—genetic algorithm or simulated annealing—but it is much faster. This is why these default settings are kept unchanged (Figure 10.2). |

*(Continued)*

**Instructions**                    **Comments**

Figure 10.2 Interface of the attribute selection mode of Weka Explorer. An attribute selection process is a combination of an attribute evaluator and a search method.

- Click on the new line in the *Result list* with the right mouse button.
- From the pop-up menu, select the item *Save reduced data...*
- Switch to the **Preprocess** submode of the **Explorer** mode.
- Click on *Open file* and open the file *preselected_descr.arff*.
- Switch to the *Classify* submode.
- Click on *Choose.*

From the hierarchical list of machine learning methods choose *weka-> classifiers -> functions -> LinearRegression*.

- To change options for the MLR method, click on **LinearRegression.**
- Change **attributeSelectionMethod** to *No attribute selection*.

The modeling will continue with a dataset reduced to the sole descriptors selected by the algorithm (Figure 10.3). Save the dataset with 30 selected descriptors to a file named *preselected_ descr.arff*.

Selected descriptors are supposed to be used in the Multiple Linear Regression model.

The loaded dataset contains only the previously selected descriptors. Therefore no further selection is needed. Since the selected descriptors are uncorrelated by construction, the value of the parameter *eliminateColinearAttributes* is inconsequential. All settings are shown on the snapshot below (Figure 10.4).

| Instructions | Comments |
|---|---|

Figure 10.3  The Attribute Selection interface while saving the dataset using only the attributes selected by the combination of the CfsSubsetEval evaluator with the BestFirst attribute search method.

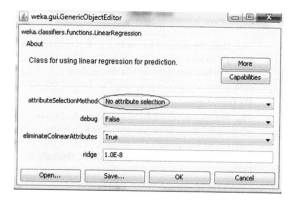

Figure 10.4  Configuration interface for the Multi-linear regression algorithm using the previously selected descriptors.

*(Continued)*

| Instructions | Comments |
|---|---|
| • Press the **OK** button in the MLR configuration window.<br><br>• Press the **Start** button in order to test MLR on the subset with selected descriptors. | The default setting is to build a model on the entire dataset, which is reported in the output widows, then to report the 10-fold cross-validated measures of success of the model (Figure 10.5).<br><br>Thus, this study shows that the cross-validated correlation coefficient between estimated values and their experimental property value is 0.8214. This lures us into accepting a statistical significance of the model based on random numbers. A similar situation is observed for some other popular machine learning methods (PLS, kNN, etc.). Hence, the whole procedure of building the model using a set of pre-selected descriptors is erroneous. |

```
Correlation coefficient              0.8214
Mean absolute error                 22.2335
Root mean squared error             27.3909
Relative absolute error             67.1857 %
Root relative squared error         58.9268 %
Total Number of Instances           74
```

Figure 10.5 Cross-validated results of the MLR model built on the previously selected descriptors.

2) *Internal cross-validation using descriptors selected in the course of model building*
In this section, the *k* Nearest Neighbor approach will be used both to optimize the *k* parameter and to select an optimal set of descriptors by minimizing the cross-validation error of the method (Figure 10.1b).

| Instructions | Comments |
|---|---|
| • Start or restart *Weka*.<br><br>• Select the item **Explorer** from the **Applications** menu.<br><br>• Click on **Open file…** and load the file *alkan-bp-louse100.arff*.<br><br>• Click on the label **Select attributes**.<br><br>• Click on the **Choose** button under the label **Attribute Evaluator**.<br><br>• Choose menu item *WrapperSubsetEval*.<br><br>• Click on *WrapperSubsetEval*.<br><br>• Click on the **Choose** button near the **classifier** label.<br><br>• Select from the hierarchical list of machine learning methods **weka -> classifiers -> lazy -> IBk**. | This time the dataset has only 100 descriptors taking random values, that is 10 times less than before. Thus the chances of finding their correlations with the property (the boiling point) are reduced.<br><br>The wrapper method for descriptor selection is prepared. The default method *ZeroR* (which consists in computing the average value of the property in the training set) should be changed to kNN.<br><br>Under these settings (Figure 10.6), the minimization of the five-fold cross-validation error of kNN models is used to guide the descriptors selection. |

| Instructions | Comments |
|---|---|
| <ul><li>Click on **IBk**.</li><li>In order to search an optimal value of $k$ in the range from 1 to 10:</li><li>Change **kNN** to 10.</li><li>Select *True* for **crossValidate**.</li></ul> | A window with default parameters of the *kNN* method (fixed value of $k = 1$) appears. The method must be configured to optimize the number of neighbours to use (parameter $k$) in the range from 1 to 10 using a cross-validation procedure (Figure 10.7). |

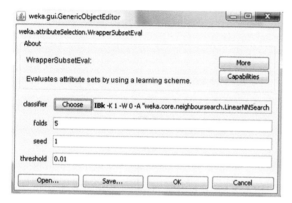

**Figure 10.6** Configuration of the WrapperSubsetEval method, which consists in using the performances of a classification or regression model based on a subset of descriptors to assess their usefulness for modeling.

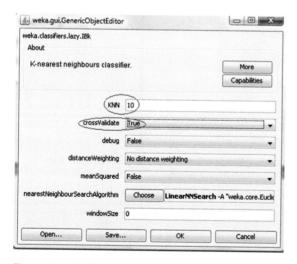

**Figure 10.7** Configuration interface for the WrapperSubsetEval method.

*(Continued)*

| Instructions | Comments |
|---|---|
| • Click on the **OK** button in the window with *k*-Nearest Neighbors classifier.<br>• Click on the **OK** button in the window with WrapperSubsetEval.<br>• In the remaining window click on the **Start** button. | Start the calculations. Results are reported in the Figure 10.8. There are six selected attributes (descriptors). |
| • Click on corresponding line in the *Result list* (left side, on the bottom) with the right mouse button.<br>• From the pop-up menu, select the item **Save reduced data…**<br>• Save the dataset with the selected descriptors to file *knn_descr.arff*. | Save the selected descriptors into a file *knn_descr. arff*. These descriptors will be used to build a kNN model. |

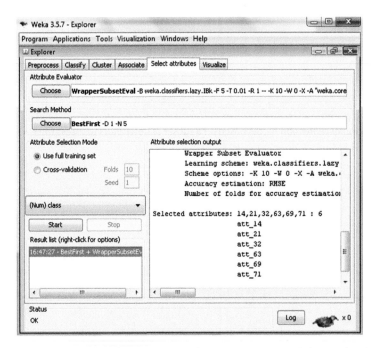

**Figure 10.8** Results of the descriptor selection using WrapperSubsetEval with a kNN model. The descriptor selection is embedded into the cross-validation procedure.

| Instructions | Comments |
|---|---|
| • Switch to the **Preprocess** submode.<br>• Click on the **Open file...** button and open the file *knn_descr.arff*.<br>• Switch to the **Classify** submode.<br>• Click on the **Choose** button near the **classifier** label.<br>• Select from the hierarchical list of machine learning methods **weka/ classifiers/lazy/IBk**.<br>• Click on **IBk**.<br>• Change **kNN** to 10.<br>• Select **True** for **crossValidate**.<br>• Click on the **OK** button in the window with K-nearest neighbors classifier.<br>• Click on the **Start** button. | Load the file *knn_desc.arff* and set up a kNN model using up to 10 nearest neighbors. The optimal number of neighbors is obtained through five-fold cross-validation. The results are reported in Figure 10.9. As in previous case, the correlation coefficient significantly deviates from zero. Hence, the descriptor selection bias takes place in the case of internal cross-validation procedures. |

```
Correlation coefficient          0.548
Mean absolute error             26.6253
Root mean squared error         39.8131
Relative absolute error         80.4571 %
Root relative squared error     85.6509 %
Total Number of Instances       74
```

Figure 10.9 Cross-validated performances of the kNN model using the cross-validated descriptor set by the WrapperSubsetEval method.

### 3) *External cross-validation*

Here, a separate set of selected descriptors is formed in each fold of the cross-validation procedure (Figure 10.1c).

| Instructions | Comments |
|---|---|
| • Start the Weka program.<br>• Select the item **Explorer** from the **Applications** menu.<br>• Click on the button **Open file...** and load the file *alkan-bp-louse100.arff*.<br>• Switch to the **Classify** submode by clicking on label **Classify**.<br>• Click on the **Choose** button near the **classifier** label.<br>• Select from the hierarchical list of machine learning methods **weka -> classifiers -> meta -> AttributeSelectedClassifier**.<br>• Click on the word **AttributeSelectedClassifier**. | The same dataset is loaded once again. The classifier meta-method AttributeSelectedClassifier is used (Figure 10.10). It can be embedded into a cross-validation loop, so, for each fold of the loop, the attribute selection and the regression model are applied to the training set and only the final model with selected descriptors accesses the test set. |

*(Continued)*

| Instructions | Comments |
|---|---|

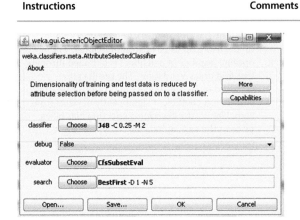

Figure 10.10  Configuration interface of the AttributeSelectedClassifier.

```
Correlation coefficient              -0.2858
Mean absolute error                  38.4063
Root mean squared error             53.6368
Relative absolute error             116.057  %
Root relative squared error         115.3901 %
Total Number of Instances            74
```

Figure 10.11  Cross-validated results of kNN models based on optimal sets of descriptors when only the final model with selected descriptors is used for making predictions in the corresponding test set in each fold of the external cross-validation loop.

- Click on the **Choose** button near the **classifier** label.
- Select from the hierarchical list of machine learning methods **weka -> classifiers -> lazy -> IBk.**
- Click on **IBk.**
- Change **kNN** to 10.
- Select *True* for **crossValidate.**
- Click on the **OK** button in the window with k-Nearest Neighbors classifier.
- Click on the **OK** button in the window with parameters of *AttributeSelectedClassifier*.
- In the remaining window click on *Start*.

The *AttributeSelectClassfier* is configured to use the kNN algorithm optimizing the number $k$ of nearest neighbors in the range from 1 to 10 using a cross-validation loop. The same setup for variable selection as in the first part of the tutorial is used: the descriptor set used for the kNN is an ensemble of the descriptors most collinear to the target property while avoiding being inter-correlated.

Results are reported in Figure 10.11. These results show that the developed models are not predictive at all. This looks reasonable since those models involve random numbers as descriptors.

## Conclusion

External cross-validation should be used to assess predictive performance of QSAR models based on automatically selected descriptors. The larger the dimension of the descriptor vector, the higher the chance of a spurious correlation between the explained variable and some of the descriptor element. Therefore, cross-validation is absolutely necessary (albeit, unfortunately, not sufficient) to avoid falling into the trap of accidental correlations.

# References

1 Needham, D.E., I.C. Wei, and P.G. Seybold, *Molecular modeling of the physical properties of alkanes*. Journal of the American Chemical Society, 1988, **110**(13), 4186–4194.

2 Stone, M., *Cross-validatory choice and assessment of statistical predictions*. Journal of the Royal Statistical Society. Series B (Methodological), 1974, 111–147.

3 Tetko, I.V., et al., *Benchmarking of linear and nonlinear approaches for quantitative structure-property relationship studies of metal complexation with ionophores*. Journal of Chemical Information and Modeling, 2006, **46**(2), 808–819.

4 Tetko, I.V., et al., *Critical assessment of QSAR models of environmental toxicity against Tetrahymena pyriformis: focusing on applicability domain and overfitting by variable selection*. Journal of Chemical Information and Modeling, 2008, **48**(9), 1733–1746.

5 Hawkins, D.M., *The problem of overfitting*. Journal of Chemical Information and Computer Sciences, 2004, **44**(1), 1–12.

6 Freedman, D.A., *A note on screening regression equations*. The American Statistician, 1983, **37**(2), 152–155.

7 Hall, M.A., *Correlation-based feature selection for machine learning*. 1999, The University of Waikato.

8 Rich, E. and K. Knight, *Artificial intelligence*. McGraw-Hill, New York, 1991.

9 Kohavi, R. and G.H. John, *Wrappers for feature subset selection*. Artificial Intelligence, 1997, **97**(1), 273–324.

11

## Classification Models

*Igor I. Baskin, Gilles Marcou, Dragos Horvath, and Alexandre Varnek*

*Goal*: Illustrate the use of simple classification methods and introduce measures of success for classification.
*Software*: WEKA, ISIDA/ModelAnalyzerC
*Data*:

- `A2AC_train.arff` – training set fragment descriptors
- `A2AC_test.arff` – test set fragment descriptors
- `A2AC_external.arff` – external test set fragment descriptors
- `A2AC_train.sdf` – training set molecular structures
- `A2AC_test.sdf` – test set molecular structures
- `A2AC_external.sdf` – external test set molecular structures
- `NB_SMO_kNN.txt` - classes estimations of individual NB, SMO, and kNN models for the test set formatted for the ModelAnalyzerR software.

The dataset contains ligands of the human adenosine receptor (A2A), a member of purinergic receptor family, crystalized recently.[1] A2A is one of the targets of caffeine. Two approved drugs currently target A2A, one agonist and one antagonist. The agonist is used as a coronary vasodilator and the antagonist is used to cure movement disorders induced by Parkinson treatments. There were, in May 2015, 14 entries for the A2A receptor in the Protein Data Bank.

Data for this tutorial were collected from the IUPHAR/BPS, ChEMBL, and PubChem BioAssay in 2012. A compound was classified as active if the pKi > 5. Decoys were randomly chosen from PubChem and have no reported affinity for A2A. The number of decoys was set so that the balance between active and inactive compound is about 15%. The classification data set contains 4948 compounds (756 actives; 4192 inactives), evenly split into a training (385 actives; 2089 inactives) and a test set (371 actives; 2103 inactives). Activity is reported in the SDF field *Activity*: 1 for active compounds and 0 for inactive compounds. The training set and the test set files are named respectively `train.sdf` and `test.sdf` in the folder Classification.

*Tutorials in Chemoinformatics*, First Edition. Edited by Alexandre Varnek.
© 2017 John Wiley & Sons Ltd. Published 2017 by John Wiley & Sons Ltd.
Companion website: www.wiley.com/go/varnek/chemoinformatics

## Theoretical Background

Three methods are illustrated in this tutorial. The first one, the Naïve Bayes algorithm, focuses on a statistical description of the data. The second one, the Support Vector Machine, provides a geometric view of the classification problem. The third one, the *k*-Nearest Neighbor algorithm, relies on the assumption that similar objects should have similar properties.

Then, the tutorial illustrates some measures of success that are recommended for classification problems. There are quite a few measures of success of classifiers.[2] Most of them are based on the confusion matrix. The confusion matrix is a double entry table: one for experimental class and the other for predicted class. Each cell $C_{ij}$ of the table counts the number of instances that are estimated as belonging to a given class $i$ while they are from the class $j$. When estimation and experience coincide for an instance, the corresponding value on the diagonal of the confusion matrix is increased by one— the estimation is accurate in this case, otherwise it is a classification error.

A very common measure of success for solving classification problems is the *accuracy*: the rate of successful classification. Formally, considering $N$ instances distributed over the cells $C_{ij}$ of the confusion matrix for K classes problem, the accuracy is:

$$Accuracy = \frac{1}{N} \sum_{i=1}^{K} C_{ii}$$

As the unbalanced populations of each class affect this measure, it is considered as good practice to look at the *balanced accuracy*: it is the average over all classes of the success rate of classification per class. Therefore, the balanced accuracy (*BA*) is:

$$BA = \frac{1}{K} \sum_{i=1}^{K} \frac{C_{ii}}{\sum_{j=1}^{K} C_{ji}}$$

A more detailed view of the classification problem is obtained by considering the K classes as $K$ binary classification problems. Each binary classification problem consists in classifying items as belonging to one of the $K$ classes or to any others.

The $N$ classified instances are distributed, according to a given class $i$, into four populations which sizes are reported as:

- True Positives (*$TP_i$*), the number of compounds of class $i$ correctly classified as such:

  $$TP_i = C_{ii}$$

- False Positives (*$FP_i$*), the count of compounds erroneously attributed to class $i$:

  $$FP_i = \sum_{j \in \{1..K\} \backslash \{i\}} C_{ji}$$

- True Negatives (*$TN_i$*), the number of compounds that are correctly classified as non-members of class $i$:

  $$TN_i = \sum_{\substack{j \in \{1..K\} \backslash \{i\} \\ k \in \{1..K\} \backslash \{i\}}} C_{jk}$$

- False Negatives ($FN_i$), the count of compounds that were erroneously attributed to class $i$:

$$FN_i = \sum_{j \in \{1..K\} \setminus \{i\}} C_{ij}$$

From those counts, some common measures are synthesized.

The True Positive Rate ($TPR$), Recall ($Rec$), or Sensitivity ($Sen$), measures the proportion of instances of a given class $i$ that are correctly predicted as the members of class $i$:

$$TPR_i = Rec_i = Sen_i = \frac{TP_i}{TP_i + FN_i}$$

The False Positive Rate ($FPR$) reports the proportion of actual non-members of class $i$ erroneously predicted to be in class $i$:

$$FPR_i = \frac{FP_i}{TN_i + FP_i}$$

The True Negative Rate ($TNR$), or Specificity ($Spe$), quantifies the proportion of instances not from class $i$, correctly assigned outside of class $i$:

$$TNR_i = Spe_i = 1 - FPR_i = \frac{TN_i}{TN_i + FP_i}$$

The False Negative Rate ($FNR$) measures the fraction of instances of class $i$ that failed to be recognized as such by the classifier:

$$FNR_i = 1 - TPR_i = \frac{FN_i}{TP_i + FN_i}$$

The Precision ($Prec$) can be considered as a measure of the reliability of the estimation of the classifier. It is the proportion of true predictions, out of all predictions assigning a compound to the class $i$:

$$Prec_i = \frac{TP_i}{TP_i + FP_i}$$

Some indicators are even more elaborated, in an attempt to characterize the performances of the models combining several points of views.

Hence the $F$-measure is the harmonic mean of the precision and the recall, which tends to be closer to the smaller of the two numbers:

$$F_i = \frac{2}{\dfrac{1}{Prec_i} + \dfrac{1}{Rec_i}}$$

The Matthews Correlation Coefficient ($MCC$) is a practical way to compute an analogue of the Pearson's correlation coefficient between the estimated classes and the experimental ones:

$$MCC_i = \begin{cases} \dfrac{TP_i.TN_i - FP_i.FN_i}{\sqrt{(TP_i + FN_i).(TP_i + FP_i).(TN_i + FP_i).(TN_i + FN_i)}}, & \begin{aligned} &(TP_i + FN_i > 0) \, and \\ &(TP_i + FP_i > 0) \, and \\ &|(TN_i + FP_i > 0) \, and \\ &(TN_i + FN_i > 0) \end{aligned} \\[2em] 0, & otherwise \end{cases}$$

The *MCC* is equal to *0* if the denominator of its expression is null.

If instances can be ranked according to *a score*, a number that reflects their likelihood to belong to class *i*, it is recommended to compute the Receiver Operator Curve (ROC) and its Area Under the Curve (ROC AUC).[3] This curve is obtained by using a threshold *t* value taking all possible values for the score. For each value, instances with a score above *t* are attributed to class *i*, thus a confusion matrix and all the aforementioned indicators can be computed for this threshold. The matrix and the indicators become a function of *t*. The ROC is a parametric plot of the *TPR* as a function of the *FPR* using the possible values of *t*. The ROC curve always starts at the point *(0,0)* and ends at the point *(1,1)*. The former is the case when *t* is chosen such as no instance has a score higher than *t* and the later is the situation where all instances have a score higher than *t*.

The ROC is discrete in essence. To compute an area under it, it is necessary to interpolate between two successive points. The preferred way to do is to interpolate linearly. However, it is possible to interpolate in stepwise manner: either by moving parallel to the *FPR* axis or by moving parallel to the *TPR*. In the former case the resulting ROC AUC is pessimistic and in the later, it is optimistic.

Finally, the ROC is equivalent to the Precision-Recall Curves (*PRC*).[4,5] The *PRC* is the parametric plot obtained by plotting the Precision as a function of the Recall for all possible values of the threshold *t*. However, the areas under the curve, the *PRC AUC* and the ROC *AUC*, are different, and optimizing the ROC *AUC* does not optimize the *PRC AUC*. The ROC are much more esthetic and intuitive to interpret. Yet the ROC *AUC* tends to be optimistic for skewed datasets. The *PRC* are much less popular. They are rather unattractive and difficult to read, but they provide interesting details that are hidden in a ROC. The *PRC AUC* is less affected by asymmetry in the data, but the interpolation procedure is not linear and the computation of *PRC AUC* is therefore much more difficult.

If the classification problem is strictly binary, it is straightforward to derive a score from a classifier. The ROC *AUC* for one class is equal to the ROC *AUC* for the other. Besides, the two ROCs are mirror images of each other through the line connecting the points *(0,1)* and *(1,0)* of the plots. In the case of multi-class problems, if a score for each class is not provided, it is impossible to reliably compute a ROC, and no relationship between the ROC curves exists *a priori*.

## Algorithms

The first classification method introduced in this tutorial is the Naïve Bayes algorithm.[6] Given a molecule descriptor vector $x$, the "probability" that the molecule belongs to a class $C$ is dependent on its structure (i.e., on its descriptor vector) and therefore noted $P(C|x)$. One may try to associate patterns $x$ often encountered among known members of a class to a high $P(C|x)$ value. However, this probability cannot be estimated from the observed frequency of the vector $x$ in $C$ because the sampling of $C$ is insufficient: the vector $x$ is often sampled only once and many possible vectors $x'$ are not present in the data set. The problem is therefore rephrased using the *Bayes* formula: $P(C|x) = P(x|C)P(C)/P(x)$. Of course, the sampling problem remains the same. But, by stating *naively* that the components of the vector $x$ are conditionally independent (i.e., they are independent for a given class membership $C$), the probability factorizes: $P(C|x) = \prod_i P(x_i|C)P(C)/\prod_i P(x_i)$. The estimation of $P(x_i|C)$, $P(C)$, and $P(x_i)$ are much better because those events are much better sampled. The problem reduces therefore to estimating the parameters of these probability distributions based on training data. Here, $P(C)$ is the "probability" to encounter a molecule of class $C$, which amounts to the total number of training examples in that class over total training set size. $P(x_i)$ is the "probability" of encountering a molecule with the $i$-th descriptor element taking the value $x_i$—that is the number of known molecules for which this is the case, over the training set size. Eventually, $P(x_i|C)$ is the conditional probability of encountering the descriptor value $x_i$ specifically within the members of class $C$.

The term "probability" is sometimes quoted in the above phrases in order to signal that the mentioned events to which those "probabilities" were associated are not really random. Strictly speaking, it makes no sense to discuss the "probability" that a molecule contains a phenyl ring (suppose $x_i$ is the phenyl ring count). Such terms should be interpreted as frequency measures— they signal how "popular" phenyl rings are in drug candidates included in the training set, and how "popular" phenyl rings are among the actives from that training set. Should the second popularity rate be significantly higher than the first, this is a hint that phenyl rings may be associated with activity: their presence correlates with activity, which may imply that they are involved in binding to the active site—or not! Recall that correlation does not imply causality.

In this tutorial, the model probability distribution of ISIDA descriptors is the multinomial distribution.[7]

The second classification method of interest is the Support Vector Machine (SVM).[8,9] An SVM performs classification using a hyper-plane dividing the chemical space (defined by the high-dimensional descriptor vectors $x$) so that *(i)* it is as far as possible from any molecule and *(ii)* all molecules of one class are on one side of the hyper-plane, which separates them from (almost) all others. The hyper-plane is therefore the optimum of the function $y^t\left(w^T x^t + w_0\right) > 1 - \xi^t$, where $y^t$ is an indicator function so that it is equal to 1 if the molecule $t$ is in the class $C$ and $-1$ otherwise, $x^t$ is the column molecule descriptor vector, $w^T$ is the row vector defining the separating hyper-plane, $w_0$ is the hyper-plane offset and $\xi^t$ is a slack variable of the molecule $t$. The slack variable is 0 if the molecule is on the correct side of the plane and far enough of the plane (that is the molecule is separated from the hyper-plane by the distance exceeding a certain *margin*), while the value of the slack variable exceeds 1 if the molecule is miss-classified and it belongs to the range [0,1] otherwise—on the right side of the hyper-plane, but too close to it. The optimization problem is solved under the constraint of the $\sum_t \xi^t$ being minimum. The optimization problem introduces a Lagrange parameter, called the cost parameter, which defines a tradeoff between following the aforementioned conditions *(i)* and *(ii)*. The higher its value, the more important it is to have a minimum sum of slack variables. This is why generally a high cost value is associated with over-fitting and a low cost value is associated with under-fitting. The problem can be expressed in an abstract feature space using a kernel.[9]

In this tutorial, only a linear kernel, corresponding to the geometric optimization problem described above, is used.

The last method presented is the $k$-Nearest Neighbor (kNN) algorithm.[10,11] Such method is considered *lazy* because it does not learn any model during the training stage. Instead, the dataset is stored. For a query compound, the $k$ most similar compounds in the training set are detected, by calculating their (here, Euclidean) distances in the chemical space defined by the descriptor vectors $x$. The most abundant class represented by those nearest-neighbor compounds is attributed to the query. A possible variation of the algorithm is to weight the decision of each neighbor by a function of the distance.

In Weka, the method *NaiveBayesMultinomial* in the *bayes* methods implements the Naïve Bayes Multinomial algorithm as described in [7]. The method *SMO* in the *functions* method implements a Support Vector Machine algorithm as described in [12–14]. The method *IBk* in the methods *lazy* implements a k-Nearest Neighbor algorithm as described in [10].

## Step-by-Step Instructions

| Instructions | Comments |
|---|---|
| • In the **Preprocess** tab, click on the button **Open File**. In the file selection interface, select the directory A2AC, then the file `train.arff`.<br>• In the **Classify** tab, select **Use training set** inside the **Test options** frame. | Load the training data set of the human adenosine receptor classification problem. Model performances will be assessed with respect to the compounds used to create the model. These are the *fitting* performances. |
| • Click on the button **More options...** Then, near the label *Output predictions*, click the button **Choose** and select the format *CSV*. | The estimated class of each test set compound will be printed in the *Classifier output* frame of the interface. These results can be saved in a file and then reanalyzed with dedicated software. The Comma Separated Value (CSV) format is a convenient exchange format for this purpose. |
| • Click on the **Choose** button in the panel **Classifier**.<br>• Choose the method **weka->classifiers->bayes->NaiveBayesMultinomial** from the hierarchical tree.<br>• Click on the **Start** button to run the Naïve Bayes method. | Results are reported in Figure 11.1. Although the model was trained on the same data on which is was applied, there are some errors. These molecules have an unexpected behavior from the point of view of the model. |
| • Right click on the line of the **Result list** corresponding to your model's report. Then select **Save model**.<br>• Save the model in the current working directory as `A2AC_NB.model`. | The model is saved and can be used later on. |

| Instructions | Comments |
|---|---|
| • Right click on the line of the **Result list** corresponding to your model's report. Then select **Save result buffer**. | The report contains the experimental and estimated class for each of the training instance. This information must be linked with the chemical structures in order to be analyzed. |
| • Save the file in the current working directory as A2AC_fit_NB.out. | |
| • Open the file A2AC_fit_NB.out using the software **ModelAnalyzerC**. Use the file train.sdf as structural source (zone **1** in Figure 11.2). | Analyzing the fitting errors of the model reveals that the phenyl xanthine family (Figure 11.3) is erroneously considered as non-binder of the human adenosine A2A receptor whereas this scaffold is patented for as such.[15] |
| • Then click the **Start** button. Select a class for the **Statistics per class** list box (zone **2** in Figure 11.2). | |
| • Click the confusion matrix cell corresponding to compounds experimentally reported as class 1 but attributed to class 0 (zone **2** in Figure 11.2). | |
| • Search for a point of the ROC curve to decreasing the number of errors of this kind (zone **5** in Figure 11.2). | |
| • In the **Classify** tab, within the **Test options** frame, select **Supplied test set** and click **Set...**. Then click the **Open file...** button and in the directory A2AC select the test.arff file. Then click **Close**. | The same model is applied to the test set. The results are essentially the same as for the fit (Figure 11.4) and are very good. |
| • Right click on the line of the **Result list** corresponding to your model's report. Select the line **Re-evaluate model on current test set**. | A similar analysis reveals that the model incorrectly estimates the class of the same chemical families—in particular phenyl xanthine. |
| • Click on the **Choose** button in the panel **Classifier**. | Select the SVM algorithm. |
| • Choose the method **weka->classifiers->functions->SMO** from the hierarchical tree. | |
| • Click on the word **SMO**. The weka.gui.GenericObjectEditor window to configure the SVM method appears on the screen (Figure 11.5). | Set the cost parameter *c* to 0.1 and do *not* apply any transformation to the dataset. The default configuration generates a linear model. |
| • Edit the value of the parameter *c* and set its value to 0.1. | |
| • Set the *filterType* parameter to *No normalization/standardization*. | |
| • Click on the **Start** button to run the SVM algorithm. | |

```
=== Summary ===

Correctly Classified Instances        2168               89.2181 %
Incorrectly Classified Instances       262               10.7819 %
Total Number of Instances             2430

=== Detailed Accuracy By Class ===

            TP Rate  FP Rate  Precision  Recall  F-Measure  MCC    ROC Area  PRC Area  Class
            0,886    0,076    0,984      0,886   0,933      0,689  0,966     0,991     0
            0,924    0,114    0,602      0,924   0,729      0,689  0,963     0,830     1
Weighted Avg.  0,892  0,082    0,924      0,892   0,901      0,689  0,966     0,966

=== Confusion Matrix ===

    a    b    <-- classified as
 1816  233 |   a = 0
   29  352 |   b = 1
```

**Figure 11.1** Fitting performances of a Naïve Bayes model. The model is applied on the training data set. Note the errors in the confusion matrix.

*(Continued)*

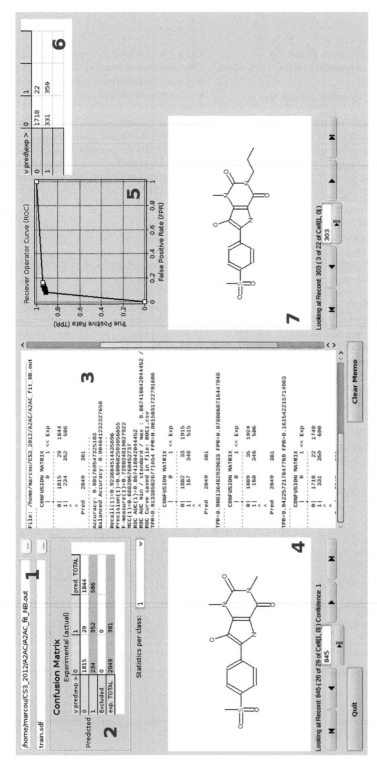

Figure 11.2 Interface of the software **ModelAnalyzerC.** The Weka output is loaded together with a structural file in area **1**. Then the confusion matrix is displayed in **2**. General statistical parameters and specific ones for a selected class are displayed in **3**. The associated ROC is displayed in **5**. The plot is interactive: selecting a point displays the measures of success associated with the selected threshold. It includes an interactive confusion matrix in area 6. The chemical content of any cell of a confusion matrix is displayed in area **4** and **7**.

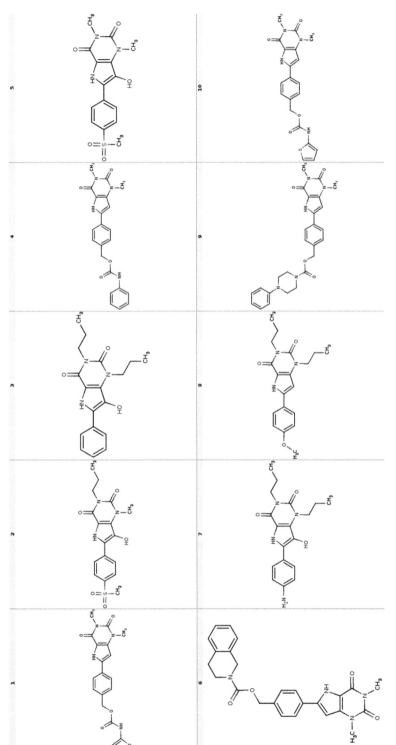

**Figure 11.3** The phenyl xanthine family is incorrectly predicted as non-binders of the human adenosine A2A receptor.

*(Continued)*

```
=== Summary ===

Correctly Classified Instances        2155              88.2835 %
Incorrectly Classified Instances       286              11.7165 %
Total Number of Instances             2441

=== Detailed Accuracy By Class ===
```

| | TP Rate | FP Rate | Precision | Recall | F-Measure | MCC | ROC Area | PRC Area | Class |
|---|---|---|---|---|---|---|---|---|---|
| | 0,875 | 0,074 | 0,985 | 0,875 | 0,927 | 0,664 | 0,961 | 0,990 | 0 |
| | 0,926 | 0,125 | 0,565 | 0,926 | 0,702 | 0,664 | 0,958 | 0,801 | 1 |
| Weighted Avg. | 0,883 | 0,082 | 0,923 | 0,883 | 0,894 | 0,664 | 0,960 | 0,961 | |

```
=== Confusion Matrix ===

   a    b   <-- classified as
1818  259 |   a = 0
  27  337 |   b = 1
```

Figure 11.4  Performances of a Naïve Bayes model on the test set.

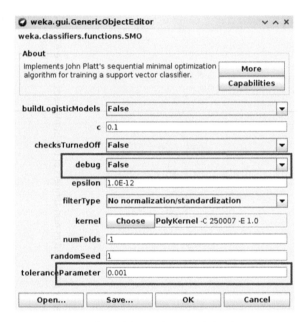

Figure 11.5  Interface of the SMO method. Set the cost parameter c to 0.1 and do not apply any transformation to the dataset.

```
=== Summary ===

Correctly Classified Instances        2421              99.6296 %
Incorrectly Classified Instances         9               0.3704 %
Total Number of Instances             2430

=== Detailed Accuracy By Class ===
```

| | TP Rate | FP Rate | Precision | Recall | F-Measure | MCC | ROC Area | PRC Area | Class |
|---|---|---|---|---|---|---|---|---|---|
| | 0,999 | 0,016 | 0,997 | 0,999 | 0,998 | 0,986 | 0,991 | 0,997 | 0 |
| | 0,984 | 0,001 | 0,992 | 0,984 | 0,988 | 0,986 | 0,991 | 0,979 | 1 |
| Weighted Avg. | 0,996 | 0,014 | 0,996 | 0,996 | 0,996 | 0,986 | 0,991 | 0,994 | |

```
=== Confusion Matrix ===

   a    b   <-- classified as
2046    3 |   a = 0
   6  375 |   b = 1
```

Figure 11.6  Fitting performance of the SVM model. Note that the number of incorrectly classified instances is very small.

| Instructions | Comments |
|---|---|
| ● Right click on the line of the **Result list** corresponding to your SVM model report. Then select **Save model**. | The model is saved and can be used later on. |
| ● Save the model in the current working directory as `A2AC_SVM.model`. | |
| ● Right click on the line of the **Result list** corresponding to your SVM model report. Then select **Save result buffer**. | The report contains the experimental and estimated class for each of the training instance. This information must be linked with the chemical structures in order to be analyzed. |
| ● Save the file in the current working directory as `A2AC_fit_SVM.out`. | |
| | The fitting performances are considerably improved, with respect to the NB model (Figure 11.6). |
| ● Open the file `A2AC_fit_SVM.out` using the software **ModelAnalyzerC**. Use the file `train.sdf` as structural source (zone **1** in Figure 11.2). | Incorrectly estimated structures in the training data are reported in Figure 11.7 and Figure 11.8. |
| ● Then click the **Start** button. Select a class for the **Statistics per class** list box (zone **2** in Figure 11.2). | Classification errors in the decoy set (Figure 11.7) are expected since the binding activity of these compounds on A2A is not measured. They are assumed inactive compounds. On the other hand, the errors on the set of active compounds highlight intriguing situations such as for the compound 4 in Figure 11.8, for which all analogues are inactive compounds. |
| ● Click the confusion matrix cell corresponding to compounds experimentally reported as class 1 but attributed to class 0 (zone **2** in Figure 11.2). | |
| ● Click the confusion matrix cell corresponding to compounds experimentally reported as class 0 but attributed to class 1 (zone **2** in Figure 11.2). | |
| ● In the **Classify** tab, inside the **Test options** frame, select **Supplied test set** and click **Set…**. Then click the **Open file…** button and in the directory A2AC select the `test.arff` file. Then click **Close**. | The same model is applied to the test set. The results are reported in Figure 11.9. They are a bit worse than for the fitting stage. Yet they remain much better than for the NB model. |
| ● Right click on the line of the **Result list** corresponding to your model's report. Select the line **Re-evaluate model on current test set**. | |
| ● Click on the **Choose** button in the panel **Classifier**. | Select the kNN algorithm. |
| ● Choose the method **weka->classifiers->lazy->IBk** from the hierarchical tree. | |

Figure 11.7 Instances classified as active while they are inactive. For compounds 1 and 3, the fused cycles scaffold are often met in active compounds. The most related compounds to compound 2 are all actives. Inactive compounds are decoys: there is no proof that they are not active.

(*Continued*)

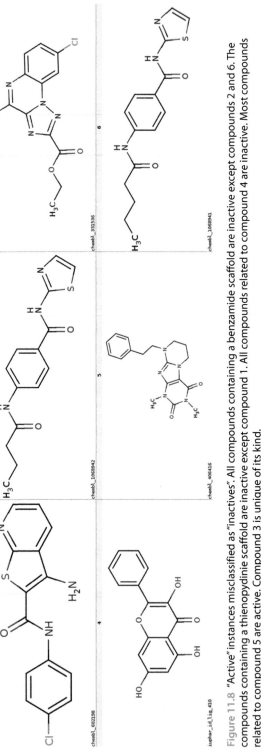

**Figure 11.8** "Active" instances misclassified as "inactives". All compounds containing a benzamide scaffold are inactive except compounds 2 and 6. The compounds containing a thienopydinie scaffold are inactive except compound 1. All compounds related to compound 4 are inactive. Most compounds related to compound 5 are active. Compound 3 is unique of its kind.

| Instructions | Comments |
|---|---|

```
=== Summary ===

Correctly Classified Instances      2401        98.3613 %
Incorrectly Classified Instances      40         1.6387 %
Total Number of Instances           2441

=== Detailed Accuracy By Class ===

               TP Rate  FP Rate  Precision  Recall  F-Measure  MCC    ROC Area  PRC Area  Class
               0,989    0,049    0,991      0,989   0,990      0,936  0,970     0,990     0
               0,951    0,011    0,940      0,951   0,945      0,936  0,970     0,901     1
Weighted Avg.  0,984    0,044    0,984      0,984   0,984      0,936  0,970     0,977

=== Confusion Matrix ===

    a    b   <-- classified as
 2055   22 |   a = 0
   18  346 |   b = 1
```

**Figure 11.9** Performances of SVM method on the test set. Note the decrease of performances compared to the fitting stage.

- Click on the word **SMO**. The configuration window appears on the screen (Figure 11.5).
- Keep all parameters to their default values.
- Click on the **Start** button to run the SVM algorithm.

The default values are fine. The number of neighbors can be set manually as a parameter **KNN**. It can be set automatically based on a leave-one-out strategy by setting the parameter **crossValidate** to **True**. The optimum value is determined by minimizing the MAE, however RMSE can be used instead of MAE if the parameter **meanSquared** is set to **True**. It is also possible to weight the vote of each neighbor by its distance to the instance by setting the parameter **distanceWeighting** to **True**. The parameter **nearestNeighbourSeachAlgorithm** allows trying several algorithms to improve the speed of the algorithm. If the data set is large and cannot fit into the computer's memory, it is possible to use only a subsample of it and this is controlled by the value of the parameter **windowSize**.

- Right click on the line of the **Result list** corresponding to your kNN model's report. Then select **Save model**.

The model is saved and can be used later on (Figure 11.10).

- Save the model in the current working directory as A2AC_kNN.model.
- Right click on the line of the **Result list** corresponding to your kNN model's report. Then select **Save result buffer**.

A kNN model (with k=1) cannot make mistakes on training data (Figure 11.11).

- Save the file in the current working directory as A2AC_fit_kNN.out
- In the **Classify** tab, inside the **Test options** frame, select **Supplied test set** and click **Set....** Then click the **Open file...** button and in the directory A2AC select the test.arff file. Then click **Close**.
- Right click on the line of the **Result list** corresponding to your model's report. Select the line **Re-evaluate model on current test set**.

The same model is applied to the test set. The results are reported in Figure 11.12. They are worse than those of the SVM. Yet, they are comparable to the Naïve Bayes classifier in terms of ROC AUC and PRC AUC. This means that as a classifier, the kNN outperforms the NB model, but the ranking based on the NB score is of comparable quality with kNN.

*(Continued)*

Figure 11.10 Interface of the kNN algorithm. Keep parameters to default values.

```
=== Summary ===

Correctly Classified Instances        2430                 100      %
Incorrectly Classified Instances         0                   0      %
Total Number of Instances             2430

=== Detailed Accuracy By Class ===

                 TP Rate  FP Rate  Precision  Recall  F-Measure  MCC    ROC Area  PRC Area  Class
                 1,000    0,000    1,000      1,000   1,000      1,000  1,000     1,000     0
                 1,000    0,000    1,000      1,000   1,000      1,000  1,000     1,000     1
Weighted Avg.    1,000    0,000    1,000      1,000   1,000      1,000  1,000     1,000

=== Confusion Matrix ===

    a     b   <-- classified as
 2049     0 |   a = 0
    0   381 |   b = 1
```

Figure 11.11 Fitting performances of the kNN. By definition, a kNN (k=1) model cannot make mistakes on the training set.

```
=== Summary ===

Correctly Classified Instances        2383              97.6239 %
Incorrectly Classified Instances        58               2.3761 %
Total Number of Instances             2441

=== Detailed Accuracy By Class ===

                 TP Rate  FP Rate  Precision  Recall  F-Measure  MCC    ROC Area  PRC Area  Class
                 0,981    0,052    0,991      0,981   0,986      0,909  0,965     0,988     0
                 0,948    0,019    0,898      0,948   0,922      0,909  0,965     0,859     1
Weighted Avg.    0,976    0,047    0,977      0,976   0,976      0,909  0,965     0,969

=== Confusion Matrix ===

    a     b   <-- classified as
 2038    39 |   a = 0
   19   345 |   b = 1
```

Figure 11.12 Performances of the kNN model on test data. Performances are a bit worse than the SVM model. According to the ROC AUC or PRC AUC, performances are close to those of Naive Bayes.

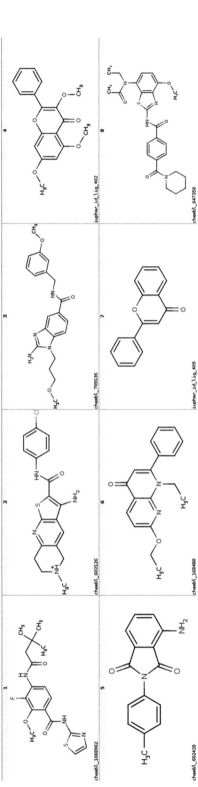

**Figure 11.13** Compounds of the test set incorrectly predicted by the NB, SVM and kNN models as inactive.

*(Continued)*

| Instructions | Comments |
|---|---|
| • Use several instances of the *ModelAnalyzerC* software to compare the errors of the different models on the same test set. | The set of compounds that NB, SVM, and kNN incorrectly predict as inactive is given in Figure 11.13. The three models tend to make different mistakes which reflects the principles on which they are built. |
| • In the *Classify* tab, within the *Test options* frame, select *Supplied test set* and click *Set...*. Then click the *Open file...* button and in the directory A2AC select the external.arff file. Then click *Close*. | The cost parameter in the SVM or the number of neighbors in the kNN can be optimized to best fit the test data. Testing the models on an external test set can highlight a learning bias toward a specific test set. The external test set is also essential to test the models in situation, as they are provided to the end user. |
| • Right click on a blank space bellow the last line of the *Result list*. Select *Load model* and chose the file A2AC_NB.model. | Load the previously saved models. |
| • Right click on a blank space bellow the last line of the *Result list*. Select *Load model* and chose the file A2AC_SVM.model.<br><br>Right click on a blank space below the last line of the *Result list*. Select *Load model* and chose the file A2AC_kNN.model. | |
| • Right click on each line of the *Result list* corresponding to a loaded model and select *Re-evaluate model on current test set*. | Apply the loaded models to the external test set. Results are provided in Figure 11.14, Figure 11.15 and Figure 11.16.<br><br>    The kNN and SVM models both have exactly one error, but on different instances. The NB model has 8 errors, including those made by the kNN and SVM models. |
| • Use the *ModelAnalyzerC* software to open the file NB_SMO_kNN.txt.<br>• Click on the button *Vote*.<br>On the updated interface select a class in the *Statistics per class* list box and observe the ROC AUC. | The file NB_SMO_kNN.txt contains a summary of the results of the NB, SVM, and kNN models. When the three models are combined in a vote, the resulting model has a ROC AUC of *1*. The ranking of the combined models is perfect; all active compounds are at the beginning of the list ranked using the votes of the classification models. |

```
=== Summary ===

Correctly Classified Instances         69               89.6104 %
Incorrectly Classified Instances        8               10.3896 %
Total Number of Instances              77

=== Detailed Accuracy By Class ===

               TP Rate  FP Rate  Precision  Recall  F-Measure  MCC     ROC Area  PRC Area  Class
               0,879    0,000    1,000      0,879   0,935      0,713   0,975     0,996     0
               1,000    0,121    0,579      1,000   0,733      0,713   0,976     0,835     1
Weighted Avg.  0,896    0,017    0,940      0,896   0,907      0,713   0,975     0,973

=== Confusion Matrix ===

  a   b   <-- classified as
 58   8 |  a = 0
  0  11 |  b = 1
```

Figure 11.14 Performances of the NB model on the external test set.

```
=== Summary ===

Correctly Classified Instances          76              98.7013 %
Incorrectly Classified Instances         1               1.2987 %
Total Number of Instances               77

=== Detailed Accuracy By Class ===

                 TP Rate  FP Rate  Precision  Recall   F-Measure  MCC     ROC Area  PRC Area  Class
                 0,985    0,000    1,000      0,985    0,992      0,950   0,992     0,998     0
                 1,000    0,015    0,917      1,000    0,957      0,950   0,992     0,917     1
Weighted Avg.    0,987    0,002    0,988      0,987    0,987      0,950   0,992     0,986

=== Confusion Matrix ===

  a  b    <-- classified as
 65  1 |  a = 0
  0 11 |  b = 1
```

Figure 11.15 Performances of the SVM model on the external test set.

```
=== Summary ===

Correctly Classified Instances          76              98.7013 %
Incorrectly Classified Instances         1               1.2987 %
Total Number of Instances               77

=== Detailed Accuracy By Class ===

                 TP Rate  FP Rate  Precision  Recall   F-Measure  MCC     ROC Area  PRC Area  Class
                 0,985    0,000    1,000      0,985    0,992      0,950   0,992     0,998     0
                 1,000    0,015    0,917      1,000    0,957      0,950   0,992     0,917     1
Weighted Avg.    0,987    0,002    0,988      0,987    0,987      0,950   0,992     0,986

=== Confusion Matrix ===

  a  b    <-- classified as
 65  1 |  a = 0
  0 11 |  b = 1
```

Figure 11.16 Performances of the kNN model on the external test set.

## Conclusion

The tutorial exposed different classification methods. The NB algorithm is clearly the weakest and the SVM the most promising. The kNN and SVM performed equally well.

The statistical performances assessed by the different measures of success allow comparing the models on the test set. An external test set is available to test the models in this case.

Yet, it is very useful to have a closer look at the errors of the models, especially during the fit or when combining the models. This highlights potential problems with the training set (wrong data, insufficient sampling of the cases), or intrinsic limitations of the models themselves.

# References

1 Jaakola, V.-P., et al., *The 2.6 angstrom crystal structure of a human A2A adenosine receptor bound to an antagonist*. Science, 2008, **322**(5905), 1211–1217.

2 Baldi, P., et al., *Assessing the accuracy of prediction algorithms for classification: an overview*. Bioinformatics, 2000, **16**(5), 412–424.

3 Provost, F.J., T. Fawcett, and R. Kohavi. *The case against accuracy estimation for comparing induction algorithms*. in *ICML*. 1998.

4 Davis, J. and M. Goadrich. *The relationship between Precision-Recall and ROC curves*. in *Proceedings of the 23rd international conference on Machine learning*. 2006, ACM.

5 Raghavan, V., P. Bollmann, and G.S. Jung, *A critical investigation of recall and precision as measures of retrieval system performance*. ACM Transactions on Information Systems (TOIS), 1989, **7**(3), 205–229.

6 Alpaydin, E., *Introduction to machine learning*. 2014, MIT press.

7 McCallum, A. and K. Nigam. *A comparison of event models for naive bayes text classification*. in *AAAI-98 workshop on learning for text categorization*. 1998, Citeseer.

8 Boser, B.E., I.M. Guyon, and V.N. Vapnik. *A training algorithm for optimal margin classifiers*. in *Proceedings of the fifth annual workshop on Computational learning theory*. 1992, ACM.

9 Smola, A.J. and B. Schölkopf, *Learning with kernels*. 1998, Citeseer.

10 Aha, D.W., D. Kibler, and M.K. Albert, *Instance-based learning algorithms*. Machine Learning, 1991, **6**(1), 37–66.

11 Witten, I.H. and E. Frank, *Data Mining: Practical machine learning tools and techniques*. 2005, Morgan Kaufmann.

12 Keerthi, S.S., et al., *Improvements to Platt's SMO algorithm for SVM classifier design*. Neural Computation, 2001, **13**(3), 637–649.

13 Platt, J., *Fast training of support vector machines using sequential minimal optimization*. Advances in Kernel Methods: Support Vector Learning, 1999, **3**.

14 Hastie, T. and R. Tibshirani, *Classification by pairwise coupling*. The Annals of Statistics, 1998, **26**(2), 451–471.

15 Daluge, S.M. and H.L. White, *Phenyl xanthine derivatives*. 2001, Google Patents.

# 12

## Regression Models

*Igor I. Baskin, Gilles Marcou, Dragos Horvath, and Alexandre Varnek*

*Goal*: Illustrate the use of common regression methods and introduce performance measures for regression.
*Software*: WEKA, ISIDA/ModelAnalyzerR
*Data*:

- `A2AR_train.arff` – training set fragment descriptors
- `A2AR_test.arff` – test set fragment descriptors
- `A2AR_train.sdf` – training set molecular structures
- `A2AR_test.sdf` – test set molecular structures

The dataset contains ligands of the human adenosine receptor (A2A), a member of purinergic receptor family, crystallized recently.[1] A2A is one of the targets for caffeine. Two approved drugs currently target A2A, an agonist and an antagonist. The agonist is used as a coronary vasodilator, while the antagonist is used to cure movement disorders induced by Parkinson treatments. There were, in May 2015, 14 entries for the A2A receptor in the Protein Data Bank.

Data for this tutorial were collected from the IUPHAR/BPS, ChEMBL, and PubChem BioAssay in 2012.

The regression problem consists in estimating ligand affinity (expressed as $pK_i$ = negative log of the complex instability constant expressed in mol/l) to A2A, as a function of the ligand structure. Ligand structures and their known $pK_i$ values were collected from the IupharDB, ChEMBL, and PubChem BioAssay databases. If several values were reported for a given ligand, the retained value is the median of the collected values, if the range of values did not exceed 2 log units (otherwise, that ligand was deleted from the present set). The final data set for building regression models contains 766 compounds, out of which 384 were assigned to training, while 383 were kept aside in an external test set. The affinity value of each compound is stored in the SDF field pKi. The training set and the test set files are named respectively `train.sdf` and `test.sdf` in the folder Regression.

*Tutorials in Chemoinformatics*, First Edition. Edited by Alexandre Varnek.
© 2017 John Wiley & Sons Ltd. Published 2017 by John Wiley & Sons Ltd.
Companion website: www.wiley.com/go/varnek/chemoinformatics

## Theoretical Background

A Regression model is a function $f(\mathbf{x})$ of a vector of *independent*, or *explaining variables* (descriptors, attributes) $\mathbf{x}$ with the property that, for any given instance $i$ characterized by its descriptor vector $\mathbf{x}_i$, $f(\mathbf{x}_i)$ is an estimator of an *explained variable (property)* $y$ which depends on $\mathbf{x}$. In other words, $f(\mathbf{x}_i) = y_i^{pred}$, the calculated/predicted property of instance $i$, should closely match the actually observed property $y_i$ associated to that instance. Formally, a function $f(\mathbf{x}_i)$ is searched/parameterized in order to minimize the sum of squared errors $\sum_i \left( y_i^{pred} - y_i^{exp} \right)^2$ between calculated and observed properties, over all instances in the training set. In this tutorial, instances are A2A ligands, the independent variables are the structural descriptors (fragment counts) which numerically encode the chemical structure, while the explained property $y$ is the A2A affinity, which is determined by the ligand structure—thus should depend on molecular descriptors if these encode the relevant structural aspects.

Three methods are illustrated in this tutorial: Ridge Regression (RR),[2] the $\epsilon$-insensitive Support Vector Machine ($\epsilon$-SVM),[3] and Artificial Neural Networks (ANN).[4] Ridge Regression is a method producing least-square errors multi-linear regression models while controlling the length of the parameter vector. The method is related to the classical least-square error algorithm but can be used for problems involving (maybe large number of) collinear (linearly correlated) predictors.

The $\epsilon$-SVM uses an "$\epsilon$-insensitive loss function," which is proportional to the fitting error absolute value where this value is larger than $\epsilon$ and equal to zero otherwise. This loss-function is more robust to outliers than that used in least-square error methods. The objective function to be minimized in the course of model building contains, along with the loss-function, an additional term controlling the complexity of the model, while the optimization problem can be formulated in terms of kernels. As a consequence, the method resists overfitting.

Finally, the ANN is based on the back-propagation algorithm. This method takes the form of a directional network capable of approximating any continuous function when the network is processed in one direction and the gradient of the function when the network is processed in the reverse order. It is a classical example of non-linear model.

Several performance measures are usually computed for regression models. The most widely used of them is the Root Mean Squared Error (RMSE). This measure originates from the discovery of linear regression by K. Pearson[5,6] following the seminal works of his colleague F. Galton.[7] Considering a set of $N$ pairs of experimental observations $y_i^{exp}$ and the corresponding estimates $y_i^{pred}$, the RMSE has the following form:

$$RMSE = \sqrt{\frac{1}{N} \sum_{i=1}^{N} \left( y_i^{exp} - y_i^{pred} \right)^2} \tag{12.1}$$

In the same works K. Pearson introduced a correlation measure that took his name, the Pearson's correlation. The measure is usually represented by the letter $R$ standing for "regression":

$$R = \frac{\sum_{i=1}^{N} \left( y_i^{exp} - \langle y^{exp} \rangle \right)\left( y_i^{pred} - \langle y^{pred} \rangle \right)}{\sqrt{\sum_{i=1}^{N} \left( y_i^{exp} - \langle y^{exp} \rangle \right)^2 \cdot \sum_{i=1}^{N} \left( y_i^{pred} - \langle y^{pred} \rangle \right)^2}} \tag{12.2}$$

where the terms $\langle y^{exp}\rangle=(1/N)\sum_{i=1}^{N}y_{i}^{exp}$ and $\langle y^{pred}\rangle=(1/N)\sum_{i=1}^{N}y_{i}^{pred}$ represent the arithmetic means of the experimental and estimated observations, respectively. It is customary to report the squared coefficient $R^2$ for the Pearson's correlation. The equation (12.2) is geometrically interpreted as the cosine between two centered vectors containing the experimental and estimated observations respectively. Its value is 1 if experiment and estimation are in agreement, –1 if they are in reversed agreement, and 0 if they are linearly unrelated.

Another very popular performance measure is the coefficient of determination:

$$R^2 = 1 - \frac{\sum_{i=1}^{N}\left(y_{i}^{exp}-y_{i}^{pred}\right)^2}{\sum_{i=1}^{N}\left(y_{i}^{exp}-\langle y^{exp}\rangle\right)^2} \tag{12.3}$$

The coefficient of determination is in fact the square of the Pearson correlation coefficient for a simple linear regression with a single explaining factor. It can also be shown that the determination coefficient is equal to the square of the Pearson's correlation coefficient if the vector formed by the residual errors is orthogonal to the vector of centered estimates. This is why the notations of the coefficient of determination and the square of the Pearson's correlation coefficient are the same, while in practical situations their numerical values are different. While the two values are usually similar, the coefficient of determination is favored because it is a qualitative comparison of the explained variance of data by a regression model compared to the no-model situation when the average of the property $\langle y^{exp}\rangle$ is used as an oracle to estimate the property of any instance.

A different viewpoint to the problem can be based on estimating the quality of a model as its ability to reproduce the experimental observation. As stated, the Concordance Correlation Coefficient (CCC)[8] was introduced as a normalized distance of the points of coordinate $\left(y_{i}^{exp},y_{i}^{pred}\right)$ to the line of slope 1 and origin 0 to which these points shall ideally belong. The CCC is formulated as follows:

$$CCC = \frac{2\cdot\sum_{i=1}^{N}\left(y_{i}^{exp}-\langle y^{exp}\rangle\right)\left(y_{i}^{pred}-\langle y^{pred}\rangle\right)}{\sum_{i=1}^{N}\left(y_{i}^{exp}-\langle y^{exp}\rangle\right)^2+\sum_{i=1}^{N}\left(y_{i}^{pred}-\langle y^{pred}\rangle\right)^2+N\cdot\left(\langle y^{exp}\rangle-\langle y^{exp}\rangle\right)^2} \tag{12.4}$$

The values of this coefficient are in the range from –1 to 1. For anti-correlated data the CCC value is –1, for perfectly correlated data it is 1, and it is 0 if the experimental and estimated values do not match in a linear manner.

It is possible to apply the Fisher's z-transformation[9,10] to each of these measures:

$$z = \text{arctanh}(x) = \frac{1}{2}\left(\ln(1+x)-\ln(1-x)\right) \tag{12.5}$$

where $x$ can be either the Pearson's correlation coefficient, the squared root of the coefficient of determination, or the CCC. The new variable $z$ follows approximately the normal law, especially if the estimated and experimental values are normally distributed. The $z$ variable has the standard deviation $1/\sqrt{N-3}$, so the confidence interval of the $z$ value can be computed using a critical value $z_{crit}$ from normal distribution tables as $z_{crit}/\sqrt{N-3}$. It is easy to compute p-values for the correlation coefficient, the determination coefficient or the CCC using this transform.

The $l_2$ norm (the RMSE) is the most widely used to assess the errors of a model. However, the $l_1$ norm was considered first by Eddington back in 1914[11] while reporting stellar velocities. This measure is usually called the Mean Absolute Error (MAE):

$$MAE = \frac{1}{N}\sum_{i=1}^{N}\left|y_i^{exp} - y_i^{pred}\right|$$

(12.6)

As reported by Eddington, the MAE is much less sensible to exaggerated errors than the RMSE and thus it is more resilient to outliers.

Since the RMSE and MAE are expressed in the same units as the experimental properties, it is common to report them in a unit-less format, the Relative RMSE (%RMSE) and Relative MAE (%MAE):

$$\%RMSE = \frac{\sum_{i=1}^{N}\left(y_i^{exp} - y_i^{pred}\right)^2}{\sum_{i=1}^{N}\left(y_i^{exp} - \left\langle y^{exp}\right\rangle\right)^2} \times 100$$

(12.7)

$$\%MAE = \frac{\sum_{i=1}^{N}\left|y_i^{exp} - y_i^{pred}\right|}{\sum_{i=1}^{N}\left|y_i^{exp} - \left\langle y^{exp}\right\rangle\right|} \times 100$$

(12.8)

Graphical plots are often used to assess the quality of regression models. The most frequently used "scatter" plot represents the estimated values $y_i^{pred}$ as a function of the experimental ones $y_i^{exp}$, or *vice versa*. Some additional elements can be added to help analyzing the plot. The first one is the line of equation $y = x$ representing the ideal case, when all estimates are in agreement with experimental values. In ISIDA/ModelAnalyzerR used in this tutorial, two lines of equation $y = x \pm 3s_Y$ are also plotted, where $s_Y$ is the standard deviation of residuals defines as:

$$s_Y = \sqrt{\frac{\sum_i\left(y_i^{exp} - y_i^{pred}\right)^2}{N-2}}$$

(12.9)

These lines are useful to detect outliers, since the area between those lines should contain about 99% of the instances. Those instances outside the lines should be considered as extreme value points and critically examined.

The correlation line of equation $y = a_0 + a_1 x$ with

$$a_0 = \left\langle y\right\rangle - \left\langle x\right\rangle \times \frac{\sum_i x_i y_i - \sum_i x_i\left\langle y\right\rangle}{\sum_i x_i^2 - \sum_i x_i\left\langle x\right\rangle}$$

(12.10)

$$a_1 = \frac{\sum_i x_i y_i - \sum_i x_i\left\langle y\right\rangle}{\sum_i x_i^2 - \sum_i x_i\left\langle x\right\rangle}$$

(12.11)

gives some indications on the learning bias of the model, by comparison to the ideal line $y = x$.

A parabolic domain centered on the correlation line is mainly present to better identify the dense population zones, where the regression error is reduced, from peripheral estimation regions where the estimations are clearly extrapolated and the estimation errors are expected to be larger. The equations of the parabolic curves are:

$$y = a_0 + a_1 x \pm 3\sqrt{V(x)} \tag{12.12}$$

$$V(x) = s_Y^2 \left( 1 + \frac{1}{n} + \frac{\left(x - \langle x \rangle\right)^2}{\sum_i \left(x_i - \langle x \rangle\right)^2} \right) \tag{12.13}$$

The latter curves are closely related to prediction intervals for simple linear regression.

A less popular plot, which is nonetheless useful for model comparison, is the Regression Error Characteristic curve (REC).[12] The y-axis of the plot is the percentage of a data set instances that are estimated better than a threshold value, and the threshold value is reported on the x-axis. It is analogous to the concept of ROC for classification. Whenever two regression models are compared, the one dominating the other is considered to be better. The dominating models estimates better a larger number of instances. The area over the curve (AOC) of a REC ranges from 0 to the maximum observed error on the data set. It can be regarded as an optimistic estimate of the expected error of the model.

## Step-by-Step Instructions

| Instructions | Comments |
| --- | --- |
| • In the **Preprocess** tab, click on the button **Open File**. In the file selection interface, select the directory A2AR, then the file `train.arff`. In the **Classify** tab, inside the **Test options** frame, select **Use training set**. | Load the training data set for building regression model for binding of ligands to the adenosine receptor. The model performances will be assessed on the same data as those used to create the model. These are the *fitting* performances. The estimated pK$_i$ of each test set compound will be printed in the *Classifier output* frame of the interface. These results can be saved to a file and then reanalyzed with dedicated software. The Comma Separated Value (CSV) format is a convenient exchange format. It uses the comma to separate columns. |
| • Click on the button **More options...** Then, near the label *Output predictions*, click the button **Choose** and select the format **CSV**. | |
| • Click on the **Choose** button in the panel **Classifier**. | The configuration interface of the Ridge Regression method is presented in Figure 12.1. No attribute selection algorithm is needed because the ridge regression algorithm is robust to collinear attributes. The ridge parameter needs to be optimized in order to maximize the model predictive performance. |
| • Choose the method **weka->classifiers->functions->LinearRegression** from the hierarchical tree. | |
| • Set the **attributeSelectionMethod** to **No attribute selection**. | |
| • Set the **eliminateColinearAttributes** to **False**. | |
| • Set the **ridge** value to 250. | |
| • Click on the **Start** button to run the Ridge Regression algorithm. | |
| • Right click on the line of the **Result list** corresponding to your model's report. Then select **Save model**. | The model is saved and can be used later on. |
| • Save the model in the current working directory as `A2AR_RR.model`. | |

(*Continued*)

| Instructions | Comments |
|---|---|

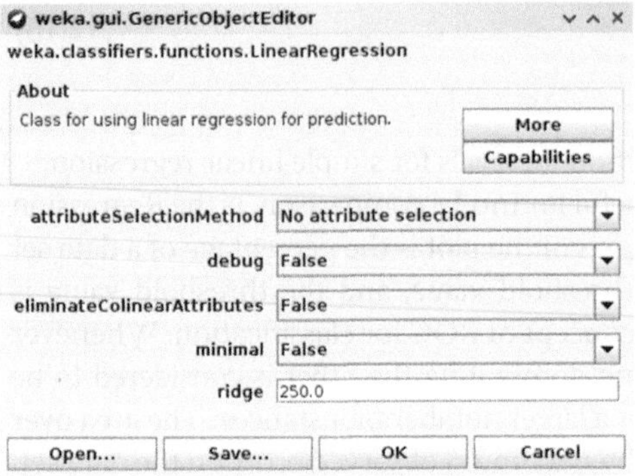

**Figure 12.1** Configuration interface of the Ridge Regression.

- Right click on the line of the **Result list** corresponding to your model's report. Then select **Save result buffer**.

Save the file in the current working directory as A2AR_fit_RR.out.

The reported performances of the fit are good (Figure 12.2). It also contains the experimental and estimated pK$_i$ values for each of the training instance. This information must be linked to the information on the chemical structures in order to be analyzed in terms of chemical structures.

```
=== Summary ===

Correlation coefficient          0.8765
Mean absolute error              0.4532
Root mean squared error          0.579
Relative absolute error          45.7262 %
Root relative squared error      49.2895 %
Total Number of Instances        384
```

**Figure 12.2** Ridge Regression performances on training data. The fitting performances are very good.

- Open the file A2AR_fit_RR.out using the software **ModelAnalyzerR**. Use the file train.sdf as the source of the information on chemical structures (zone **1** in Figure 12.3).
- Then click the **Start** button.
- Click the experimental versus estimated values plot (zone **5** in Figure 12.3). Select those points that are outside the red lines.
- Search for a point of the REC (zone **6** in Figure 12.3). The rightmost points are the worst estimated compounds. Observe where they are located in the experimental/estimated plot.

The worst predicted compound is a xanthine derivative (Figure 12.14a). Interestingly, it is the only one of the xanthine derivatives reported in [13] that is not active on A2A. The authors reported their finding as anomalous without explanations.

The second outlier is furan-2-yl-pyrazolo[3,4-d] pyrimidine, N-substituted with a carboxybenzyl (Figure 12.14c), which is reported in [14]. All related compounds are nanomolar, except this one. Besides, the methoxycarbonylbenzyl analogue is also nanomolar.

The third outlier (Figure 12.14b) is reported in [15]. The A2A activity of pyrimidine carboxamides is very sensitive to the nature of substituent at position C2 (the carbon atom between two nitrogen atoms in the ring) of the pyrimidine. The methylamine substituted compound is indeed in two orders of magnitude more potent than the dimethyl amine substituted compound (the outlier). In fact, the amine substituted compound is even more potent (Figure 12.3).

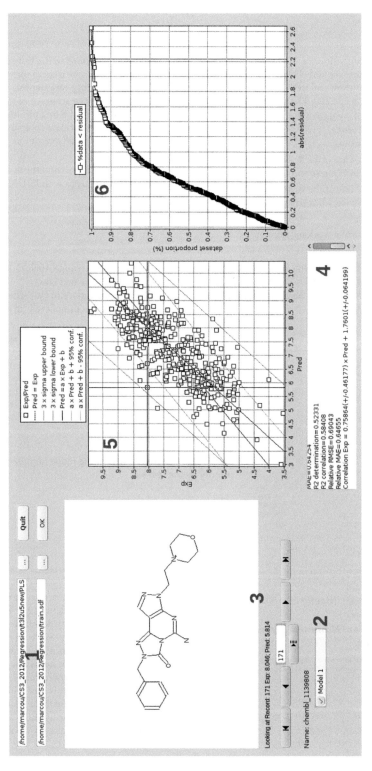

Figure 12.3 Interface of the software ISIDA/ModelAnalyzerR. The names of input files are provided in area 1. If multiple models are available, they can be selected in area 2. Molecules are depicted in area 3. Statistical parameters are reported in the text box 4. The experimental versus estimated values are plotted in area 5. The REC is plotted in area 6.

(Continued)

| Instructions | Comments |
|---|---|

- In the **Classify** tab, inside the **Test options** frame, select **Supplied test set** and click **Set...**. Then click the **Open file...** button and in the directory A2AR select the test.arff file. Then click **Close**.
- Right click on a blank line of the **Result list** and then select **Load model**. Load the model A2AR_RR.model previously saved.
- Right click on the last line of the **Result list** and select the line **Re-evaluate model on current test set**.
- Right click on the line of the **Result list**.
- **t** corresponding to your model's report. Then select **Save result buffer**.

Save the file in the current working directory as A2AR_test_RR.out.

The same model is applied to the test set. The results are a bit worse than in the previous case with fitting parameters (Figure 12.4): the average accuracy of the estimations ranges from 1.5 to 2 pKi units, which is worse on the test data compared to the train data. This is a quite common behavior.

Save the statistical report and the estimates to a file for further analysis with ISIDA/ModelAnalyzerR.

- Open the file A2AC_test_RR.out using the software **ModelAnalyzerR**. Use the file test.sdf as the source of information on chemical structures. Then click the **Start** button.

An analysis with ISIDA/ModelAnalyzerR reveals two outliers. The first one (Figure 12.15a) is mentioned in [16]. The 4,6-amino-triazyl scaffold is present only in the test set.

The second outlier (Figure 12.15b) is mentioned in [17]. Among the triazolopyrimidine derivatives mentioned in the article, the outlier is the most potent one, while only the weakest analogues are present in the training set. This analysis therefore reveals some flaws in the data collection.

```
=== Summary ===

Correlation coefficient              0.7429
Mean absolute error                  0.5945
Root mean squared error              0.758
Relative absolute error             63.1935 %
Root relative squared error         66.9524 %
Total Number of Instances            383
```

Figure 12.4 Performances of the Ridge Regression model on the test set.

- Click on the **Choose** button in the panel **Classifier**.
- Choose the method **weka->classifiers->functions->SMOreg** from the hierarchical tree.
- Set the **filterType** to **No normalization/standardization**.

Set the **c** value to 0.01.

Prepare for the second regression method, the SVM.

The main configuration interface of the SVM method is shown in Figure 12.5. The linear kernel is already set by default, and it does not require any change. Since all descriptors represent counts of fragments in chemical structures, it is not necessary to use scaling factors. Therefore, molecular descriptors can be used in this case without any standardization or normalization. The cost parameter $c$ needs to be optimized, but the recommended value is already a good choice.

| Instructions | Comments |
|---|---|
| • Click on the *regOptimizer* text field, on the word *RegSMOImproved*.<br><br>In *RegSMOImproved* configuration interface, set the *epsilonParameter* to 0.01 and the *tolerance* to 0.05. | This action opens the configuration interface of the algorithm solving the regression SVM problem (Figure 12.6). The *RegSMOImproved* refers to an improvement of the Sequential Minimal Optimization (SMO) algorithm to solve the SVM problem.[18,19]<br><br>The $\epsilon$ parameter represents the width of the $\epsilon$-insensitive loss function. Fitting errors in the range $[-\varepsilon, \varepsilon]$ are ignored. The value of 0.01 is representative of the precision of the values of the pKi values in the dataset that are given up to two digits. Therefore smaller fitting errors are meaningless. Two alternative approaches would be to set this value to actual experimental precision, or to use it as an adjustable parameter to be optimized.<br><br>The tolerance parameter controls the convergence of iterative procedures. A smaller value will improve the fit but require more computations.<br><br>The *epsilon* parameter manages the numerical round-off errors. It should not be changed and not be confused with the $\epsilon$ parameter already discussed. |
| • Click on the *OK* button on all forms.<br>Click the *Start*. | Validate all settings and start calculations. The results should be close to those reproduced in Figure 12.7. |

Figure 12.5 Main configuration interface of the SVM method.

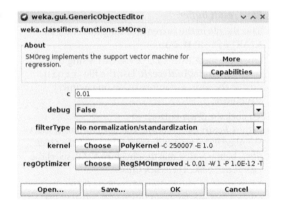

| • Right click on the line of the *Result list* corresponding to your model's report. Then select *Save model*.<br><br>Save the model in the current working directory as A2AR_SVM.model. | The model is saved and can be used later on. |

*(Continued)*

| Instructions | Comments |
|---|---|

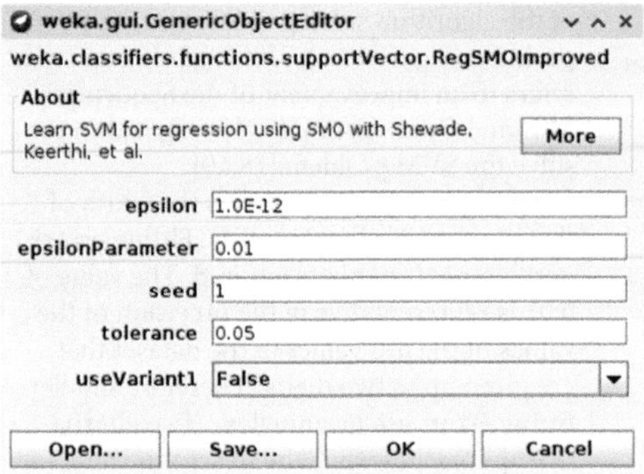

Figure 12.6 Configuration interface of the optimization algorithm of the SVM, including the loss function parameter.

Figure 12.7 Performance measures of the regression SVM algorithm on training data.

- Right click on the line of the **Result list** corresponding to your model's report. Then select **Save result buffer**.
- Save the file in the current working directory as A2AR_fit_SVM.out.
- Open the file A2AC_fit_SVM.out using the software **ModelAnalyzerR**. Use the file train.sdf as the source of information on chemical structures.

Then click the **Start** button.

- In the **Classify** tab, inside the **Test options** frame, select **Supplied test set** and click **Set...**. Then click the **Open file...** button and, within the directory A2AR, select the test.arff file. Then click **Close**.
- Right click on a blank line of the **Result list** then select **Load model**. Load the model A2AR_SVM.model previously saved.
- Right click on the last line of the **Result list** and select the line **Re-evaluate model on current test set**.
- Right click on the line of the **Result list** corresponding to your model's report. Then select **Save result buffer**.

Save the file in the current working directory as A2AR_test_RR.out.

The reported performances of the fit are worse for SVM than for the Ridge Regression. However, the REC curve highlights that the SVM model fit is worse only for large residual errors. This is expected since the loss function used for the SVM model is linear and not quadratic as for the Ridge Regression.

The outliers are the same for the SVM model and the Ridge Regression model.

The same model is applied to the test set. The results are a bit worse than the performances previously reported for fitting (Figure 12.8). The predictive performances of the SVM model are very close to those of the Ridge Regression.

The statistical report and the estimates can be saved to a file for further analysis with ISIDA/ModelAnalyzerR. However, the outliers are the same for SVM and Ridge Regression models, and the performances of both models are similar.

| Instructions | Comments |
|---|---|

Figure 12.8 Performance measures of the SVM model on the test set.

```
=== Summary ===

Correlation coefficient                  0.7453
Mean absolute error                      0.5819
Root mean squared error                  0.7594
Relative absolute error                 61.8521 %
Root relative squared error             67.0822 %
Total Number of Instances              383
```

- Click on the **Preprocess** tab.
- Click the button **Choose** and select the method *filters- > unsupervised- > attribute- > NumericToNominal*.
- Click on the word *NumericToNominal*.
- Configure the **attributeIndices** to **last**.
- Set the **invertSelection** to **True**.

Click the button **Apply**

- Click the button **Save...**
- Save the dataset as `trainNML.arff`.
- Click the button **Open file...**
- Load the file `test.arff`.
- Click the button **Apply**.
- Click the button **Save...**
- Save the dataset as `testNML.arff`.
- Click the button **Open file...**

Load the file `trainNML.arff`.

The third method to be applied is a neural network. The method performs much better if the ISIDA descriptors are not considered as numerical.

The first step to preprocess the data is to transform the integer-valued descriptors into nominal attributes (Figure 12.9). Only the ISIDA descriptors must be transformed, but not the $pK_i$, which must remain numerical (Figure 12.10).

Save the training set with converted values as the file `trainNML.arff`. Then, the test set is loaded and the same conversion is applied. The resulting dataset is saved as `testNML.arff`. The training set is then reloaded as `trainNML.arff`

Figure 12.9 Location of the filter converting numeric values to nominal ones.

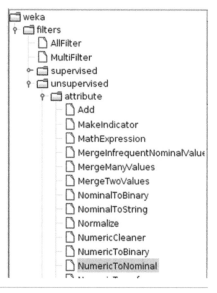

| Instructions | Comments |
|---|---|

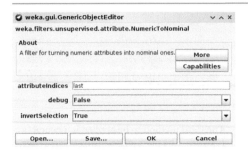

**Figure 12.10** Configuration interface of the filter converting numerical values to nominal ones.

- Click on the **Choose** button in the panel **Classifier**.
- Choose the method **weka->classifiers->functions->MultilayerPerceptron** from the hierarchical tree.
- Set the **decay** to **True**.
- Set the **hiddenLayers** value to 20.
- Set the **nominalToBinaryFilter** to **True**.
- Set the **normalizeAttributes** to **False**.
- Set the **normalizeNumericClass** to **True**.
- Set the **validationSetSize** to 20.

Select a neural network algorithm with back propagation of errors.

The configuration interface is represented in Figure 12.11. The complexity of the neural networks methods requires a large number of parameters to be specified.

First, the neural network performs better with normalized property values. Hence, for the training of the network, the $pK_i$ values are transformed.

Second, for a proper management of the nominal nature of the ISIDA descriptors, they must be converted to binary values internally. In other words, each possible value of an ISIDA descriptor is replaced with a bit that is set on if the descriptor takes this value and off otherwise.

Third, to accelerate network convergence, the decay method is applied. This means that each time the training set is presented to the network (this is termed an *epoch*), the learning rate decreases.

Fourth, the default architecture of the network implies the use of a single hidden layer with a very large number of neurons. However, the higher the number of neurons in the hidden layer, the more non-linear and complex is the model. This is very bad practice if the number of molecular descriptors is large, as is very often the case in Chemoinformatics. Here, the proposed number of neurons in the hidden layer is 20. This is a reasonably small number in comparison to the 451 nominal attributes that are converted to binary attributes for the learning. Yet, it implies some non-linearity in the model, which is important for good fitting capacities of the network.

Last, an early stopping criterion is used. The model is trained on 80% of the training set. The remaining 20% are used to check whether the network predictive performances are improving. If the predictive performances are dropping 20 times in a sequence, then the training stops. This is a very common way to prevent the tendency of neural networks to overfit the data.

| Instructions | Comments |
|---|---|

Figure 12.11  Configuration interface of the neural network.

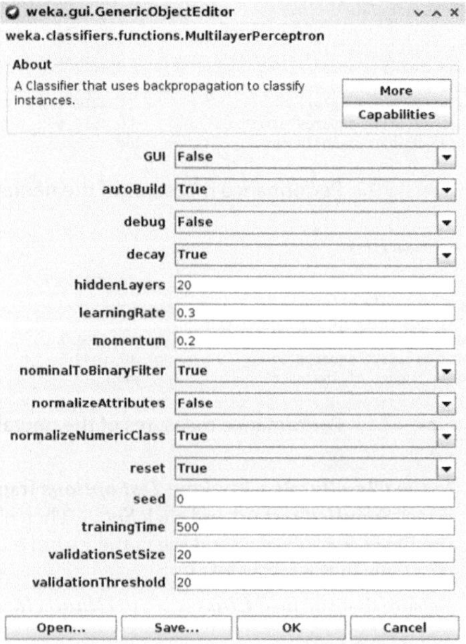

The following is the content shown in the configuration interface:

weka.gui.GenericObjectEditor

weka.classifiers.functions.MultilayerPerceptron

**About**

A Classifier that uses backpropagation to classify instances.

More

Capabilities

| | |
|---|---|
| GUI | False |
| autoBuild | True |
| debug | False |
| decay | True |
| hiddenLayers | 20 |
| learningRate | 0.3 |
| momentum | 0.2 |
| nominalToBinaryFilter | True |
| normalizeAttributes | False |
| normalizeNumericClass | True |
| reset | True |
| seed | 0 |
| trainingTime | 500 |
| validationSetSize | 20 |
| validationThreshold | 20 |

Open...   Save...   OK   Cancel

- Click the **Start** button.
- Right click on the line of the **Result list** corresponding to your model's report. Then select **Save model**.
- Save the model in the current working directory as A2AR_NN.model.
- Right click on the line of the **Result list** corresponding to your model's report. Then select **Save result buffer**.
- Save the file in the current working directory as A2AR_fit_NN.out.
- Open the file A2AC_fit_NN.out using the software **ModelAnalyzerR**. Use the file train.sdf as structural source.
- Then click the **Start** button.

The fit performances of the neural network are reported in Figure 12.11. They are, as expected, extremely high because of the non-linearity of the model and a big number of adjustable parameters (Figure 12.13).

The model is saved to the file A2AR_NN.model, and the performances estimates on the training set are saved to the file A2AR_fit_NN.out for further analysis.

The analysis with **ModelAnalyzerR** reveals the same outlies, that is, furan-2-yl-pyrazolo[3,4-d]pyrimidine, N-substituted with a carboxybenzyl (Figure 12.14c) reported above for the Ridge Regression model and the SVM model. The case is difficult because the scaffold is associated with very potent inhibitors. The presence of a carboxylic acid on the contrary, deteriorates completely the activity. However, there are too few examples in the dataset to learn this effect.

The pyrimidine carboxamide outlier (Figure 12.14b) previously reported is still not well fitted. However, there are seven other compounds that are fitted worse. The xanthine outlier (Figure 12.14a) is this time rather well fitted.

*(Continued)*

| Instructions | Comments |
|---|---|

```
=== Summary ===

Correlation coefficient            0.915
Mean absolute error                0.3117
Root mean squared error            0.4755
Relative absolute error            31.4416 %
Root relative squared error        40.4814 %
Total Number of Instances          384
```

Figure 12.12 Performance measures of the neural network on training data.

```
=== Summary ===

Correlation coefficient            0.7559
Mean absolute error                0.5866
Root mean squared error            0.759
Relative absolute error            62.3491 %
Root relative squared error        67.0444 %
Total Number of Instances          383
```

Figure 12.13 Performance measures of the neural network on test data.

| Instructions | Comments |
|---|---|
| • In the **Classify** tab, within the **Test options** frame, select **Supplied test set** and click **Set...**. Then click the **Open file...** button and select the testNML. arff file in the folder A2AR. | The test set with attributes converted to nominal values must be loaded to be compatible with the current training set and the trained model. |
| • Be sure that the field **Class** is set to **(Num) class**. | The same neural network model is applied to the test set. The results are reported in Figure 12.13. The performances are almost identical to those of the Ridge Regression and the SVM model. Despite very good fit performances, the model does not generalize better than the Ridge Regression and the SVM models. |
| • Then click **Close**. | |
| • Right click on a blank line of the **Result list** then select **Load model**. Load the model A2AR_ NN.model saved previously. | |
| • Right click on the last line of the **Result list** and select the line **Re-evaluate model on current test set**. | |
| • Right click on the line of the **Result list** corresponding to your model's report. Then select **Save result buffer**. | The statistical report and the estimates are written to a file for further analysis with the *ModelAnalyzerR* software. |
| • Save the file in the current working directory as A2AR_test_NN.out. | Interestingly, the same main outliers are found with the neural network as with two other methods (Figure 12.15). |

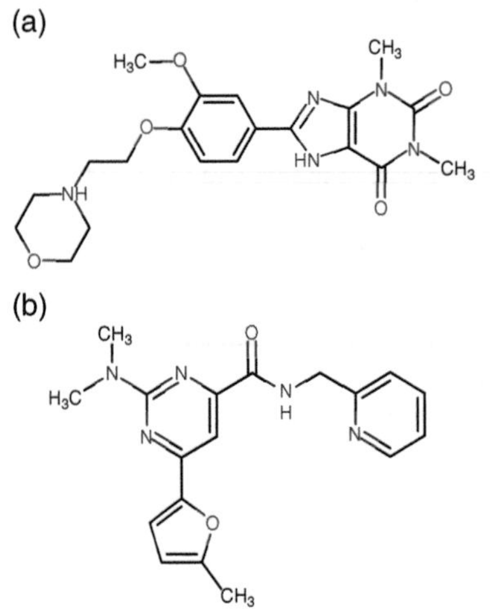

(a)

(b)

Figure 12.14 Outliers identified in the training set with the Ridge Regression and the SVM models.

| Instructions | Comments |
|---|---|

Figure 12.14 (Continued)

(c)

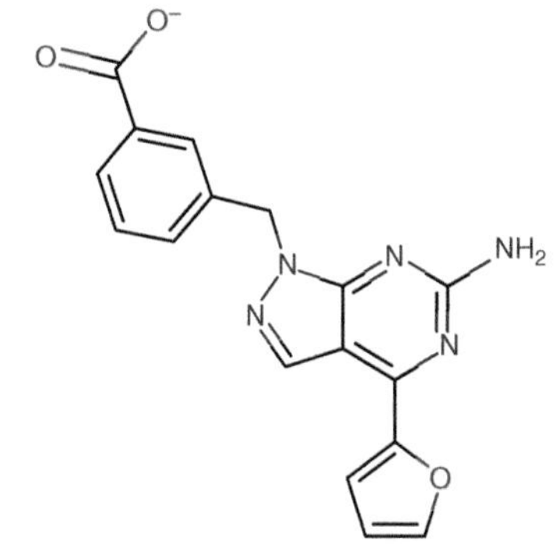

Figure 12.15 Outliers identified in the test set with the Ridge Regression model.

(a)

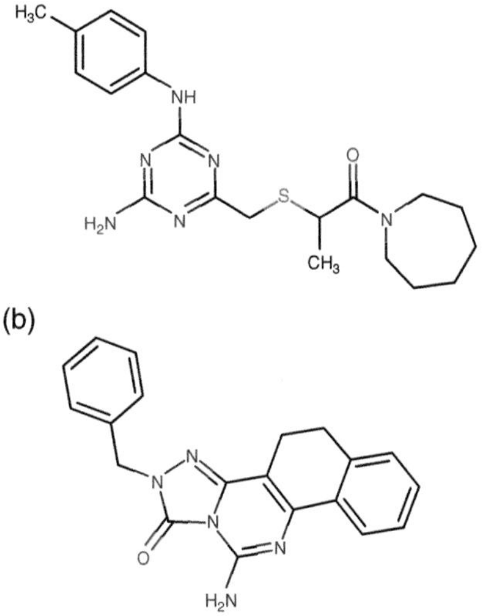

(b)

## Conclusion

The tutorial deals with different regression methods. Although the Ridge Regression, SVM, and Neural Network methods have different behaviors on the training data, yet the models generalize similarly on the test set.

The analysis of the fitting results must not be disregarded. First, the particularities of the modeling strategies are more visible. Besides, the fitting errors of the models may reflect some intrinsic "conflicts" between pairs of instances that are very similar in terms of their structural descriptors and therefore must be predicted to display similar activity levels, in contradiction to observation (activity cliffs[20]). This gives some precious hints about the limitations of the models. This kind of analysis can efficiently be made only on the training set, where predictions depend on the peculiar collection of training examples used to fit the model, as well as on the similarity between the compounds in the training and the test sets.

The use of external test sets and cross-validation is, however, essential to assess the predictive performances of the model.

## References

1 Jaakola, V.-P., et al., *Science*, 2008, **322**(5905), 1211–1217.

2 Hoerl, A.E. and R.W. Kennard, *Technometrics*, 1970, **12**(1) 55–67.

3 Vapnik, V., *The Nature of Statistical Learning Theory*. 2013: Springer Science & Business Media.

4 Williams, D.R.G.H.R. and G. Hinton, *Nature*, 1986, **323**, 533–538.

5 Pearson, K., *The London, Edinburgh, and Dublin Philosophical Magazine and Journal of Science*, 1901, **2**(11), 559–572.

6 Pearson, K., *Proceedings of the Royal Society of London*, 1895, **59**(353–358), 69–71.

7 Galton, F., *Journal of the Anthropological Institute of Great Britain and Ireland*, 1886, 246–263.

8 Lawrence, I. and K. Lin, *Biometrics*, 1989, 255–268.

9 Fisher, R.A., *Biometrika*, 1915, 507–521.

10 Fisher, R.A., *Metron*, 1924, **3**, 329–332.

11 Eddington, A.S., *Stellar Movements and the Structure of the Universe*. 1914: Macmillan and Co., Limited.

12 BIJ, J.B., R. Edu, and K.P.B. BENNEK. *Regression error characteristic curves.* in *Twentieth International Conference on Machine Learning (ICML-2003). Washington, DC.* 2003.

13 Bansal, R., et al., *European Journal of Medicinal Chemistry*, 2009, **44**(5), 2122–2127.

14 Gillespie, R.J., et al., *Bioorganic & Medicinal Chemistry Letters*, 2008, **18**(9), 2924–2929.

15 Gillespie, R.J., et al., *Bioorganic & Medicinal Chemistry*, 2009, **17**(18), 6590–6605.

16 Katritch, V., et al., *Journal of Medicinal Chemistry*, 2010, **53**(4), 1799–1809.

17 Harris, J.M., et al., *Bioorganic & Medicinal Chemistry Letters*, 2011, **21**(8), 2497–2501.

18 Smola, A.J. and B. Schölkopf, *Statistics and Computing*, 2004, **14**(3), 199–222.

19 Shevade, S.K., et al., *Neural Networks, IEEE Transactions on*, 2000, **11**(5), 1188–1193.

20 Maggiora, G. M., *J. Chem. Inf. Model.*, 2006, **46**, 1535–1535.

# 13

## Benchmarking Machine-Learning Methods

*Igor I. Baskin, Gilles Marcou, Dragos Horvath, and Alexandre Varnek*

*Goal*: Nowadays, there exist hundreds of different machine learning methods. Can one suggest "the best" approach for QSAR/QSPR studies? The aim of this tutorial is to compare different regression methods. However, the methodology is analogous for classification tasks.
*Software*: WEKA
*Data*: The following structure-activity/property datasets are analyzed in the tutorial:

- `alkan-bp-connect.arff` – the boiling points for 74 alkanes;[1]
- `alkan-mp-connect.arff` – the melting points for 74 alkanes;[1]
- `selwood.arff` – the Selwood dataset of 33 antifilarial antimycin analogs;[2]
- `shapiro.arff` – the Shapiro dataset of 124 phenolic inhibitors of oral bacteria.[3,4]

Obviously, `alkan-bp` and `alkan-mp` are structure-property datasets, while `selwood` and `shapiro` are structure-activity ones. It is rather easy to build QSPR models for the `alkan-bp` set, while the `alkan-mp` set poses a serious challenge for QSPR modeling.

The Kier-Hall connectivity topological indexes ($0\chi$, $1\chi$, $2\chi$, $3\chi^P$, $3\chi^c$, $4\chi^P$, $4\chi^{PC}$, $5\chi^P$, $5\chi^c$, $6\chi^P$)[5,6] are used as descriptors for the *alkan-bp* and *alkan-mp* datasets. Compounds from the Selwood[2] dataset are characterized by means of 52 physico-chemical descriptors.[7] The Shapiro dataset[3,4] is characterized using 14 TLSER (Theoretical Linear Solvation Energy Relationships) descriptors.[8]

Additionally, the file `compare1.exp` is a configuration file for the Weka Explorer software and the file `compare1-results.arff` provides some results to compare to.

## Theoretical Background

The following machine learning methods for performing regression are considered in the tutorial:

1) Zero Regression (ZeroR)—pseudo-regression method that always builds models with cross-validation coefficient $Q^2 = 0$. In the framework of this method the value of

*Tutorials in Chemoinformatics*, First Edition. Edited by Alexandre Varnek.
© 2017 John Wiley & Sons Ltd. Published 2017 by John Wiley & Sons Ltd.
Companion website: www.wiley.com/go/varnek/chemoinformatics

a property/activity is always predicted to be equal to its average value on the training set. This method is usually used as a reference point for comparing with other regression methods.

2) Multiple Linear Regression (MLR) with the M5 descriptor selection method and a fixed small ridge parameter ($\gamma = 10^{-8}$). In the M5 method, a MLR model is initially built on all descriptors, and then descriptors with the smallest standardized regression coefficients are step-wisely removed from the model until no improvement is observed in the estimate of the average prediction error given by the Akaike information criterion.[9]

3) Partial Least Squares (PLS)[10–13] with 5 latent variables.

4) Support Vector Regression (SVR)[14] with the default value 1.0 of the trade-off parameter *C*, the linear kernel and using the Shevade *et al.* modification of the SMO algorithm.[15]

5) *k* Nearest Neighbors (*k*NN) with automatic selection of the optimal value of parameter *k* through the internal cross-validation procedure and with the Euclidean distance computed with all descriptors.[16] Contributions of neighbors are weighted by the inverse of distance.

6) Back-Propagation Neural Network (BPNN)[17] with one hidden layer with the default number of sigmoid neurons trained with 500 epochs of the standard generalized delta-rule algorithms with the learning rate 0.3 and momentum 0.2.

7) Regression Tree M5P (M5P) using the M5 algorithm.[18]

8) Regression by Discretization based on Random Forest (RD-RF).[19] This is a regression scheme that employs a classifier (random forest, in this case) on a copy of the data which have the property/activity value discretized with equal width. The predicted value is the expected value of the mean class value for each discretized interval (based on the predicted probabilities for each interval). The random forest classification algorithm[20] is used here.

## Step-by-Step Instructions

This part of the tutorial requires the installation of the Partial Least Square method, which is provided in Weka as an additional package. The installation is done in two steps. First, in the Weka main interface, in the menu *Tools*, choose the **Package Manager** option (Figure 13.1). Then in a new window, find the *partialLeastSquares* package (Figure 13.2). It is categorized as a preprocessing tool. Finally, click the *Install* button.

Figure 13.1 Weka main interface. The package manager is accessible through the Tools in the menu bar.

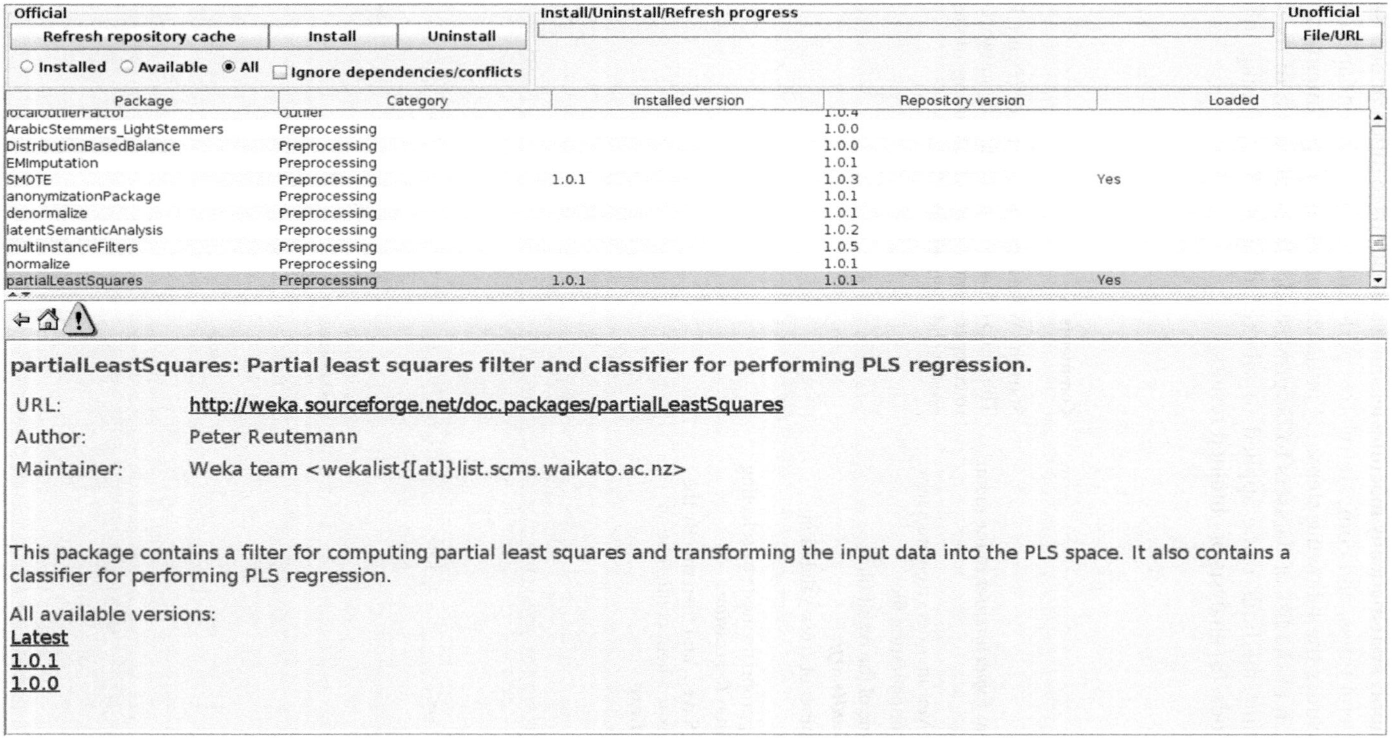

Figure 13.2 Interface of the package manager. It facilitates installation of contributed methods to Weka. Here the Partial Least Square method is to be installed.

In this tutorial, the **Experimenter** mode of the Weka program is used. This mode allows one to apply systematically machine learning algorithms to different data sets, to repeat this several times, and compare relative performances of different approaches. This study includes the following steps: (1) initialization of the **Experimenter** mode, (2) specification of the list of datasets to be processed, (3) specification of the list of machine learning methods to be applied to selected datasets, (4) running machine learning methods, (5) analysis of obtained results.

| Instructions | Comments |
|---|---|
| • Start **Weka**. | Start the Weka *Experimenter* mode (Figure 13.1). The setup consists in preparing the software to compare regression methods using 5-fold cross-validation (Figure 13.3). |
| • Select the item **Experimenter** in the menu **Applications**. | |
| • Press button **New** in order to create a new experiment configuration file. | |
| • Enter the name of the result file: **compare1-results.arff**. | |
| • Enter the number of cross-validation folds: **5**. | |
| • Choose the appropriate mode by selecting the radio-button **Regression**. | |
| • Press button **Save...** and then specify the name of the experiment configuration file **compare1.exp**. | |

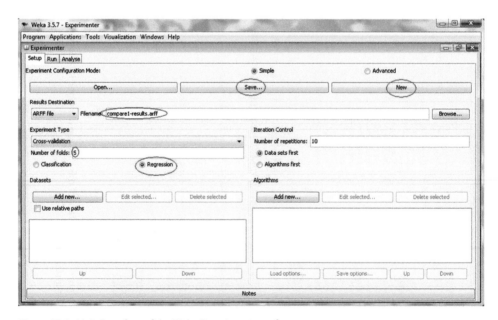

Figure 13.3 Main interface of the Weka Experimenter software.

| Instructions | Comments |
|---|---|
| • Press the button *Add new…* in the **Datasets** panel (at the left side of the window) and select file with the first dataset *alkan-bp-connect.arff*.<br>• Similarly, add the files *selwood.arff* and *shapiro.arff* files to the list *alkan-mp-connect.arff*.<br>• Press button **Save…** and then specify the name of the configuration file *compare1.exp*. | Compose the list of the datasets to which all machine learning methods will be applied (Figure 13.4). |
| • Click on the *Add new…* button in the **Algorithm** panel. | Since default parameters are used, click on the **OK** button. After that the method *ZeroR* appears in the list of currently selected methods in the **Algorithms** panel (at the right side of the main program window).<br>**Optionally**: button *Choose* selects another machine learning method; button *More* shows short description of the currently selected method; button *Capabilities* lists its capabilities (for example, whether it supports regression); button *Save…* saves specifications of the currently selected method; and *Open…* reads previously saved specifications of method. |
| • Click on the *Add new…* button in the **Algorithm** panel.<br>• Click on the *Choose* button in the window entitled **weka.gui.GenericObjectEditor**.<br>• Choose the method *weka - > classifiers - > functions - > LinearRegression* from the hierarchical tree.<br>• Since the default parameters are used here, click on the **OK** button. | Add the MLR algorithm with default settings to the benchmark list (Figure 13.4 and 13.5).<br>The method *LinearRegression* with all its parameters appears in the list of currently selected methods in the **Algorithms** panel (Figure 13.6). |
| • Click on the *Add new…* button in the **Algorithm** panel.<br>• Click on the *Choose* button and select the *weka - > classifiers - > functions - > PLSClassifier* method from the hierarchical tree.<br>• Click on *PLSFilter*. A window containing different parameters of the PLS method appears.<br>• Set the number of components to 5.<br>• Click on the **OK** button. | The PLS implementation is based on the wrapper *PLSClassifier* (Figure 13.7), which applies filter *PLSFilter* to perform calculations (Figure 13.8). The default settings for *PLSFilters* imply 20 components (latent variables). In this tutorial, we will set the number of components to be 5.<br>If the PLS method is not installed yet, refer to the beginning of the step-by-step instruction to do so.<br>Then the *PLSClassifier* method with all its parameters appears in the list of currently selected methods in the **Algorithms** panel.<br>**Optionally**, Weka allows user to find the "optimal" number of latent variables by means of a special *meta*-procedure (Figure 13.7). |

*(Continued)*

| Instructions | Comments |
|---|---|

**Figure 13.4** State of the Weka Experimenter when the list of data sets for the benchmark is setup.

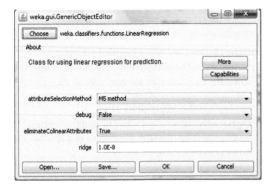

**Figure 13.5** GenericObjectEditor, the interface to select and configure machine learning methods.

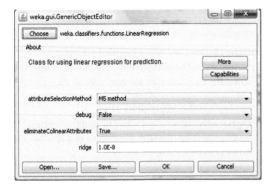

**Figure 13.6** Configuration interface for the Multi-linear regression algorithm with default setup.

| Instructions | Comments |
|---|---|

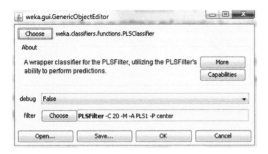

Figure 13.7 The configuration interface of the PLS classifier.

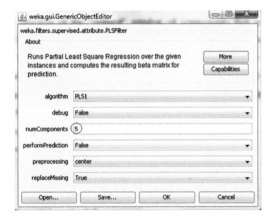

Figure 13.8 Configuration interface of the PLSFilter preprocessing tool.

- Click on the ***Add new...*** button in the **Algorithm** panel
- Click on the ***Choose*** button in the window entitled **weka.gui.GenericObjectEditor** and choose the ***weka- > classifiers - > functions - > SMOreg*** method from the hierarchical tree.
- Click on the ***OK*** button

Setup the SVM machine learning method (Figure 13.9). In this tutorial, the default values for all parameters are used: the error-complexity tradeoff parameter $C$ is 1.0, the type of kernel is polynomial with degree 1 (i.e., the linear kernel), the value of the $\varepsilon$ parameter for the $\varepsilon$-insensitive loss function (key -L) is 0.001.

Then the *SMOreg* method with all its parameters appears in the list of currently selected methods in the **Algorithms** panel.

**Optionally**, one can either apply a special wrapper procedure for finding the optimal values of parameter $C$, $\varepsilon$ and a kernel-specific parameter (degree for the polynomial kernel and $\gamma$ for the Gaussian [RBF – Radial Basis Function] kernel), or one can define the SVR algorithm with a certain set of parameter values as a separate method, and run all of them sequentially.

*(Continued)*

| Instructions | Comments |
|---|---|

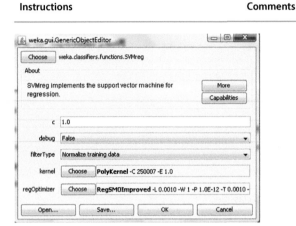

Figure 13.9 Configuration interface of the SVM algorithm with default settings.

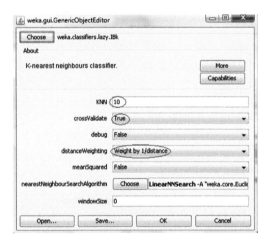

Figure 13.10 Configuration interface of the kNN method.

- Click on the **Add new...** button in the **Algorithm** panel.
- Click on the **Choose** button in the window entitled **weka.gui.GenericObjectEditor** and choose the **weka - > classifiers - > lazy - > IBk** method from the hierarchical tree.
- Type value **10** for **kNN**.
- Choose value **True** for **crossValidate**.
- Choose value **Weight by 1/distance** for **distanceWeighting**.

Add the k Nearest Neighbors (kNN) regression method. The default number of neighbors is 1. The default distance is Euclidean, which is computed using all descriptors. In this tutorial, we optimize the number of neighbors by means of a leave-one-out cross-validation procedure varying k from 1 to 10. The weighted version of the kNN regression with weights inversely proportional to distances is used.

The kNN regression (*IBk* in Weka) method with all its parameters appears in the list of currently selected methods in the Algorithms panel (Figure 13.10).

| Instructions | Comments |
|---|---|
| • Click on the ***Add new...*** button in the **Algorithm** panel.<br>• Click on the ***Choose*** button in the window entitled **weka.gui.Generic ObjectEditor** and choose the ***weka - > classifiers - > functions - > MultilayerPerceptron*** method from the hierarchical tree. The default values of all parameters are used in the tutorial.<br>• Click on the ***OK*** button. | Adding Backpropagation Neural Networks (MultilayerPerceptron). The MultilayerPerceptron method (backpropagation neural networks) with all its default parameters appears in the list of currently selected methods in the **Algorithms** panel. |
| • Click on the ***Add new...*** button in the **Algorithm** panel.<br>• Click on the ***Choose*** button in the window entitled **weka.gui.GenericObjectEditor** and choose the ***weka - > classifiers - > trees - > M5P*** method from the hierarchical tree.<br>Click on the ***OK*** button. | Adding the Regression Trees method (M5P). At this stage, a window containing numerous options for building and running regression trees pops up. The default values are used in this tutorial. The M5P regression trees method with all its default parameters appears in the list of currently selected methods in the **Algorithms** panel. |
| • Click on the ***Add new...*** button in the **Algorithm** panel.<br>• Click on the ***Choose*** button in the window entitled **weka.gui.GenericObjectEditor** and choose the method ***weka - > classifiers - > meta - > RegressionByDiscretization*** from the hierarchical tree.<br>• Click on the ***Choose*** button.<br>• Choose the method ***weka/classifiers/trees/ RandomForest*** from the hierarchical tree.<br>• Click on ***RandomForest***.<br>• Change the number of trees (numTrees) from 10 to 100.<br>• Click on the ***OK*** button to save new parameters of the method.<br>Click on the ***OK*** button. | Adding Regression by Discretization based on Random Forest (RD-RF) (Figure 13.11). The **RegressionByDiscretization** meta-method can be linked to any classification algorithm to produce numerical property/activity estimation. The **RandomForest** classifier is used in this tutorial instead of the default classifier for the **RegressionByDiscretization** *meta*-method, which is **J48** (a kind of classification trees, known also as **C4.5**[18]). The number of trees in the forest is changed from 10 (default number) to 100 (Figure 13.12).<br><br>The *RegressionByDiscretization* regression meta-method based on the Random Forest classification algorithm with all its default parameters appears in the list of currently selected methods in the **Algorithms** panel. |

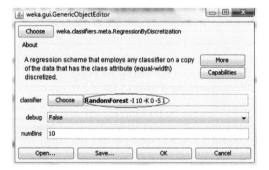

**Figure 13.11** Default configuration interface of the RegressionByDiscretization method.

*(Continued)*

| Instructions | Comments |
|---|---|

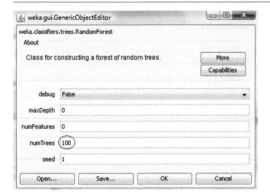

Figure 13.12  Configuration interface of the Random Forest algorithm.

Figure 13.13  Final state of the Weka Experimenter interface when all data sets and all machine learning methods are setup.

- Save the whole setup to the experiment configuration file **compare1.exp** by clicking on the **Save...** button in this window.
- Switch to the **Run** mode by clicking on the **Run** label.
- Click on the **Start** button.

Save the setup of the benchmark to the file *compare1.exp* (Figure 13.13).

Since the setup for the experimenter mode has been prepared, one can apply selected machine learning methods to the set of data sets. The current status of the process is indicated at the bottom side of the window (Figure 13.14). The history is written to the central log window. The selected machine learning methods are applied to each of the selected data sets 10 times, each time a data set being randomized. Successful termination of the job is indicated on the log window.

Instructions                                    Comments

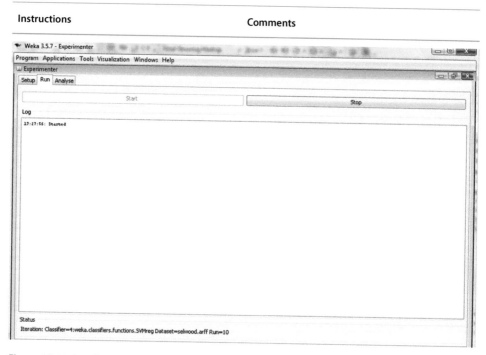

**Figure 13.14** Interface of the Weka Experimenter dedicated to monitoring the benchmark run.

- Switch to the **Analyse** panel.
- Load the result file **compare1-result.arff** by clicking on the *File...* button and selecting the appropriate file.

The analysis interface (Figure 13.15) compares the results of different runs of each machine learning method on each of the dataset and performs Student's *t*-tests. The **Configure test** panel (on the left) contains the options for running comparison test, while the test output (on the right) contains the list of machine learning methods just executed against the data sets. The **Configure test** panel has different options to assess relative performance of machine learning methods. The most important ones are the following:

- **Testing with** - statistical method used to compare performances of machine learning methods. Default is **Paired T-tester (corrected)** based on Student's t-criterion.
- **Comparison field** - performance measure. The default value is **Root_relative_squared_error** (which is related to $Q^2$). Other options include **Relative_absolute_error** and **Correlation_coefficient** (between predicted and experimental values). All performance measures are computed using the cross-validation procedure!
- **Significance** - Significance value used for the *t*-test.

*(Continued)*

| Instructions | Comments |
|---|---|

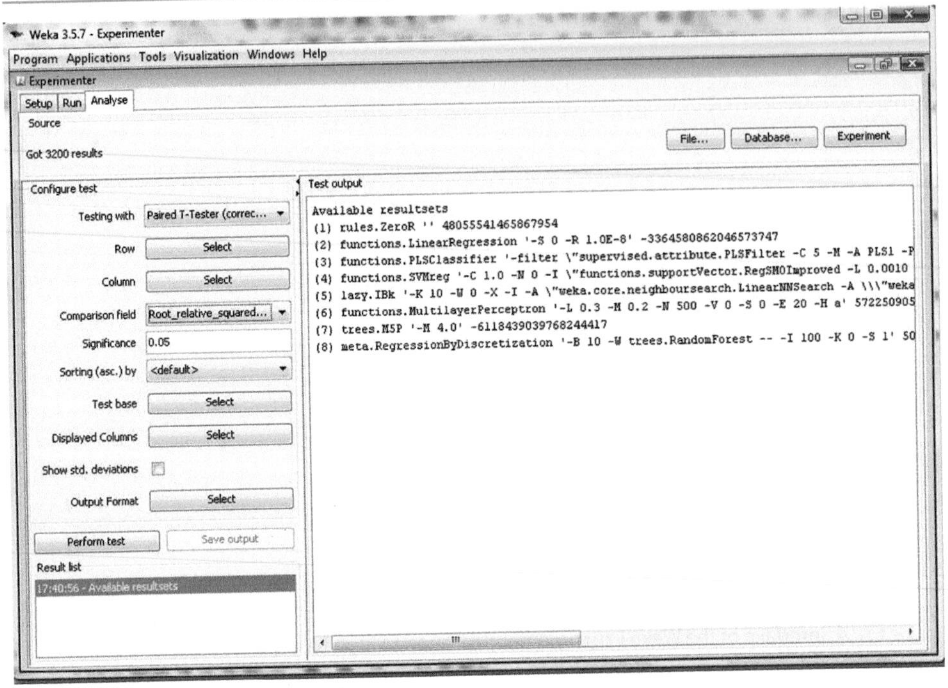

Figure 13.15 Analysis interface of the Weka Experimenter.

- Run test by clicking on the **Perform test** button.
- Read and analyze the content of the **Test output** panel.

Use the Relative Mean Squared Error during all 10 experiments to compare the machine learning algorithm performances on the different data sets. The obtained results are represented in Table 13.1. The obtained results can slightly deviate from those depicted in the table because of the stochastic nature of randomization.

Table 13.1 Root relative squared errors obtained by applying each machine learning method to each dataset. The rank is in the parenthesis, values not passing the t-test are shown in italic.

|  | alkan-bp | alkan-mp | selwood | shapiro |
|---|---|---|---|---|
| *ZeroR* | 100.00 (8) | 100.00 (6) | 100.00 (5) | 100.00 (8) |
| *MLR* | 20.38 (6) | *112.64 (8)* | *221.18 (8)* | 48.86 (5) |
| *PLS* | 13.40 (4) | *105.06 (7)* | *91.17 (3)* | 43.00 (1) |
| *SVR* | 10.16 (2) | *95.81 (3.5)* | 99.01 (4) | 47.06 (3) |
| *kNN* | 21.45 (7) | *95.81 (3.5)* | 88.45 (2) | 50.95 (6) |
| *BPNN* | 8.79 (1) | 90.76 (2) | *117.95 (6)* | 64.26 (7) |
| *M5P* | 10.29 (3) | *102.24 (5)* | *127.48 (7)* | 47.69 (4) |
| *RD-RF* | 19.70 (5) | 89.67 (1) | *77.39 (1)* | 45.59 (2) |

| Instructions | Comments |
|---|---|
| • Select the ***Relative_absolute_error*** option for **Comparison field**.<br>• Run test by clicking on the ***Perform test*** button.<br>• Read and analyze the content of the **Test output** panel. | Use the Relative Absolute Error during all 10 experiments to compare the machine learning algorithm performances on the different data sets. Results are summarized in the Table 13.2. |
| • Select the ***Correlation_coefficient*** option for **Comparison field**.<br>• Run test by clicking on the ***Perform test*** button.<br>• Read and analyze the content of the **Test output** panel. | Use the Correlation coefficient measure of success during all 10 experiments to compare the machine learning algorithm performances on the different data sets. Results are summarized in Table 13.3. |

**Table 13.2** Relative absolute errors obtained by applying each machine learning method to each data set. Rank is in the parenthesis, the values not passing the t-test (i.e., not significantly different from those of ZeroR) are shown in italic.

|  | alkan-bp | alkan-mp | selwood | shapiro |
|---|---|---|---|---|
| *ZeroR* | 100.00 (8) | 100.00 (5) | 100.00 (5) | 100.00 (8) |
| *MLR* | 20.43 (7) | *117.27 (8)* | *197.31 (8)* | 47.46 (5) |
| *PLS* | 13.28 (4) | *107.69 (7)* | *95.89 (3)* | 40.77 (1) |
| *SVR* | 9.30 (2) | *95.31 (2)* | *96.48 (4)* | 45.97 (4) |
| *kNN* | 18.69 (6) | *98.52 (4)* | *87.84 (2)* | 48.57 (6) |
| *BPNN* | 8.52 (1) | *95.52 (3)* | *119.12 (6)* | 60.36 (7) |
| *M5P* | 9.72 (3) | *106.20 (6)* | *119.73 (7)* | 45.45 (3) |
| *RD-RF* | 18.38 (5) | 91.35 (1) | 74.58 (1) | 42.59 (2) |

**Table 13.3** Correlation coefficient of the estimated values compared to the experimental ones obtained using each machine learning method for each data set. Rank is in the parenthesis, the values not passing the t-test (i.e., not significantly different from those of ZeroR) are shown in italic.

|  | alkan-bp | alkan-mp | selwood | shapiro |
|---|---|---|---|---|
| *ZeroR* | 0.00 (8) | 0.00 (8) | 0.00 (8) | 0.00 (8) |
| *MLR* | 0.98 (5.5) | *0.25 (7)* | *0.38 (5)* | 0.87 (5) |
| *PLS* | 0.99 (3.5) | 0.40 (4) | *0.40 (6)* | 0.90 (1) |
| *SVR* | 1.00 (1.5) | *0.30 (5)* | 0.60 (1) | 0.88 (3.5) |
| *kNN* | 0.97 (7) | 0.43 (3) | *0.42 (4)* | 0.86 (6) |
| *BPNN* | 1.00 (1.5) | 0.71 (1) | 0.47 (3) | 0.82 (7) |
| *M5P* | 0.99 (3.5) | *0.29 (6)* | *0.30 (7)* | 0.88 (3.5) |
| *RD-RF* | 0.98 (5.5) | 0.47 (2) | 0.56 (2) | 0.89 (2) |

## Conclusion

The relative performance of machine learning methods sharply depends on datasets and on comparison methods. In order to achieve increased predictive performance of the QSAR/QSPR models, it is suggested to apply different machine learning methods, to compare results and to select the most appropriate one.

## References

1 Needham, D.E., I.C. Wei, and P.G. Seybold, *Journal of the American Chemical Society*, 1988, **110**(13), 4186–4194.

2 Selwood, D.L., et al., *Journal of Medicinal Chemistry*, 1990, **33**(1), 136–142.

3 Shapiro, S. and B. Guggenheim, *Quantitative Structure-Activity Relationships*, 1998, **17**(04), 327–337.

4 Shapiro, S. and B. Guggenheim, *Quantitative Structure-Activity Relationships*, 1998, **17**(04), 338–347.

5 Kier, L., *Molecular connectivity in chemistry and drug research*. Vol. **14**. 2012, Elsevier.

6 Kier, L.B. and L.H. Hall, *Molecular connectivity in structure-activity analysis*. 1986, Wiley.

7 Kubinyi, H., *Journal of Chemometrics*, 1996, **10**(2), 119–133.

8 Famini, G. and L. Wilson, *Theor. Comput. Chem.*, 1994, **1**, 213–241.

9 Akaike, H., *Automatic Control, IEEE Transactions on*, 1974, **19**(6), 716–723.

10 Wold, H., *Multivariate Analysis*, 1966, **1**, 391–420.

11 Geladi, P. and B.R. Kowalski, *Analytica Chimica Acta*, 1986, **185**, 1–17.

12 Höskuldsson, A., *Journal of Chemometrics*, 1988(2), 211–228.

13 Helland, I.S., *Scandinavian Journal of Statistics*, 1990, 97–114.

14 Smola, A.J. and B. Schölkopf, *Statistics and Computing*, 2004, **14**(3), 199–222.

15 Shevade, S.K., et al., *Neural Networks, IEEE Transactions on*, 2000, **11**(5), 1188–1193.

16 Aha, D.W., D. Kibler, and M.K. Albert, *Machine Learning*, 1991, **6**(1), 37–66.

17 Williams, D.R.G.H.R. and G. Hinton, *Nature*, 1986, **323**, 533–538.

18 Quinlan, J.R., *Learning with continuous classes*. in *5th Australian Joint Conference on Artificial Intelligence*. 1992, World Scientific.

19 Frank, E. and R.R. Bouckaert, *Conditional density estimation with class probability estimators*, in *Advances in Machine Learning*. 2009, Springer. p. 65–81.

20 Breiman, L., *Machine Learning*, 2001, **45**(1), 5–32.

# 14

# Compound Classification Using the scikit-learn Library

*Jenny Balfer, Jürgen Bajorath, and Martin Vogt*

*Goal*: Demonstrate the use of different machine learning models for binary classification problems including naïve Bayes, decision trees, and support vector machines. Furthermore, their applicability to nonlinear data, interpretability, use for multi-class problems, and computational complexity are discussed.

*Software/Code*: Python with packages NumPy,[1] scikit-learn,[2] and optionally pydot.

The packages can be installed using pip and the following commands:

```
pip install -U numpy
pip install -U scikit-learn
pip install -U pydot
```

We used Python 2.7 with NumPy 1.8.2, scikit-learn 0.16.1, and pydot 1.0.29 in the preparation of this tutorial. The complete code for the step by step instructions is provided:

- `classification_models.py` contains utility methods for reading the provided data files, accessing model performance, and cross-validation of the models.
- `naive_bayes_classification.py` takes one or more input files as command line parameters, trains a naïve Bayes model, and reports its performance.
- `decision_tree_classification.py` takes one or more input files as command line parameters, trains a decision tree, and reports its performance.
- `support_vector_classification.py` takes one or more input files as command line parameters, trains a support vector machine, and reports its performance.

*Data*: `CDC4.dat`, `MCL1.dat`.

Both data sets were assembled from PubChem's bioassay database.[3] The first data set contains 1333 ligands for the CDC-like kinase 4 (CDC4). They were assembled from three different assays, in which they were categorized as active (376 compounds) or inactive (957 compounds).

*Tutorials in Chemoinformatics*, First Edition. Edited by Alexandre Varnek.
© 2017 John Wiley & Sons Ltd. Published 2017 by John Wiley & Sons Ltd.
Companion website: www.wiley.com/go/varnek/chemoinformatics

The second data set consists of 1550 ligands that were tested against induced myeloid leukemia cell differentiation protein MCL-1 in two different assays. Of those compounds, 786 and 764 were found to be active or inactive, respectively.

Both data sets were randomly split into training (80% of the available ligands) and test compounds (the remaining 20%). A list of 196 real-valued compound descriptors available from RDKit[4] was computed for each of the ligands. These contain several atom and bond count descriptors, as well as fragment, shape index, or MOE-type surface area contribution descriptors.

Both files are tabulator-separated text files. The first line contains a header with descriptor information, and each following line represents one molecule. Each molecule is represented by a compound ID (CID), its class label ("active" or "inactive"), its SMILES string, and whether it belongs to the training or test set. Then, the values of the descriptors are given.

## Theoretical Background

Classification refers to the division of input objects into a finite number of different categories or classes. In chemoinformatics, input objects are often molecules and the different classes can be "active" or "inactive," describing the compounds' abilities to activate or inhibit a certain target. All three algorithms described in this chapter belong to the class of supervised classification models. The basic workflow of supervised machine learning methodology is illustrated in Figure 14.1. During training, the parameters of the model are learned from a set of training compounds, which are given together with their known class labels. Once a model and its parameters are obtained, it can be used to predict the class labels of new and previously unseen test compounds. The following tutorials will show how to derive different models from labeled training data, and to apply these models on the test data.

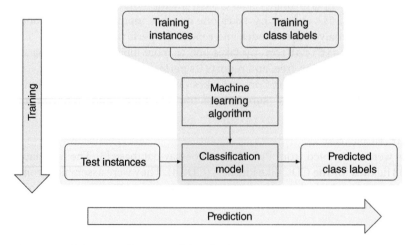

Figure 14.1 Schematic of supervised classification.

Some of the models in this chapter require careful setting of parameters. However, determining the best parameter setting for a certain data set is not trivial. Therefore, we will use $k$-fold cross-validation to find preferred parameter settings. This means that the training data is split into $k$ equally-sized folds. Each fold is once used as a test set, and a model is trained using the different parameter variations on the remaining folds. The parameter set that yields the best mean performance on the test folds is then chosen for the final model.

In this chapter, we use the F1-score balancing precision and recall to assess the performance of the parameter sets, and automatically choose the one with the best mean over all folds. However, for practical applications, it often makes sense to use a different performance metric, a choice that depends on the problem at hand and the relative importance of precision versus recall. Furthermore, a careful analysis of the training and validation set errors in different folds should be carried out. For many algorithms, the learning curve will be steep for some simpler parameter settings and further increase only slightly for more complex ones. In this case, it is meaningful not to select the setting that performs maximally, but the one separating the steep section of the learning curve from more shallow ones.

## Algorithms

The **naïve Bayes** classifier uses Bayes' theorem to predict the probability of an object $x$ to belong to a class $y$:[5]

$$P(y|x) \propto P(x|y)P(y)$$

Here, the parameters $P(x|y)$ and $P(y)$ are estimated from the training data. A new test object is then assigned to the class with the highest posterior probability $P(y|x)$. As a simplifying assumption, the naïve Bayes classifier assumes that the different descriptors of $x$ are conditionally independent from each other. It is possible to estimate the complex probability $P(x|y)$ as a product of individual descriptor contributions:

$$P(x|y) = \prod P(x_i|y)$$

In practical applications, descriptor independence is usually not given; however, a minimally correlated set of descriptors can be chosen prior to modeling. Furthermore, it has been shown that the naïve Bayesian classifier can work well even on correlated descriptor sets.[6]

A **decision tree** is a tree of rules derived from a training set in which at each branching point one feature is used to split the data depending on its values.[7] To classify a new instance, the tree is traversed from the root, and the descriptor values are used to determine which branch to follow. Once a tree leaf is reached, the class label associated with that leaf is assigned to the test instance. An exemplary decision tree is shown in Figure 14.2. Using this tree a hypothetical molecule with molecular weight 350, one hydrogen donor, and two aromatic rings would be predicted as active.

Decision trees can be recursively built from a training set. Starting at the root, the feature that best splits the training set into the different classes is determined. Here, the decision which feature is best is computed by an information measure, for example,

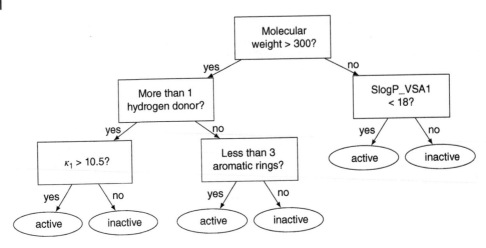

Figure 14.2 An exemplary decision tree.

the information gain or Gini index. The training data is then split at the given feature, and for each feature subset, a child tree is built. Tree construction is stopped if all leaf nodes only contain data from one class, if the data cannot be split any further, or if a maximum depth is reached.

The idea underlying **support vector machines** is the derivation of a plane in descriptor space that linearly separates instances of two classes.[8] This plane is a linear combination of the descriptors and described by a weight vector **w** and a bias $b$:

$$\mathcal{H}(\mathbf{x}) = \mathbf{w} \cdot \mathbf{x} - b$$

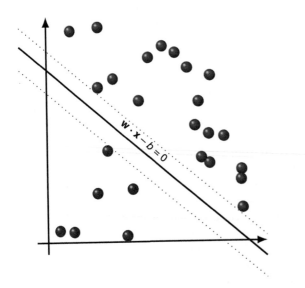

Figure 14.3 Schematic visualization of a linear Support Vector Machine (SVM).

Because infinitely many of such hyperplanes exist, the one with the maximum margin, that is, distance to the nearest training instance on each side, is chosen. Figure 14.3 illustrates this concept for a two-dimensional space. In this case, the hyperplane is a one-dimensional line with maximum distance to the training instances.

To find this hyperplane, the training class labels are expressed numerically as $-1$ and $+1$, and the following optimization problem is solved:

$$\min \frac{1}{2} \mathbf{w} \cdot \mathbf{w} \text{ subject to } y_i \left( \mathbf{x}_i \cdot \mathbf{w} - b \right) \geq 1$$

If this problem is reformulated using Lagrangian multipliers $\alpha$, the weight vector $\mathbf{w}$ can be expressed as a linear combination of a subset of training examples:

$$\mathbf{w} = \underset{\text{support vectors}}{\sum} \lambda_i y_i x_i$$

The training examples used to describe the weight vector are called the "support vectors"; they are the ones that fall exactly on the margin. However, such a hyperplane can only be found if the training data is indeed linearly separable in the descriptor space. In practice, this is usually not the case: either there is noise in the data, and some training instances fall on the wrong side of the hyperplane, or the data is not linearly separable in the descriptor space. To enable the use of support vector machines for "real-world" data, two adjustments are made. First, so-called slack variables are introduced to allow a certain amount of misclassified training examples.[9] This changes the primal optimization problem:

$$\min \frac{1}{2} \mathbf{w} \cdot \mathbf{w} + C \sum_{i=1}^{n} \xi_i \text{ subject to } y_i \left( \mathbf{x}_i \cdot \mathbf{w} - b \right) \geq 1 - \xi_i, \xi_i \geq 0$$

Second, the dot product can be replaced by a kernel function that computes a similarity between two instances and corresponds to a dot product of vectors projected into some high-dimensional space. Theoretically, this kernel space can even be infinite dimensional, because only the similarity of two instances is computed, not their explicit mapping to the feature space.[10] This in turn requires the weight vector $\mathbf{w}$ to lie in kernel space, which means it cannot be explicitly computed. Instead, the hyperplane is now given implicitly in kernel space:

$$\mathcal{H}(\mathbf{x}) = \underset{\text{support vectors}}{\sum} \lambda_i y_i K \left( \mathbf{x_i}, \mathbf{x} \right) - b$$

New training instances are then classified according to the side of the hyperplane they fall on, alternatively, they can be ranked by their signed distance to said hyperplane.

|  | Naïve Bayes | Decision Tree | Support Vector Machine (SVM) |
|---|---|---|---|
| **Complexity** | The NB classifier can be trained in linear time. Its complexity is $O(ndc)$, where $n$ is the number of training examples, $d$ the number of descriptors, and $c$ the number of classes. | The complexity to construct a binary decision tree is quadratic with respect to the number of samples, $O(n^2 d \log(n))$, where $n$ is the number of training examples and $d$ the number of descriptors. Scikit-learn claims that by pre-sorting descriptors, it is possible to train in linear time of $O(nd\log(n))$. However, this assumes that the tree is fairly balanced. | Since support vector machine training involves the solution of a complex quadratic programming problem, its complexity is at least $O(n^2)$ and in practice between $O(n^2)$ and $O(n^3)$, where $n$ is the number of training examples. Furthermore, the kernel determines the complexity in terms of the descriptor space. For the linear kernel, algorithms have been proposed that enable training in linear time ($O(nd)$). |
| **Interpretability** | Statistical measures such as the log-odds ratios or information gain can be used to access the importance of descriptors; however, an intuitive "rule" for classification cannot be derived. | Single decision trees can be easily visualized and are interpretable in terms of rules from root to leaf nodes. For very large and high-dimensional data sets, tree visualization might become too complex and unclear; however, it is possible to assess the importance of each descriptor and directly reconstruct individual predictions. | Only linear SVMs can be interpreted in terms of descriptor contributions. It is usually not possible to generate a back projection of kernel functions of nonlinear SVMs into descriptor space; therefore, only the support vectors and their coefficients are known. Hence, SVMs are often referred to as "black box" methods. |

| | | | |
|---|---|---|---|
| **Multi-class classification** | The NB classifier can be used with any number of classes and is not restricted to the binary case. | Decision trees are not restricted to binary classification problems and can be applied to any number of different classes. | Support vector machines were designed for binary classification. Hence, multi-class problems cannot be directly addressed but require specialized application protocols involving multiple SVM classifiers. Prominent strategies for multi-class SVM modeling include the one-vs-all or one-versus-one approach, which involve the training of $c$ or $c(c-1)/2$ separate SVMs, respectively. |
| **Nonlinearity** | The NB classifier is linear in feature space and modeling of nonlinearity depends on the choice of a model variant. Using the Gaussian formulation, data that roughly follows a Gaussian in feature space can be modeled. However, data with inherent substantial nonlinearity cannot be modeled. | In principle, decision trees can capture nonlinearity of any kind. Due to the nature of the splitting procedure, decision boundaries are orthogonal. Decision boundaries that are diagonal, for example, can thus not be modeled. | If a suitable kernel is chosen, even extreme cases of nonlinearity can be modeled. However, both the kernel function and its parameters have to be selected carefully. |

## Step-by-Step Instructions

### Naïve Bayes

- **Select a variant of naïve Bayes.** Scikit-learn offers three different variants of the naïve Bayes classifier, depending on the nature of the input data: Gaussian, multinomial, or Bernoulli. These three variants differ in the way they model the probability $P(x_i \mid y)$, and hence, in the parameters for each distribution that need to be estimated from the training data. If the training compounds are represented as real-valued descriptors, the Gaussian variant is used. In the case of binary fingerprints, the Bernoulli formulation is preferred, and if categorical descriptors or count fingerprints are used, one should select the multinomial formulation. Since the provided training data contains real-valued descriptors, we will use the Gaussian naïve Bayes variant throughout this chapter.
- **Choose a parameter set for the model.** Some of the model variants described above are further parameterized. The multinomial and Bernoulli formulations determine relative frequencies of descriptor values, and typically use a smoothing parameter to prevent ill-defined probabilities. If one of these variants is chosen, the smoothing parameter $\alpha$ has to be given. However, in the case of the Gaussian formulation, no parameterization is required, and the model can be instantiated without any parameters:

```
1    import sklearn.naive_bayes
2
3    model = sklearn.naive_bayes.GaussianNB()
```

- **Train the model and access its parameters.** After the training data have been loaded and the model has been trained using the `fit` method, the parameter set of the model can be accessed for further analysis. In the case of Gaussian naïve Bayes, the classes, their prior probabilities, mean and standard deviation for each feature and class are stored. While it is not essential to analyze the model parameters, it can be helpful for model interpretation.

```
1    cids, smiles, Xtrain, ytrain, Xtest, ytest = load_data
(infile)
2    model.fit(Xtrain, ytrain)
3    print infile
4    for i,cls in enumerate(model.classes_) :
5        print "\tP({0}) = {1:.2%}".format(cls,
model.class_prior_[i])
```

For the two data sets provided, this should give the following output:

```
../data/CDC4.dat
    P(active)  = 27.12%
    P(inactive) = 72.88%
../data/MCL1.dat
    P(active)  = 50.64%
    P(inactive) = 49.36%
```

Here, we can see that the prior probability of inactivity is almost three times higher than the one of activity for the CDC-like kinase 4 ligands. For MCL1, there were roughly the same number of active and inactive compounds in the training set, leading to almost equal class priors.

- **Predict the class labels for the test data and evaluate prediction quality.** The naïve Bayes classifier predicts for each test instance the probabilities to belong to one or the other class. The test instances are then assigned to the class with the maximum probability. Scikit-learn therefore offers the possibility to predict not only the class labels, but also the class probabilities for each test instance:

```
1    ypred = model.predict(Xtest)
2    yprob = model.predict_proba(Xtest)
```

Here, ypred is a vector containing the class labels "active" or "inactive" for each of the test instances. In contrast, yprob is a matrix, where each row contains the two probabilities for the "active" and "inactive" class, and both add up to 1. The probabilities can be used to access the confidence of the predictions. Since there are only two classes, we will only use the probability scores of the "active" class (the probabilities of the other class are then given by subtracting the "active" probabilities from one).

```
1    yprob = yprob[:, model.classes_ == "active"]
2    roc, auc, f1 = access_performance(ytest, ypred, yprob)
3    print infile
4    print "\tAUC: {0:.2f} \n\tF1-score: {1:.2f}".
format(auc, f1)
```

The output obtained for the two data files is the following:

```
../data/CDC4.dat
  AUC: 0.82
  F1-score: 0.69
../data/MCL1.dat
  AUC: 0.68
  F1-score: 0.68
```

While the F1-score is almost the same for both activity classes, the area under the ROC curve (AUROC) differs by 14%. This mirrors one major difference between both performance metrics: while the F1-score focuses on true positives, the AUROC also takes true negatives into account. In the first data set, there are more inactive than active compounds, a fact that is directly reflected by the higher AUROC value.

Decision Tree

- **Load the data and train a model with default parameters.**

```
1    from sklearn.tree import DecisionTreeClassifier
2
3    cids, smiles, Xtrain, ytrain, Xtest, ytest =
load_data(infile)
4    model = DecisionTreeClassifier()
5    model.fit(Xtrain, ytrain)
```

• **Visualize the tree.** The intuitive concept of decision trees lends itself well to the visualization of the trained model. Using scikit-learn and pydot, it is possible to export the trained decision tree into a graph format and print it to a figure file. The final model can then easily be analyzed, and rules can be inferred for new compound classification. Also, the visualization can aid in the analysis of the decision tree model, for example, to find out whether the tree is imbalanced or overfit.

```
1    import os.path
2    import pydot
3    from sklearn.externals.six import StringIO
4    from sklearn.tree import export_graphviz
5
6    descriptors = np.loadtxt(infile, delimiter="\t",
comments=None, dtype=str) [0,4:]
7    dot_data = StringIO()
8    export_graphviz(model, out_file=dot_data,
feature_names=descriptors)
9    graph = pydot.graph_from_dot_data(dot_data.getvalue
())
10   graph.write_png(os.path.splitext (infile)
[0] + "_graph.png")
```

Exercise: Investigate the image files that were created. Which descriptors are chosen to split the data? Which classification rules are derived by traversing the trees from top to bottom?

• **Predict the class labels and access the prediction quality.** In contrast to the naïve Bayes classifier, the decision tree does not directly produce probability estimates. Here, the computed probabilities correspond to the fraction of samples per class in the respective leaf node.

```
1    ypred = model.predict(Xtest)
2    yprob = model.predict_proba(Xtest)
3    yprob = yprob[:, model.classes_ == "active"]
4    roc, auc, f1 = access_performance(ytest, ypred, yprob)
5    print "\tAUC: {0:.2f} \n\tF1-score: {1:.2f}".
format(auc, f1)
```

Exercise: Compare the prediction statistics of the decision tree to those of the naïve Bayes classifier. Which approach performs better?

• **Choose a parameter set for the model via cross-validation.** While the principles of the decision tree algorithm are easy to comprehend, there are a number of parameters that control the way in which the data is split. Carefully selecting these parameters is important to yield a model that is neither over- nor underfitting the data. In scikit-learn, the following parameters have to be taken into account:

• The **split criterion** measures the quality of a possible split. For example, this can be the Gini impurity or the information gain. Usually, at each node the descriptor is chosen that splits the data best; to determine the meaning of "best," the split criterion is used. The Gini impurity measures how often a randomly chosen element from the current training subset would be misclassified if it was labeled randomly, with probabilities drawn from the current training label distribution. It is maximal if all labels occur equally often in each split subset, and minimal if a "pure" subset is created by a split. The information gain measures the expected change in information entropy when a split is created at a certain descriptor. Information entropy is maximal for a subset with equally frequent labels, and minimal for "pure" subsets.

- The **split strategy** can be used to induce randomness into the splitting process. Here, one can choose between the "best" split determined by the splitting criterion or the best "random" split of a number of random splits.
- The **maximum number of descriptors** to consider at each node. To speed up the training process, it can be reasonable to only consider a subset of descriptors when looking for a split.
- The **maximum depth or maximum number of tree leaves** can be limited to avoid a possible overfitting of the training data. If the maximum depth is limited, each subtree can only grow until that depth is reached. If a maximum for the number of leaf nodes is provided, the tree is grown "best-first" until the given number of leaf nodes is reached. Then, the training is terminated, even if there are impure leaf nodes left.
- The **minimum number of samples** required to consider another split. Splitting a decision tree further even if there are only very few examples left may lead to overfitting. Therefore it can make sense to require a minimum number of examples in each node that is split further. Analogously, the minimum number of samples required in a leaf node can be given.
- **Class weighting** can be used if the training set is imbalanced, that is, if one class occurs considerably more often than the other one. In many virtual screening applications, there are more inactive than active training compounds. Here, the active class can be weighted higher than the inactive one. If classes are weighted, it is also possible to define the minimum weighted fraction of training examples that are required at each leaf node.
- In practice, it is hard to predict which parameter combination yields the best model. For this tutorial, we use the best split determined via Gini impurity and allow all features to be tested. Furthermore, automatic class weighting according to the inverse observed frequencies is used for CDC4, and no weighting for MCL1. It is also possible to assign an initial random state to the model to make all computations reproducible.

```
1    cids, smiles, Xtrain, ytrain, Xtest, ytest = load_data
(infile)
2    freq = [sum(ytrain == cls) for cls in np.unique (ytrain)]
3    weight = "auto" if 1.0 * max(freq)/min (freq) > 2 else None
4    base_params = {
5        "criterion" : "gini",
6        "splitter" : "best",
7        "max_features" : Xtrain.shape[1],
8        "class_weight" : weight,
9        "random_state" : 1
10    }
```

To determine the maximum depth of the tree and the minimum fraction of training examples per leaf, we will perform 5-fold cross-validation. The following code shows how to create a grid of values for the maximum tree depth (from 1 to 10) and the minimum weighted fraction of training examples (1%, 5%, 10%, 25%, or 50%). Then, cross-validation is performed:

```
1    param_sets = []
2    for maxdepth in range(1, 11) :
3        for minleaf in [0.01, 0.05, 0.1, 0.25, 0.5] :
4    params = dict(base_params,
max_depth=maxdepth, min_weight_fraction_leaf=minleaf)
5    param_sets.append(params)
6    params = cross_validate (Xtrain,
ytrain, DecisionTreeClassifier, param_sets, 5, "f1_score")
7    model = DecisionTreeClassifier (**params)
8    print infile
9    print "\tchosen parameters:", params
```

For the two provided data sets and a random seed of 1, the following parameters are determined:

```
../data/CDC4.dat
    chosen parameters: {'splitter': 'best', 'random_state':
1, 'criterion': 'gini', 'min_weight_fraction_leaf': 0.05,
'max_features': 196, 'max_depth': 4, 'class_weight':
'auto'}
../data/MCL1.dat
    chosen parameters: {'splitter': 'best', 'random_state':
1, 'criterion': 'gini', 'min_weight_fraction_leaf': 0.01,
'max_features': 196, 'max_depth': 8, 'class_weight': None}
```

**Train the model and visualize the tree.** Once the validated parameters have been stored in the variable params, the model can be trained:

```
1    model = DecisionTreeClassifier (**params)
```

Exercise: Use the code shown before to visualize the models. Compare the cross-validated decision trees to the ones obtained with default parameters. Are the descriptors and rules the same?

● **Compare the prediction quality to the default models.** Compare the prediction statistics of the cross-validated to the default models. What can one observe?

## Support Vector Machine

● **Normalize the data.** Support vector machines are not scale-invariant, that is, descriptors with value ranges differing in their magnitude influence the results. Therefore, the training and test data have to be normalized. Here, we use scikit-learn to subtract the mean of each feature and scale it to unit variance. Furthermore, the categorical labels are transformed into numerical labels in {−1, +1}.

```
1    from sklearn.preprocessing import StandardScaler
2
3    cids, smiles, Xtrain, ytrain, Xtest, ytest = load_data(infile)
```

```
4    scaler = StandardScaler().fit(Xtrain)
5    Xtrain = scaler.transform(Xtrain)
6    Xtest = scaler.transform(Xtest)
7    ytrain = 2 * (ytrain == "active") - 1
8    ytest = 2 * (ytest == "active") - 1
```

- **Choose a model variant.** For support vector classification, scikit-learn offers two variants: SVC and NuSVC. SVC implements the formulation described above, which means that the parameter *C* penalizing the amount of training error has to be chosen. On the other hand, NuSVC uses a parameter, $v$, controlling the number of training errors and support vectors. Both formulations are mathematically equivalent, but slightly different in their use. Here, we select the original formulation using the parameter *C*.
- **Train the model using default parameters.** In scikit-learn, the parameter *C* is set to 1 per default, and the radial basis function kernel is used with a width inverse to the number of input descriptors. Our first model will be trained using these default parameters:

```
1    from sklearn.svm import SVC
2
3    model = SVC()
4    model.fit(Xtrain, ytrain)
```

- **Access the model parameters.** Once the SVM has been trained, the support vectors, their dual coefficients, and the bias of the hyperplane can be accessed. In case of the linear kernel, it is also possible to derive the primal coefficients in descriptor space, which can be seen as feature importance weights. However, this is not possible for most other kernels. The number of support vectors can give an intuition about the generalization performance of the model: if most or all of the training data end up as support vectors, the model has most likely to just memorized the data without learning a generalized pattern. On the other hand, if there are only very few support vectors, the model might be underfitting, that is, too simple to characterize the data. Overall, non-linear SVMs are hard to interpret using only their trained parameters.

```
1    print "\tnumber of support vectors:"
2    for cls in model.classes_ :
3        nsvs = int(model.n_support_[model.classes_ == cls])
4        print "\t", cls, nsvs, '/', sum(ytrain == cls)
5    print "\tbias:", float (model.intercept_)
6    svectors = Xtrain[model.support_, :]
7    lambda_y = model.dual_coef_
8    if model.kernel == "linear" :
9        feature_weights = model.coef_
```

- **Predict the class labels and assess the prediction quality.** Support vector machines are discriminative models, meaning that they aim to best distinguish instances of two classes. If the instances fall on the correct side on the hyperplane, and outside the margin, it is not important how far away they are. Hence, the distance of the instances to the hyperplane does not necessarily have a meaning in terms of classification confidence.

Neither does the magnitude of their distances have an intuitively interpretable scale, nor does the SVM produce probability estimates. Nevertheless, instances can be ranked according to their signed distance to the hyperplane to compute metrics such as the receiver operator characteristic.

```
1    ypred = model.predict(Xtest)
2    yscore = model.decision_function(Xtest)
3    roc, auc, f1 = access_performance(ytest, ypred, yscore,
labels=[-1, +1], pos=+1)
4    print "\tAUC: {0: .2f} \n\tF1-score: {1:.2f}".format (auc, f1)
```

Exercise: What is observed if you compare the prediction quality of SVMs is compared to naïve Bayes or decision trees?

- **Choose a parameter set via cross-validation.** We will now build a model with cross-validated parameters on normalized data. Depending on the chosen variant, either $C$ or $v$ have to be chosen to control the amount of permitted training error. Furthermore, a kernel function has to be chosen. Prominent kernels are the linear, radial basis function (or Gaussian), the polynomial, or the sigmoid kernel. However, any function implicitly defining a dot product in some high-dimensional space can be applied.[9] The kernel function itself might again require parameterization; for instance, the Gaussian kernel requires a value $\gamma$ controlling its size, or the polynomial kernel will require the degree and the coefficient of the polynomial. This means that not only $C$ or $v$, but also the kernel parameters have to be determined. Furthermore, it is also possible to use a class weighting for SVMs. This makes sense if the training data is imbalanced. The following code applies an automatic class weighting if the majority class occurs more than twice as often as the minority class. For the provided data sets, this means that the active CDC4 ligands will be weighted higher than the inactive ones, while the MCL1 ligands will all be assigned the same weight. Furthermore, the linear kernel with varying choices for $C$, and the Gaussian kernel with different settings for $C$ and $\gamma$ are subjected to five-fold cross-validation.

```
1    freq = [sum(ytrain == cls) for cls in np.unique (ytrain)]
2    weight = "auto" if 1.0 * max(freq)/min(freq) > 2 else None
3    base_params = {
4        "class_weight" : weight
5    }
6    param_sets = []
7    for C in [10**x for x in range (-3, 3)] :
8        params = dict(base_params, C=C, kernel="linear")
9        param_sets.append(params)
10       for g in [10**x for x in range (-3, 3)] :
11           params = dict(base_params, C=C,
kernel="rbf", gamma=g)
12           param_sets.append(params)
13   params = cross_validate(Xtrain, ytrain, SVC,
param_sets, 5, "f1_score", labels=[-1, +1], pos=+1)
14   model = SVC(**params)
15   print infile
16   print "\tchosen parameters:", params
```

- **Assess the model parameters and compare them to the default model.** Exercise: If you compare the parameters of the cross-validated to the default model, what can you observe? Which kernel function was chosen, and how many support vectors were derived?
- **Compare the prediction quality.** Exercise: How does the cross-validated model compare to the default one in terms of AUROC and F1 score?

## Notes on Provided Code

- **Loading training and test data.** In scikit-learn, data is usually represented in the form of matrices where each row corresponds to an example and each column to a descriptor. The following code uses the function `loadtxt` from the Numpy (Numeric Python) module to read the data from the provided data files:

```
1    import numpy as np
2
3    def load_data(infile) :
4        data =        np.loadtxt(infile, delimiter="\t",
     skiprows=1, dtype=str, comments=None)
5        cids      =        data[:, 0]
6        y         =        data[:, 1]
7        smiles    =        data[:, 2]
8        subset    =        data[:, 3]
9        X         =        data[:, 4:].astype(float)
10       Xtrain    =        X[subset == "training", :]
11       ytrain    =        y[subset == "training"]
12       Xtest     =        X[subset == "test", :]
13       ytest     =        y[subset == "test"]
14       return cids, smiles, Xtrain, ytrain, Xtest, ytest
```

Only those compounds that are marked as training instances are stored in the matrix `Xtrain`, and their corresponding class labels in the vector `ytrain`. The same is done for the test instances.

- **Quality assessment.** For the binary classification case, many different performance measures are available in the package `sklearn.metrics`. In this tutorial, we use the area under the receiver operator characteristic curve (AUROC) and the F1 score. To calculate the ROC curve, an ordering of test instances has to be given, while the F1-score requires binary labels. Hence, the probabilities of the active class are passed into the function `roc_curve`, while the prediction labels are passed into the method `f1_score`. Both functions also require the parameter `pos_label` indicating which of the labels corresponds to the positive class.

```
1    from sklearn.metrics import auc, f1_score, roc_curve
2
3    def access_performance(ytest, ypred, yscore,
     labels=["active", "inactive"], pos="active") :
4        assert pos in labels,
```

```
       "The positive label is not in the list of labels!"
5         roc = roc_curve(ytest, yscore, pos_label=pos)
6         auroc = auc(roc[0], roc[1])
7         f1 = f1_score(ytest, ypred, labels=labels, pos_label=pos)
8         return roc, auroc, f1
```

- **Cross-validation.** We are using the following code to determine the best parameter set among pre-defined alternatives:

```
1     from sklearn.cross_validation import KFold
2
3     def cross_validate(X, y, model_fun, param_sets, nfolds,
      quality_fun, labels=["active", "inactive"], pos="active") :
4         performance = []
5         for params in param_sets :
6           perf = []
7           for train, test in KFold(X.shape[0], n_folds=nfolds) :
8             model = model_fun(**params)
9             model.fit(X[train, :], y[train])
10            ypred = model.predict(X[test,:])
11            perf.append(quality_fun(y[test], ypred))
12          performance.append(np.mean(perf))
13        return param_sets[np.argmax(performance)]
```

Here, X and y are the training instances and labels, respectively. The function given as model_fun takes the parameters from param_sets to build a classification model. This can for instance be a DecisionTreeClassifier from the module sklearn.tree. The parameter sets are then given as a list where each entry is a dictionary with parameter names as keys and their values as values. nfolds gives the number of folds for cross-validation, and quality_fun is a function that takes true and predicted labels and returns a performance metric.

## Conclusion

This tutorial demonstrates how to build naïve Bayes, decision tree, and support vector machine models. These models require different parameterization and reach different performance levels on the compound data sets provided as illustrated in Figure 14.4.

While naïve Bayes is a generative model, meaning that it derives full probabilities for all variables, decision trees and support vector machines are discriminative models. The naïve Bayes classifier can easily be applied without many parameter choices, whereas decision trees and support vector machines require careful parameter selection. However, the decision tree is the most interpretable model in terms of rules that can be derived directly from the tree. All models can work on nonlinear data, but their ability to do so depends strongly on the data set, the chosen model variant, and its parameters. While naïve Bayes and decision trees are suitable for multi-class problems, support vector machines cannot be directly applied to non-binary class data but require

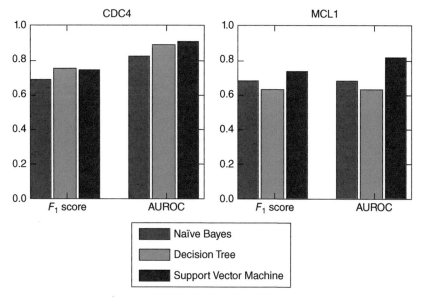

Figure 14.4 Performance comparison for the different models and data sets.

multi-model strategies. In terms of computational complexity, support vector machines usually require much longer training time than naïve Bayes and decision tree classifiers, which can be trained in linear time.

## References

1 Van der Walt, S.; Colbert, S. C.; Varoquaux, G. The NumPy Array: A Structure for Efficient Numerical Computation. *Computing in Science & Engineering* 2011, 13, 22–30.

2 Pedregosa, F. et al. Scikit-learn: Machine Learning in Python. *Journal of Machine Learning Research* 2011, 12, 2825–2830.

3 Wang, Y.; Xiao, J.; Suzek, T.; Zhang, J.; Wang, J.; Zhou, Z.; Han, L.; Karapetyan, K.; Dracheva, S.; Shoemaker, B. A.; Bolton, E.; Gindulyte, A.; Bryant, S. H. PubChem's BioAssay Database. *Nucleic Acids Research* 2012, 40, D400–412.

4 RDKit: Open-source Cheminformatics.,0000 http://www.rdkit.org.

5 Duda, R. O.; Hart, P. E.; Stork, D. G., *Pattern Classification, 2nd*; Wiley-Interscience: 2000.

6 Zhang, H. The Optimality of Naive Bayes. In *Proceedings of the 17th International Florida Artificial Intelligence Research Society Conference*, 2004, pp 562–567.

7 Breiman, L.; Friedman, J.; Stone, C. J.; Olshen, R. A., *Classification and Regression Trees*; Chapmann and Hall: 1984.

8 Vapnik, V. N., *The Nature of Statistical Learning Theory, 2nd*; Springer New York: 2000.

9 Cortes, C.; Vapnik, V. N. *Support-Vector Networks. Machine Learning* 1995, 20, 273–297.

10 Boser, B. E.; Guyon, I. M.; Vapnik, V. N. A training algorithm for optimal margin classifiers. In *Proceedings of the 5th Annual Workshop on Computational Learning Theory*, 1992, pp 144–152.

Part 5

Ensemble Modeling

# 15

# Bagging and Boosting of Classification Models

*Igor I. Baskin, Gilles Marcou, Dragos Horvath, and Alexandre Varnek*

*Goal*: This exercise illustrates two important approaches of Ensemble Learning: bagging and boosting. The methods are demonstrated by the example of building classification models based on interpretable rules. Some general behavior of these approaches are highlighted, in particular the situations where one needs to prefer one approach to another.
*Software*: WEKA
*Data*:

- `AChE/train_ache.sdf` – training set molecular structures
- `AChE/test_ache.sdf` – test set molecular structures
- `AChE/train_ache_t3ABl2u3.arff` – training set fragment descriptors
- `AChE/test_ache_t3ABl2u3.arff` – test set fragment descriptors
- `AChE/ache-t3ABl2u3.hdr` – molecular fragments definition
- `AChE/AllSVM.txt` – Estimations by SVM models built using various fragment descriptors

The task is to distinguish active compounds from inactive ones for the Acetylcholineesterase. The data set contains 27 ligands of Acetylcholinesterase (AchE) and 1000 decoy compounds chosen from the BioInfo database.[1] This data set is split into the training set (15 actives and 499 inactives) and the test set (12 actives and 501 inactives). The t3ABl2u3 ISIDA fragments are used as descriptors.

## Theoretical Background

Ensemble Learning (EL) consists in combining several base models into a more predictive one. The approach requires defining the base models and how they should be combined. There exists abundant literature devoted to EL.[2] Two main EL algorithms are concerned in this tutorial: Bagging[3] and Boosting.[4,5]

*Tutorials in Chemoinformatics*, First Edition. Edited by Alexandre Varnek.
© 2017 John Wiley & Sons Ltd. Published 2017 by John Wiley & Sons Ltd.
Companion website: www.wiley.com/go/varnek/chemoinformatics

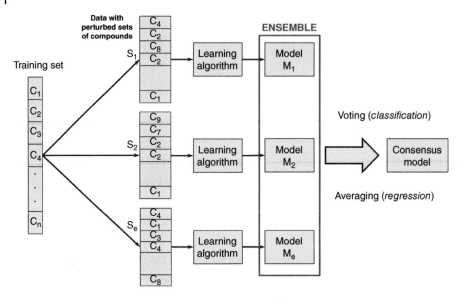

**Figure 15.1** Bagging algorithm. Resampling the training data with replacement generates an ensemble of training sets. Each of them is used to learn a base model. Predictions made by these models are combined by voting or averaging.

## Algorithm

In Weka, the EL algorithms are generally grouped into the class of *meta* methods. Both *Bagging* and *Boosting* methods belong to it. The algorithm of Bagging (Figure 15.1) consists of:

i) Training a pool of base models on the training sets sampled from the same distribution;
ii) Combine them either by majority vote (for classification tasks) or by averaging (for regression tasks).

The training sets for Bagging are obtained by resampling the original training set uniformly (i.e., all instances (chemical compounds) are drawn from the training set with the same probability) with replacement (such procedure is called *bootstrapping*).[6] In this way, each base model will be biased in a different way. Thus, taking a decision by majority voting or by averaging will tend to cancel the bias and lead to an optimal classifier. Another way to explain the Bagging is to consider prediction errors made by each base model as a random "noise"—in this case the increase in predicted performance can be attributed to the fact (known from the probability theory) that the variance of the average of two independently distributed random variables is always lower than the average of their variances.

The *AdaBoostM1* method is an implementation of Boosting. The algorithm starts by assigning equal weight to each instance in the training set. Then the algorithm iterates the following steps:

i) Train a base model on the weighted instances;
ii) Compute a model's weight based on the training errors;
iii) Increase the weight for poorly estimated instances and decrease it for well fitted ones;
iv) Repeat i-iii for the next base model.

Finally, all base models are combined using their individual weights to take a decision. The Boosting algorithm works by training a set of classifiers sequentially with each base model focusing on the prediction errors made by the previous one (Figure 15.2).

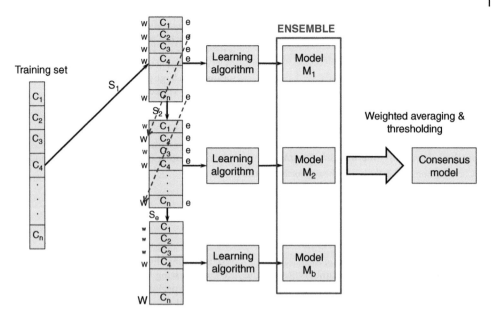

**Figure 15.2** Boosting algorithm. Each instance of the training set is assigned an equal weight w. After a training stage, the weight of those instances that were most difficult to fit is increased and the weight of the best-fitted instances is decreased. The next modeling stage uses these new weights to train the next base model. Finally, all obtained models are combined into a weighted vote (for classification) or a weighted average (for regression).

## Step by Step Instructions

| Instructions | Comments |
| --- | --- |
| • Start *Weka*. Click on the button *Explorer*. | Load the training and the test sets of the acetylcholin esterase data set. |
| • In the **Preprocess** tab, click on the button **Open File**. In the file selection dialog box, select the directory t3AB12u3, then the file `train.arff`. | |
| • In the **Classify** tab, select **Supplied test set** in the **Test options** frame and click **Set...**. The *Test Instances* dialog box pops up.Then click the **Open file...** button and in the directory t3AB12u3 select the `test.arff` file. Then click **Close** to close the *Test Instances* dialog box. | |
| • Click the **Classify** tab, then the **Choose** button. | Configure a Bagging algorithm using the *JRip* method as the base classifier and use an ensemble of exactly one model. |
| • Select **classifiers->meta->Bagging**. | |
| • Click on the name of the method to the right of the **Choose** button to open the configuration dialog box. | |
| • In the configuration dialog box, click **Choose** then select **classifiers->rules->JRip**. Set the **numIterations** to 1 and click OK. | |
| • Click on the button **Start** at the left panel on the screen | Build a model. |

*(Continued)*

| Instructions | Comments |
|---|---|

- Right-click on the last line of the **Result** list and select **Visualize threshold curve** and then **1**.

  Analyze the classification performances of the model on the test set. The performance criterion is the ROC curve and the ROC AUC. The ROC AUC value (about 0.6) is rather poor which means that a large portion of active compounds cannot be retrieved using only one rule set.

- Produce new bagging models using an increasing number of models by repeating the previous steps with different values of **numIterations**.

  One can see that ROC AUC for the consensus model increases up to 0.95 (see Figure 15.3).

- In the **Classify** tab, click **Choose** and select the method **classifiers->meta->AdaBoostM1**.

  Set up and execute a boosting consensus model consisting of exactly one Ripper*k* model.

- Click **AdaBoostM1** in the box to the right of the button. The configuration interface of the method appears.

- Click **Choose** of this interface and select the method **classifiers->meta->JRip**.

- Set the **numIterations** to 1.

- Click on the button **OK** then on **Start**.

- Repeat the experiment by setting the parameter **numIterations** to increasing numbers. Record the classification performances as measured by the ROC AUC.

  When plotting the evolution of the ROC AUC of boosting EM with the number of models, the performances are increasing (Figure 15.4).

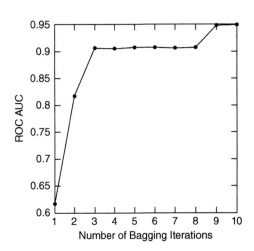

Figure 15.3 Evolution of the ROC AUC with several models in the bag. As the number increases, the classification performances reach a plateau at a ROC AUC value above 0.9.

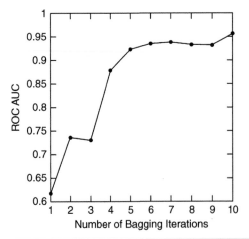

Figure 15.4 Evolution of the ROC AUC with the number of boosting iterations. The performances reach a plateau situated above 0.9.

## Conclusion

Bagging and boosting are two methods transforming "weak" individual models in a "strong" ensemble of models. In fact, Unless the accuracy of individual models per se already tends to approach optimality (i.e. the experimental precision level of the explained variable), bagging and boosting may be relied on to further reduce prediction errors towards the above-mentioned theoretical limit. However, bagging and boosting are still useful to obtain better models.

## References

1 Rognan, D. (2005). "BioinfoDB: un inventaire de molécules commercialement disponibles à des fins de criblage biologique." *La Gazette du CINES*, 1–4.
2 Brown, G. (2010). Ensemble learning. Encyclopedia of Machine Learning, Springer: 312–320.
3 Breiman, L. (1996). "Bagging Predictors." *Machine Learning*, 24(2): 123–140.
4 Freund, Y. and R. E. Schapire (1996). Experiments with a new boosting algorithm. ICML.
5 Friedman, J. H. (2002). "Stochastic gradient boosting." *Computational Statistics & Data Analysis*, 38(4): 367–378.
6 Efron, B. (1979). "Bootstrap methods: another look at the jackknife." *Annals of Statistics*, 7: 1–26.

# 16

## Bagging and Boosting of Regression Models

*Igor I. Baskin, Gilles Marcou, Dragos Horvath, and Alexandre Varnek*

*Goal*: Bagging and boosting are illustrated in the context of regression models.
*Software*: WEKA, ISIDA/Model Analyzer
*Data*:

- `LogS/train_logs.sdf` – training set molecular structures
- `LogS/test_logs.sdf` – test set molecular structures
- `LogS/train_logs_t1ABl2u4.arff` – training set fragment descriptors
- `LogS/test_logs_t1ABl2u4.arff` – test set fragment descriptors
- `LogS/LogS-t1ABl2u4.hdr` – molecular fragments definition

The task is to estimate quantitatively the aqueous solubility of organic compounds (LogS). The initial data set has been randomly split into the training (818 compounds) and the test (817 compounds) sets. A set of 438 ISIDA fragment descriptors (t1ABl2u4) was computed for each compound. Although this particular set of descriptors is not optimal for building the best possible models for this property, it however allows for high speed of all calculations and makes it possible to demonstrate clearly the effect of ensemble learning.

## Theoretical Background

Bagging and boosting are discussed in a separate tutorial dealing with their use for building classification models. However, the regression case deserves a dedicated tutorial in order to consider the specific implementation of the methods for regression and to present the weak learning algorithms recommended for bagging and boosting.

## Algorithm

The first base regression method used in this tutorial is the classical algorithm of Multiple Linear Regression (MLR) implemented in Weka in the class *classifiers/functions* with the name *LinearRegression*.

*Tutorials in Chemoinformatics*, First Edition. Edited by Alexandre Varnek.
© 2017 John Wiley & Sons Ltd. Published 2017 by John Wiley & Sons Ltd.
Companion website: www.wiley.com/go/varnek/chemoinformatics

The bagging procedure consists of [1]: *(i)* generating several samples from the original training set by drawing each compound with the same probability with replacement (so-called bootstrapping), *(ii)* building a base learner (MLR in our case) model on each of the samples, *(iii)* averaging the values predicted for test compounds over the whole ensemble of models. This procedure is implemented in Weka by means of a special "meta-classifier" in the class *classifiers/meta* with the name *Bagging*.

Additive regression is a Weka implementation of the *Gradient Boosting* ensemble learning method, which enhances the performance of a base regression base method.[2] Each iteration fits a new base model to the residuals left on the previous iteration. Prediction is accomplished by summing up the predictions of each base model. Reducing the shrinkage (learning rate) parameter, on one hand, helps to prevent overfitting and has a smoothing effect leading to enhanced predictive performance but, on the other hand, increases the learning time. Default = 1.0, that is, no shrinkage is applied. This method of ensemble learning is implemented in Weka in the class *classifiers/meta* with the name *AdditiveRegression*.

## Step-by-Step Instructions

| Instructions | Comments |
|---|---|
| • Start *Weka*. Click on the button *Explorer*. | Load the training and test sets with data on aqueous solubility. This is a dataset for building regression models. |
| • In the **Preprocess** tab, click on the button **Open File**. In the file dialog box, select the directory logS, then the file `train-logs-t1ABl2u4.arff`. | |
| • In the **Classify** tab, select **Supplied test set** in the **Test options** frame and click **Set...**. The Test Instances dialog box pops up. Then click the **Open file...** button and in the directory logS select the `train-logs-t1ABl2u4.arff` file. Then click **Close** to close the *Test Instances* dialog box. | |
| • Click **Choose** in the panel **Classifier**. | • Setting the parameters of bagging algorithm based on a single multi-linear model. |
| • Choose the method **weka- > classifiers- > meta- > Bagging** from the hierarchical tree of classifiers. | |
| • Click on the word **Bagging**. The **weka.gui. GenericObjectEditor** window related to the bagging procedure with default values of its parameters appears on the screen (Figure 16.1). | The results are summarized below (Figure 16.3). |
| • Set the number of iterations (near the label **numIterations**) to **1** (Figure 16.1). | • All statistical characteristics are worse in comparison with an individual MLR model built outside the bagging algorithm. This could be explained by the fact that after resampling the data set contains approximately 67% of unique examples, so approximately 33% of information does not take part in learning in a single bagging iteration. |
| • Click on the **Choose** button near the word *classifier*. | |
| • Choose the method **weka- > classifiers- > functions- > LinearRegression** from the hierarchical tree. | |
| • Click on the word **LinearRegression**. | |
| • Switch off the descriptor selection option by changing the option **attributeSelectionMethod** to **No attribute selection**. Press **OK** to close the window. (Figure 16.2) | |
| • Press **OK** to close the window, then the button **Start** to launch the calculations. | |

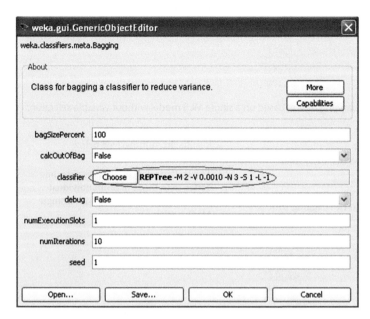

Figure 16.1 Default configuration panel for the Bagging method.

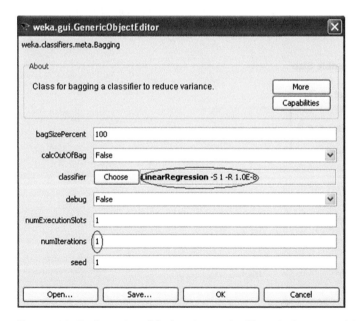

Figure 16.2 Configuration of the bagging method for a single MLR model without variable selection.

*(Continued)*

**Instructions**                                      **Comments**

```
Correlation coefficient           0.8254
Mean absolute error               0.8454
Root mean squared error           1.3627
Relative absolute error           51.0352 %
Root relative squared error       64.1859 %
```

Figure 16.3 Statistics of the bagging model based on a single MLR model without variable selection.

- Click on the word **Bagging**.
- Set the number of iterations (near the label **numIterations**) to **10**.
- Press **OK** to close the window.
- Click on the **Start** button to run bagging with 10 iterations.

- Results are summarized below (Figure 16.4).
- The statistical characteristics become better than those of both individual MLR model and bagging with a single MLR model.

```
Correlation coefficient           0.9011
Mean absolute error               0.6975
Root mean squared error           0.9503
Relative absolute error           42.1044 %
Root relative squared error       44.7614 %
```

Figure 16.4 Results of the bagging of 10 MLR models without variable selection.

- Repeat the study with the following numbers of bagging iterations: 5, 10, 15, 20, 30, 40, 50.
- Plot RMSE versus **numIterations**

- One may conclude that ensemble learning by bagging MLR models leads to decrease of prediction errors (Figure 16.5).

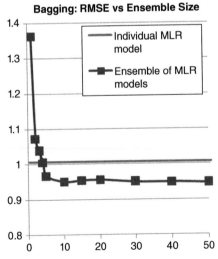

Figure 16.5 RMSE as a function of the number of models. The performance of the individual MLR model built outside the bagging algorithm is shown in blue.

| Instructions | Comments |
|---|---|

- In the **Preprocess** tab, click on the button **Open File**. In the file dialog box, select the directory logS, then the file `train-logs-t1AB12u4.arff`.
- In the **Classify** tab, select **Supplied test set** in the **Test options** frame and click **Set...**. The Test Instances dialog box pops up. Then click the **Open file...** button and in the directory logS select the `train-logs-t1AB12u4.arff` file. Then click **Close** to close the *Test Instances* dialog box.
- Click on **Choose** in the **Classifier** panel.
- On the hierarchical tree of classifiers, choose the method: **weka- > classifiers- > meta- > AdditiveRegression**.
- Click on the word **AdditiveRegression**. Notice that the default classifier (i.e., machine learning method) for the additive regression procedure is *DecisionStump* (Figure 16.6).

Load the training and test sets with data on aqueous solubility. It is useful to reload the data set before starting manipulations on the boosting algorithm.

Open the setup window of the *AdditiveRegression* method.

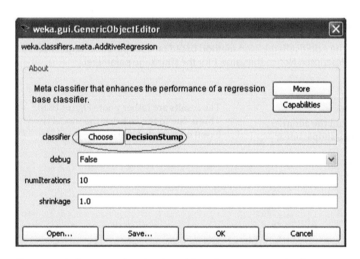

**Figure 16.6** Setup window for the AdditiveRegression method.

- Click on the **Choose** button near the word *classifier*.
- Choose the method **weka- > classifiers- > functions- > SimpleLinearRegression** from the hierarchical tree.

Set the base method to *SimpleLinearRegresssion*, which is a standard linear regression with a single molecular descriptor (Figure 16.7).

Notice the default value 1.0 for the *shrinkage* parameter. This means that we are not doing shrinkage at this stage of tutorial.

*(Continued)*

| Instructions | Comments |
|---|---|

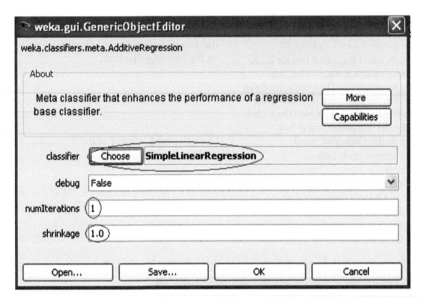

Figure 16.7 Setup window for the AdditiveRegression method based on a simple linear regression model with a single molecular descriptor. Notice the value 1 for the shrinkage parameter.

- Press **OK** to close the window.
- Click on the **Start** button to run the additive regression procedure with one simple linear regression model.

The results are rather poor (Figure 16.8). This is due to the fact that the model is based on a single descriptor.

```
Correlation coefficient        0.5863
Mean absolute error            1.323
Root mean squared error        1.7204
Relative absolute error        79.8654 %
Root relative squared error    81.0341 %
```

Figure 16.8 Results of an AdditiveRegression method based on one simple linear regression model.

- Repeat the modeling while varying the number of additive regression iterations:
- **numIterations** = 10, 50, 100, 500, and 1000.
- Change the **shrinkage** parameter to 0.5 and repeat the study for the same number of iterations.
- Build the plot RMSE versus **numIterations**

Increasing the number of models in the additive regression ensemble leads to a decrease in the prediction error. Besides, the shrinkage helps to further reduce the error at the expense of a slower convergence of the ensemble (Figure 16.9).

| Instructions | Comments |
| --- | --- |

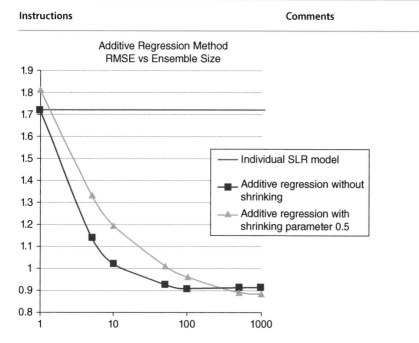

Figure 16.9 RMSE of prediction on the test set as a function of the number of models in ensemble. The prediction performance of a simple linear regression based on a single descriptor is shown in blue, while the performance of additive regression (gradient boosting) without shrinkage is shown in brown, whereas its performance with a shrinkage parameter of 0.5 is shown in green.

## Conclusion

The use of bagging and boosting is very beneficial for building both classification and regression models.

As has already been noticed for the classification case, one should use "weak" (i.e., with poor performance) base methods in order to build the strongest bagging or boosting ensemble models. Typically, bagging benefits of the methods that are "weak" because of overfitting. This is why the MLR without variable selection is a good choice for bagging. The default method, *REPtree* is also a very good choice. On the other hand, to benefit optimally from boosting, one shall use a learning algorithm that is "weak" because of underfitting. This is why the simple linear regression on a single descriptor is a good choice for boosting. For the same reason the default base method for *AdditiveRegression* in *Weka* is *DecisionStump*, which is a simple regression tree grown on a single descriptor.

## References

1 Breiman, L. (1996). *Machine Learning*, **24**(2), 123–140.
2 Friedman, J. H. (2002). *Computational Statistics & Data Analysis*, **38**(4), 367–378.

# 17

# Instability of Interpretable Rules

*Igor I. Baskin, Gilles Marcou, Dragos Horvath, and Alexandre Varnek*

*Goal*: to demonstrate the interpretable rules method. In this method selected rules are sensitive to any modification of the training data, even to the order of the data in the input file.

*Software*: WEKA, ISIDA/Model Analyzer

*Data*:

- AChE/train_ache.sdf – training set molecular structures
- AChE/test_ache.sdf – test set molecular structures
- AChE/train_ache_t3ABl2u3.arff – training set fragment descriptors
- AChE/test_ache_t3ABl2u3.arff – test set fragment descriptors
- AChE/ache-t3ABl2u3.hdr – molecular fragments definition
- AChE/AllSVM.txt – Estimations by SVM models built using various fragment descriptors

The task is to distinguish active compounds from inactive ones for the Acetylcholineesterase. The data set contains 27 ligands of Acetylcholinesterase (AChE) and 1000 decoy compounds chosen from the BioInfo database.[1] This data set is split into the training set (15 actives and 499 inactives) and the test set (12 actives and 501 inactives). The t3ABl2u3 ISIDA fragments are used as descriptors.

## Theoretical Background

Some machine learning methods allow users to obtain easily interpretable models involving a relatively small number of attributes. Generally, such models consist of a limited number of rules which are organized in a logical way. For small and/or unbalanced data sets (e.g., a data set used in this tutorial), these methods may produce "instable" models, which means that small changes in the training data lead to significant changes in selected rules. As a consequence, this may cause a problem with the model's interpretation. However, predictions for the same test set instances performed with different models (before and after training data variations) may be similar.

*Tutorials in Chemoinformatics*, First Edition. Edited by Alexandre Varnek.
© 2017 John Wiley & Sons Ltd. Published 2017 by John Wiley & Sons Ltd.
Companion website: www.wiley.com/go/varnek/chemoinformatics

## Algorithm

*JRip* is the Weka implementation of the algorithm Ripper*k* (**R**epeated **I**ncremented **P**runing to **P**roduce **E**rror **R**eduction)[2] for learning sets of "if...then..." classification rules. For example: *if* (LogP > 2.8) *and* (pKa < 5.0) *then* class = active. The sets of such rules can easily be interpreted by chemists. The algorithm starts with dividing a training set into *growing* and *pruning* sets (see Figure 17.1). Rules are built ("grown") on the growing set instances (see Figure 17.2) followed by their incremental reduced-error pruning using the pruning set. This procedure—rule growing and pruning—is repeated *k* times. Then the rules can be applied to an external test set for making predictions, and the content of the rules can be analyzed by users.

| Training set | | Test set |
|---|---|---|
| Growing set | Pruning set | |

Figure 17.1  Composition of data sets.

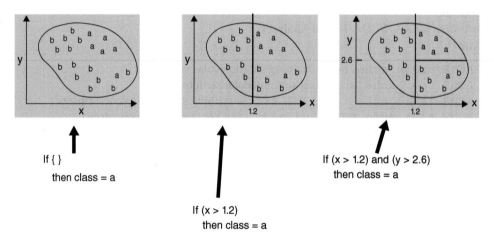

If { }

then class = a

If (x > 1.2)

then class = a

If (x > 1.2) and (y > 2.6)

then class = a

Figure 17.2  Growing a rule from data.

## Step-by-Step Instructions

| Instructions | Comments |
|---|---|
| • In the starting interface of Weka, click on the button ***Explorer***.<br>• In the ***Preprocess*** tab, click on the button ***Open File***. In the file selection interface, select the directory t3ABl2u3, then the file train.arff. | *Access training set data.*<br>The data set is characterized in the ***Current relation*** frame: the name, the number of instances, and the number of attributes (see Figure 17.3). The ***Attributes*** frame allows the user to modify the set of attributes using *select* and *remove* options. Information about the selected attribute is given in the ***Selected attribute*** frame in which a histogram depicts the attribute distribution. |

| Instructions | Comments |
|---|---|
| ● Click on the tab *Classify*. | *Access test set data.* |
| ● In the *Test options* frame, select *Supplied test set* and click *Set....* | |
| ● In the pop-up window, click the *Open file...* button and in the directory t3ABl2u3 select the test.arff file. Then click *Close*. | |
| ● Click *More options*, then in the pop-up window click the *Choose* button near *output predictions* and select *CSV*. | *Control output format to make analysis easy* |
| ● In the *classifier* frame, click *Choose*, then select the *JRip* method. | *Select and configure the implementation of the RIPPER machine learning method. Then build the model and validate it on the test set.* |
| ● Click *Start* to learn the model and apply this to the test set. | *The main parameter to configure is $k$, the number of optimization cycles of rules sets. Here, the default value is kept.* |
| ● Right click on the last line of the *Result list* frame and select *Save result buffer* in the pop-up menu. Name the file as JRip1.out. | *Analyze the estimated activities on test set compound.* |
| ● Use *ISIDA/Model Analyzer* to visualize both confusion matrix and structures of the compounds corresponding to different blocks of this matrix. Here, on the "..." button and select the JRip1.out file, then click to *Start*. | *Weka does not provide a mean to analyse the modeling results related to chemical structures. An external software is needed, for instance, the ISIDA/ModelAnalyzer (Figure 17.4 and 17.5).* |
| ● In the *Weka Classifier* output frame, check the model. | *Attributes involved in the rules can be decoded using the train.hdr file.* |

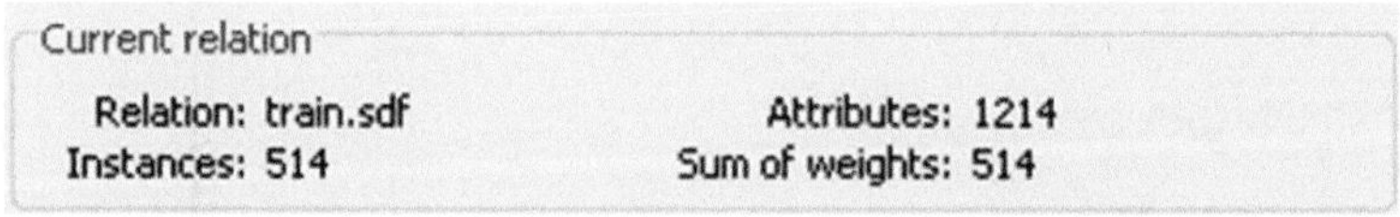

Current relation

| Relation: train.sdf | Attributes: 1214 |
|---|---|
| Instances: 514 | Sum of weights: 514 |

Figure 17.3  Current relation frame showing the parameters of loaded training set.

```
=== Classifier model (full training set) ===

JRIP rules:
===========

(att_187 >= 1) => class=1 (10.0/4.0)
(att_81 >= 3) and (att_12 <= 0) => class=1 (4.0/1.0)
=> class=0 (500.0/6.0)

Number of Rules : 3
```

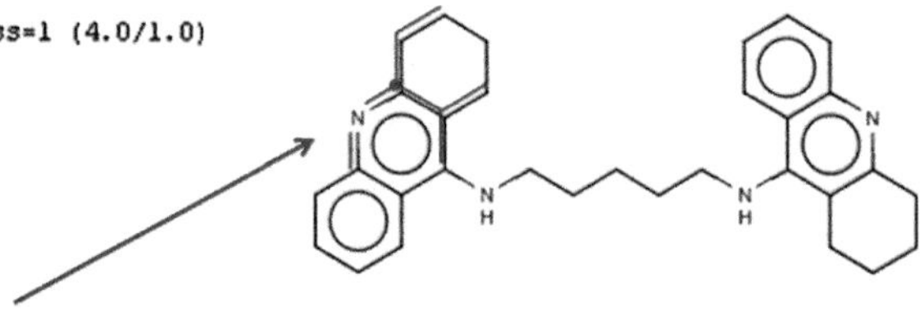

187.  (C*C),(C*C*C),(C*C-C),(C*N),(C*N*C),(C-C),(C-C-C),xC*
81.   (C-N),(C-N-C),(C-N-C),(C-N-C),xC
12.   (C*C),(C*C),(C*C*C),(C*C*C),(C*C*N),xC

Figure 17.4  RIPPER model and structural interpretation of a decision rule discovered by RIPPER. The rules involve atom-centered fragments with the neighborhood radius of two or three.

*(Continued)*

| Instructions | Comments |
|---|---|
| • In Weka, return to the **Pre-process** tab. | Apply a slight modification to the training set. Here only the order of the molecules in the dataset is modified. |
| • Click **Choose** and select **randomize** in the *filters- > unsupervised- > instance* folder. Click **Apply**. | |
| • Return to **Classify** and click **Start**. | Rebuild a model using the same machine learning method and parameterization. Then analyse the model obtained: observed rules should be changed, as exemplified in the figure 17.6. |
| • Right click on the last line of the **Result** list frame. This opens a pop-up menu, in which select **Save result buffer**. Name the file as JRip2.out. | |
| • Analyze the file JRip2.out using *ISIDA/ModelAnalyzer*. | |

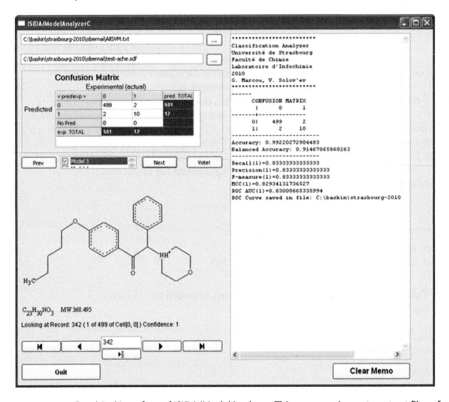

Figure 17.5 Graphical interface of ISIDA/ModelAnalyzer. This program imports output files of some data mining programs (e.g., WEKA), visualizes chemical structures, computes statistics for classification models, and builds concensus models by combing different individual models.

```
JRIP rules:
===========

(att_187 >= 1) => class=1 (10.0/4.0)
(att_831 >= 2) => class=1 (2.0/0.0)
(att_81 >= 4) => class=1 (3.0/1.0)
=> class=0 (499.0/5.0)

Number of Rules : 4
```

Figure 17.6 New set of rules obtained using the RIPPERk algorithm after reshuffling the training data.

## Conclusion

One can conclude that the data reordering is sufficient to modify the interpretable rules model. This could be explained by the fact that reshuffling of the training set leads to different composition of the growing and the pruning sets. Therefore, the model built on the training set becomes different from the previous one. This difference is more pronounced for small and/or unbalanced sets.

## References

1  http://bioinfo-pharma.u-strasbg.fr/bioinfo/
2  Cohen, William W. (1995). "Fast effective rule induction." In *Proceedings of the twelfth international conference on machine learning*, 115–123.

18

# Random Subspaces and Random Forest

*Igor I. Baskin, Gilles Marcou, Dragos Horvath, and Alexandre Varnek*

*Goal*: To introduce the concept of Random Subspace and to demonstrate the ability of the Random Forest method to produce strong predictive models.
*Software*: WEKA, ISIDA/Model Analyzer
*Data*:

- `LogS/train_logs.sdf` – training set molecular structures
- `LogS/test_logs.sdf` – test set molecular structures
- `LogS/train_logs_t1ABl2u4.arff` – training set fragment descriptors
- `LogS/test_logs_t1ABl2u4.arff` – test set fragment descriptors
- `LogS/LogS-t1ABl2u4.hdr` – molecular fragments definition

The task is to estimate quantitatively the aqueous solubility of organic compounds (LogS). The initial data set has been randomly split into the training (818 compounds) and the test (817 compounds) sets. A set of 438 ISIDA fragment descriptors (t1ABl2u4) was computed for each compound. Although this particular set of descriptors is not optimal for building the best possible models for this property, it however allows for high speed of all calculations and makes it possible to demonstrate clearly the effect of ensemble learning.

- `AChE/train_ache.sdf` – training set molecular structures
- `AChE/test_ache.sdf` – test set molecular structures
- `AChE/train_ache_t3ABl2u3.arff` – training set fragment descriptors
- `AChE/test_ache_t3ABl2u3.arff` – test set fragment descriptors
- `AChE/ache-t3ABl2u3.hdr` – molecular fragments definition
- `AChE/AllSVM.txt` – Estimations by SVM models built using various fragment descriptors

The task is to distinguish active compounds from inactive ones for the Acetylcholineesterase. The data set contains 27 ligands of Acetylcholinesterase (AchE) and 1000 decoy compounds chosen from the BioInfo database.[1] This data set is split into the training set (15 actives and 499 inactives) and the test set (12 actives and 501 inactives). The t3ABl2u3 ISIDA fragments are used as descriptors.

*Tutorials in Chemoinformatics*, First Edition. Edited by Alexandre Varnek.
© 2017 John Wiley & Sons Ltd. Published 2017 by John Wiley & Sons Ltd.
Companion website: www.wiley.com/go/varnek/chemoinformatics

## Theoretical Background

The Random Forest method is based on bagging (bootstrap aggregation, see definition of bagging) models built using the Random Tree method, in which classification trees are grown on a random subset of descriptors.[2] The Random Tree method can be viewed as an implementation of the Random Subspace (RS) method[3] for the case of classification trees. Combining two ensemble learning approaches, bagging and random space method, makes the Random Forest method a very effective approach to build highly predictive classification models.

## Algorithm

The Random Subspace (RS) algorithm consists in building base models using data sets with randomly selected molecular descriptors (Figure 18.1). Then, these models are combined using a voting or averaging procedure (Figure 18.2). In Weka, the *RandomSubSpace* method is considered as a *meta* classifier and can be found under the corresponding submenu.

The RF algorithm combines RS and bagging. The base model is a Random Tree: a fully-grown decision tree, without pruning based on a randomly selected subset of descriptors. Each tree is grown on a subsample of the data set obtained using the bootstrap procedure. The *RandomForest* method is considered in Weka as a *tree* classifier and can be found under the corresponding submenu.

Training set with initial pool of descriptors

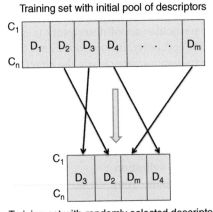

Training set with randomly selected descriptors

Figure 18.1 The Random Subspace method. Each base model is build on a randomly selected subset of descriptors.

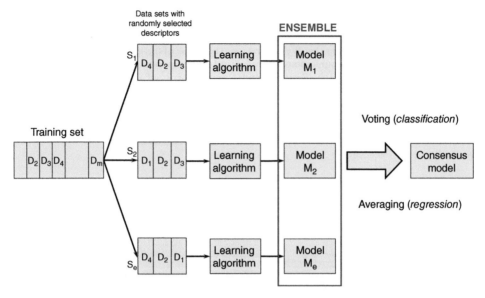

Figure 18.2 The Random Subspace method. Base models built on randomly selected descriptors are combined by voting or averaging.

## Step-by-Step Instructions

| Instructions | Comments |
|---|---|
| • Start *Weka*. Click on the button *Explorer*. | Load the training and test sets with data on |
| • In the **Preprocess** tab, click on the button **Open File**. In the file dialog box, select the directory logS, then the file `train_logs_t1ABl2u4.arff`. | aqueous solubility. They will be used to illustrate the RS approach. |
| • In the **Classify** tab, select **Supplied test set** in the **Test options** frame and click **Set....** The Test Instances dialog box pops up. Then click the **Open file...** button and in the directory logS select the `test_logs_t1ABl2u4.arff` file. Then click **Close** to close the *Test Instances* dialog box. | |
| • Click the **Classify** tab, then the **Choose** button. | Configure a Random Subspace algorithm |
| • Select **classifiers->meta->RandomSubSpace**. | using the Multiple Linear Regression method |
| • Click on the name of the method to the right of the **Choose** button to open the configuration dialog box. | *without* variable selection as the base classifier, and use an ensemble consisting of a single model. |
| • In the configuration dialog box, click **Choose** then select **classifiers->functions->LinearRegression**. | The configuration interface should look as follows (Figure 18.3). |
| • Open the configuration dialog box for the *LinearRegression* method (by clicking on the name of the method right to the Choose button) and set the **attributeSelectionMethod** to **No attribute selection**. Leave all other parameters to default values. Click **OK** to close the configuration dialog box for the LinearRegression method. | |
| • Set the **numIterations** to 1 in the configuration dialog box for the RandomSubSpace method and click **OK**. | |

*(Continued)*

| Instructions | Comments |
|---|---|
| • Clock on the button **Start**. After that Weka starts building a model. Once the computations are finished, click on the last line of the **Result** list and look at the *Root mean squared error* (RMSE) line. | The RMSE of the model should be about 1.14, corresponding to a determination coefficient, $R^2$, of about 0.74 (Figure 18.4). |

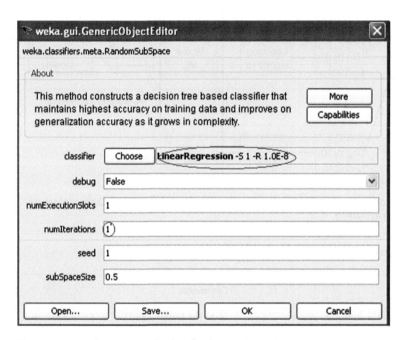

Figure 18.3 Configuration dialog box for the RandomSubSpace method using the LinearRegression as base learning method.

1 model

| | |
|---|---|
| Correlation coefficient | 0.8477 |
| Mean absolute error | 0.8415 |
| Root mean squared error | 1.1357 |
| Relative absolute error | 50.7968 % |
| Root relative squared error | 53.49 % |

10 models

| | |
|---|---|
| Correlation coefficient | 0.9024 |
| Mean absolute error | 0.6758 |
| Root mean squared error | 0.9155 |
| Relative absolute error | 40.796 % |
| Root relative squared error | 43.1219 % |

Figure 18.4 Example of the performances of a Random Subspace model using a Linear Regression method combining 1 or 10 multiple linear regression models.

| Instructions | Comments |
|---|---|
| | Produce new RS models using an increasing number of models by repeating the previous steps, setting **numIterations** value. The performances sharply improve with the number of individual models. With 10 multiple linear regression models, the RMSE is about 0.92 and the determination coefficient is about 0.83. One can follow this improvement on the curve of the RMSE as a function of the number of individual models (Figure 18.5). |
| • Start *Weka*. Click on the button *Explorer*. | Load the training and test sets of the |
| • In the **Preprocess** tab, click on the button **Open File**. In the file selection dialog box, select the file `train_ache_t3AB12u3.arff`. | acetylcholin esterase data set. This is a classification data set to illustrate the RF method. |
| • In the **Classify** tab, select **Supplied test set** in the **Test options** frame and click **Set...** The *Test Instances* dialog box pops up. Then click the **Open file...** button select the file `test_ache_t3AB12u3.arff`. Then click **Close** to close the *Test Instances* dialog box. | |
| • Click **Choose** and select the method *classifiers- > trees- > RandomForest.* | Setting the parameters of the RF method (by analogy to the Figure 18.6). |
| • Click on the word *RandomForest* to the right of the button. A configuration dialog box appears. | |
| • Set the **numTrees** to **1**, then click the button **OK**. | This setup specifies a Random Forest with a single tree. The model obtained is already rather good (Figure 18.7). |
| • Click **Start**. | |
| • Repeat the experiment by setting the parameter **numTrees** to increasing numbers. Record the classification performances as measured by the ROC AUC. | With 10 trees, all statistical characteristics become considerably stronger (Figure 18.8). When plotting the dependence of the ROC AUC of Random Forest on the number of models, the performances are increasing more and faster than with boosting or bagging (Figure 18.9). |

Figure 18.5 The dependence of the performance of ensemble models, as measured by the RMSE, on the number of base multiple linear regression (MLR) models combined using the Random Subspace method (red line). The performances of an individual MLR model with default parameters is also given (blue line) for reference.

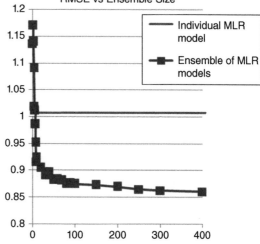

Random Subspace Method
RMSE vs Ensemble Size

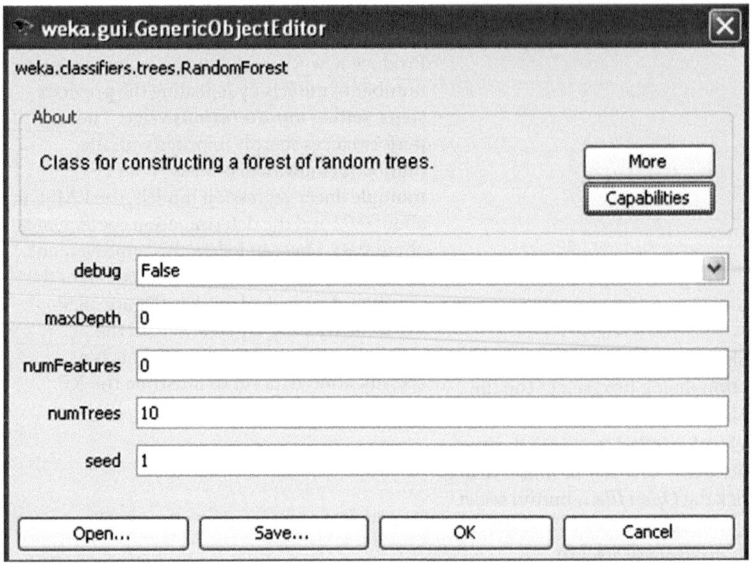

**Figure 18.6** Configuration dialog box for the Random Forest method with 10 trees.

```
Correctly Classified Instances          502              97.8558 %
Incorrectly Classified Instances         11               2.1442 %

              TP Rate  FP Rate  Precision  Recall  F-Measure  ROC Area  Class
               0.988    0.417     0.99      0.988    0.989      0.786     0
               0.583    0.012     0.538     0.583    0.56       0.786     1
Weighted Avg.  0.979    0.407     0.979     0.979    0.979      0.786

              === Confusion Matrix ===

                a    b    <-- classified as
               495   6 |   a = 0
                5    7 |   b = 1
```

**Figure 18.7** Performances of a Random Forest consisting of one tree.

```
Correctly Classified Instances          504              98.2456 %
Incorrectly Classified Instances          9               1.7544 %

              TP Rate  FP Rate  Precision  Recall  F-Measure  ROC Area  Class
               1        0.75      0.982     1        0.991      0.952     0
               0.25     0         1         0.25     0.4        0.952     1
Weighted Avg.  0.982    0.732     0.983     0.982    0.977      0.952

              === Confusion Matrix ===

                a    b    <-- classified as
               501   0 |   a = 0
                9    3 |   b = 1
```

**Figure 18.8** Performances of a Random Forest consisting of 10 trees.

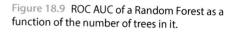

Figure 18.9 ROC AUC of a Random Forest as a function of the number of trees in it.

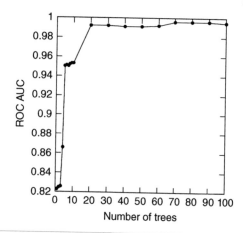

## Conclusion

The RS algorithm, which combines the bagging and the random spaces ensemble learning approaches, demonstrates that robust models with high prediction performance can be achieved without the use of sophisticated variable selection algorithms. A stochastic process can outperform them. One may observe that Random Forest models outperform the bagging and boosting models based on JRip (Ripper*k*) base models. First, a single fully-grown and unpruned random tree performs at least not worse than JRip (Ripper*k*) models based on more interpretable small set of rules. Second, the bagging and boosting models based on JRip (Ripper*k*) are saturated later, while requiring more base individual models to achieve the same performance. The resulting predictive performance of Random Forest models is very high.

## References

1 Rognan, D. (2005). "BioinfoDB: un inventaire de molécules commercialement disponibles à des fins de criblage biologique." *La Gazette du CINES*, 1–4.
2 Breiman, L. (2001). *Machine Learning*, **45**(1): 5–32.
3 Ho, T. K. (1998). "The random subspace method for constructing decision forests." *Pattern Analysis and Machine Intelligence. IEEE Transactions on*, **20**(8), 832–844.

19

## Stacking

*Igor I. Baskin, Gilles Marcou, Dragos Horvath, and Alexandre Varnek*

*Goal*: To demonstrate the ability of stacking to improve predictive performance by combining four base classifiers: (*i*) partial least squares regression (PLS), (*ii*) regression trees M5P, (*iii*) multiple linear regressions (MLR), (*iv*) a nearest neighbor model (IBk).
*Software*: WEKA
*Data*:

- `LogS/train_logs.sdf` – training set molecular structures
- `LogS/test_logs.sdf` – test set molecular structures
- `LogS/train_logs_t1ABl2u4.arff` – training set fragment descriptors
- `LogS/test_logs_t1ABl2u4.arff` – test set fragment descriptors
- `LogS/LogS-t1ABl2u4.hdr` – molecular fragments definition

The task is to estimate quantitatively the aqueous solubility of organic compounds (LogS). The initial data set has been randomly split into the training (818 compounds) and the test (817 compounds) sets. A set of 438 ISIDA fragment descriptors (t1ABl2u4) was computed for each compound. Although this particular set of descriptors is not optimal for building the best possible models for this property, however this set of descriptors allows for high speed of all calculations and makes it possible to demonstrate clearly the effect of ensemble learning.

## Theoretical Background

Stacking is historically one of the first ensemble learning methods. It combines several base models (lower-level models) built using methodologically distinct machine learning methods by means of a "meta-learner" (high-level model) that takes as its inputs the output values of the base models.[1,2]

It is expected that the diversity of the base machine learning methods will infer models that are biased differently. Therefore the combination of the models would generalize better than any individual model. Besides, the meta-learner can optimally combine the base learners so that the meta-model should perform better than voting.

*Tutorials in Chemoinformatics*, First Edition. Edited by Alexandre Varnek.
© 2017 John Wiley & Sons Ltd. Published 2017 by John Wiley & Sons Ltd.
Companion website: www.wiley.com/go/varnek/chemoinformatics

It is important to note that the mixing weights of a meta-learner used to combine base regression models should be non-negative,[3] because the estimates of each base level model should correlate with the target property. A negative weight would mean that the learning algorithm have identified the estimation of one model as efficient to compensate for the biases of another model. Since the models are independent, the correlation between the errors of a model with the estimates of a second one is likely to be fortuitous and a negative weight is likely to be a symptom of over-fitting. Yet the non-negativity constraint has been reported as irrelevant for combining classification models.[4]

Although stacking is a heuristic method and does not guarantee improvement in all cases, in many practical studies it shows excellent performance. In this tutorial we will use stacking as a high-level method (meta-learner) to combine lower-level regression models for predicting water solubility of organic compounds.

## Algorithm

The training of a stacked model is illustrated in Figure 19.1. In this case, the training consists of two stages, *Level 0* and *Level 1*. At the level 0, $k$ models based on different machine learning algorithm are trained on the training sets formed in each fold of an $n$-fold cross-validation procedure. The models are applied to the corresponding test sets,

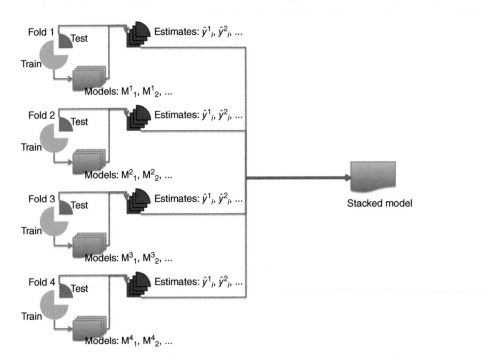

Figure 19.1 Training set up of the stacking algorithm. First, the data set is divided into non-overlapping n folds, here n=4. Each fold in turn is used as a test set while the others are fused into the corresponding training set, as for standard cross-validation. Then, on each training set, k models are trained with different machine learning algorithms and applied to the corresponding test set. Thus, finally there are k estimates for each instance. These estimates are used as input features to train a final model, the stacked model.

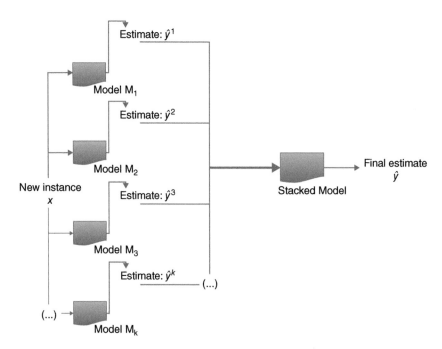

**Figure 19.2** Prediction setup of the stacking algorithm. An estimate (prediction) is made for a new instance by each of the k level-0 models. Then the stacked model combines these k estimates to produce a final estimate.

so that eventually, each instance in the full data set has $k$ estimates associated to it. At the level 1, the level 0 estimates are used as input features to train a high-level model. This is the stacked model. Finally, all level-0 models are rebuilt using the full training set.

The algorithm of applying the stacked model to new data is illustrated in Figure 19.2. The $k$ level-0 models are applied to a new instance. Then the level-1 model combines the estimates made by $k$ level-0 models to produce the final estimate.

In Weka, the method *stacking* in the *meta* methods implements the stacking algorithm as described in. [2,4]

## Step-by-Step Instructions

- Start *Weka*. Click on the button *Explorer*.
- In the **Preprocess** tab, click on the button **Open File**. In the file dialog box, select the directory `logS`, then the file `train_logs_t1AB12u4.arff`.

  Load the training and test sets with data on aqueous solubility. This is a data set for building regression models.

- In the **Classify** tab, select **Supplied test set** in the **Test options** frame and click **Set...**. The Test Instances dialog box pops up. Then click the **Open file...** button and in the directory `logS` select the `test_logs_t1AB12u4.arff` file. Then click **Close** to close the *Test Instances* dialog box.

*(Continued)*

- Click on the *Choose* button in the panel *Classifier*.
- Choose the method
  *weka- > classifiers- > functions- > PLSClassifier*
  from the hierarchical tree.
- Click on the *Start* button to run the PLS method.

Assess the predictive performance of the PLS method (with the default number of components 20). Results are reported on Figure 19.3.

```
Correlation coefficient              0.9171
Mean absolute error                  0.6384
Root mean squared error              0.8518
Relative absolute error             38.5358 %
Root relative squared error         40.12   %
```

Figure 19.3 Performances of a PLS model with 20 latent variables

- Click on the *Choose* button in the panel *Classifier*.
- Choose the method *weka- > classifiers- > trees- > M5P* from the hierarchical tree.
- Click on the *Start* button to run the M5P method.

Assess the predictive performance of the M5P method. Results are reported in Figure 19.4.
   Predictive performance of the MLR model has been assessed in another tutorial.

```
Correlation coefficient              0.9176
Mean absolute error                  0.6152
Root mean squared error              0.8461
Relative absolute error             37.1349 %
Root relative squared error         39.8532 %
```

Figure 19.4 Performance of the M5P model with default parameters.

- Click on the *Choose* button in the panel *Classifier*.
- Choose the method
  *weka- > classifiers- > meta- > Stacking*
  from the hierarchical tree of classifiers.
- Click on the word *Stacking*. The weka.gui.
  GenericObjectEditor window related to
  the stacking procedure with default
  values of its parameters appears on the screen
  (Figure 19.5).
- Click on the field containing the text "1 weka.
  classifiers.Classifier" right from the
  label *classifiers*.
- A new window (Figure 19.6) containing the list of
  currently selected classifiers pops up.
- Delete the *ZeroR* method by clicking on the *Delete*
  button.
- Click on the *Choose* button near the word
  *classifier*.
- Choose the method
  *weka- > classifiers- > functions- > PLSClassifier*
  from the hierarchical tree.
- Click on the *Add* button.

Initialize the stacking method.

- Prepare an empty list of base classifiers
  (level-0 models). Pay attention: all
  regression and classification machine
  learning algorithms are called in WEKA
  *classifiers*.

- Add the PLS classifier to the empty list of
  classifiers.

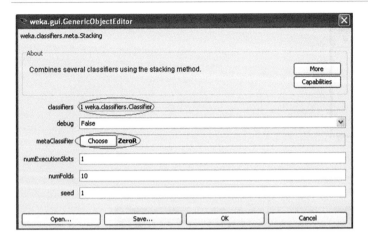

Figure 19.5 The configuration interface of the stacking method.

Figure 19.6 Interface to prepare the list of stacked machine learning methods.

- Click on the **Choose** button near the word *classifier*.
- Choose the method
  *weka->classifiers->trees->M5P* from the
  hierarchical tree.
- Click on the **Add** button.
- Click on the **Choose** button near the word *classifier*.
- Choose the method *weka->classifiers->functions->*
  *LinearRegression* from the hierarchical tree.
- Click on the word *LinearRegression*.
- Switch off the descriptor selection option by changing
  the option *attributeSelectionMethod* to *No attribute*
  *selection*.
- Press **OK** to close the window.
- Click on the **Add** button.
- Close the window by clicking at the cross.

- Click on the **Choose** button near the word
  *metaClassifier*.
- Choose the method
  *weka->classifiers->functions->*
  *LinearRegression* from the hierarchical tree.

- Add the M5P method (a kind of regression
  trees) to the list of currently selected
  classifiers.

- Add the MLR method to the list of
  currently selected classifiers.

At this stage the window should look like in
Figure 19.7.
Set the meta-classifier for the stacking method
to be the multiple linear regressions (MLR).
  At this stage the configuration interface,
the *weka.gui.GenericObjectEditor* window,
should be as in Figure 19.8.

*(Continued)*

Figure 19.7 Final state of the list of stacked machine learning methods.

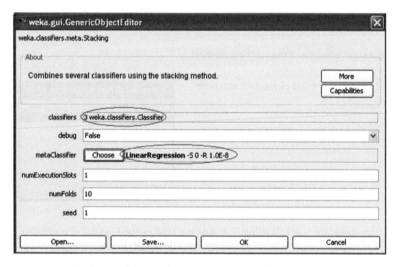

Figure 19.8 Final state of the configuration interface of the stacking method.

- Press **OK** to close the window.
- Click on the **Start** button to run the stacking method.

Run stacking of methods and assess the predictive performance of the resulting ensemble model.

Weka finds an optimal combination of the base classifiers (Figure 19.9).

The statistical results are reported in Figure 19.10.

```
 0.465   * weka.classifiers.functions.PLSClassifier +
 0.445   * weka.classifiers.trees.M5P +
 0.0644  * weka.classifiers.functions.LinearRegression +
-0.0932
```

Figure 19.9 An optimal combination of PLS, nearest neighbor and multi-linear regressions models according to the stacking method.

```
Correlation coefficient                0.9366
Mean absolute error                    0.562
Root mean squared error                0.746
Relative absolute error                33.9262 %
Root relative squared error            35.1378 %
```

Figure 19.10  Performance of the stacking model combining a PLS, nearest neighbor, and multi-linear regression models.

- Click on the field containing the text "1 weka. classifiers.Classifier" right from the label *classifiers*.
- A new window (Figure 19.11) containing the list of currently selected classifiers pops up.
- Choose the method *weka- > classifiers- > lazy- > IBk* from the hierarchical tree.

Repeat the study by adding 1-NN. The results become even better (Figure 19.11).

```
class =

      0.2368 * weka.classifiers.lazy.IBk +
      0.0384 * weka.classifiers.functions.LinearRegression +
      0.3713 * weka.classifiers.functions.PLSClassifier +
      0.3559 * weka.classifiers.trees.M5P +
     -0.0589


Time taken to build model: 99.22 seconds

=== Evaluation on test set ===
=== Summary ===

Correlation coefficient                0.9392
Mean absolute error                    0.537
Root mean squared error                0.7301
Relative absolute error                32.4193 %
Root relative squared error            34.3861 %
Total Number of Instances              817
```

Figure 19.11  Optimal weights of the stacked models combining a nearest neighbor, PLS, regression tree, and multi-linear regression, followed by the performances of the stacking model.

## Conclusion

The results for stacking are summarized below.

| Learning algorithm | R (Correlation Coefficient) | MAE | RMSE |
| --- | --- | --- | --- |
| MLR | 0.8910 | 0.7173 | 1.0068 |
| PLS | 0.9171 | 0.6384 | 0.8518 |
| M5P | 0.9176 | 0.6152 | 0.8461 |
| 1-NN | 0.8455 | 0.85 | 1.1889 |
| Stacking of MLR, PLS, M5P | 0.9366 | 0.5620 | 0.7460 |
| Stacking of MLR, PLS, M5P, 1-NN | 0.9392 | 0.537 | 0.7301 |

One may conclude that the stacking of several base classifiers has led to considerable decrease of prediction error (RMSE = 0.730) compared to that for the best base classifier (RMSE = 0.846).

## References

**1** Seewald AK. (2002). How to make stacking better and faster while also taking care of an unknown weakness. In: Proceedings of the Nineteenth International Conference on Machine Learning, Morgan Kaufmann Publishers Inc., 554–561.

**2** Wolpert DH. (1992). Stacked generalization. *Neural Networks*, **5**(2), 241–259.

**3** Breiman L. (1996). Stacked regressions. *Machine Learning*, **24**, 49–64.

**4** Ting KM, Witten IH. (1999). Issues in stacked generalization. *J Artif Intell Res (JAIR)*, **10**, 271–289.

Part 6

3D Pharmacophore Modeling

20

# 3D Pharmacophore Modeling Techniques in Computer-Aided Molecular Design Using LigandScout

*Thomas Seidel, Sharon D. Bryant, Gökhan Ibis, Giulio Poli, and Thierry Langer*

## Introduction

LigandScout enables chemists and molecular modelers to automatically derive 3D-interaction feature models starting from 1) a macromolecular ligand complex, 2) a set of ligands without active-site information, and 3) an active site where there are no ligands present. Within the platform the user can utilize these models to rapidly search very large libraries (millions of compounds) virtually to find new hit compounds, prioritize them for synthesis and biological testing, and make various 3D pharmacophore based alignment experiments to support lead optimization projects. The first version of LigandScout released in 2005, featured capabilities to automatically derive pharmacophore models from protein-ligand complexes.[1–4] Its evolution during the last ten years has expanded to include expert capabilities, such as, pharmacophore derivations from ligands, methods to explore active sites without ligands, docking, advanced filtering tools, fragment based approaches, advanced virtual screening, and alignment capacities available to users within an intuitive and friendly graphical user interface (Figure 20.1). LigandScout has been used widely by researchers in industry and academic research for early hit and lead discovery activities on a diverse number of clinically relevant targets such as, enzymes, G protein–coupled receptors (GPCRs), kinases, ion-channel, cytokine, integrin, and nuclear receptors to name a few. Recent examples in the literature cite the use of LigandScout for rapidly visualizing and deciphering key interaction features between proteins and ligands,[5,6] using 3D-pharmacophore models for virtual screening to successfully find biologically active compounds,[7,8] target fishing,[9,10] drug repurposing,[11] exploring protein-protein interfaces,[12–14] and profiling drug targets for side effects.[15–17] Furthermore, LigandScout pharmacophore models and virtual screening have been reported to outperform other virtual screening methods in a benchmarking study performed using C-X-C motif chemokine receptor type 4 (CXCR-4) antagonists.[18]

This chapter will review theory related to pharmacophore models and guide the user through six essential workflows using LigandScout: 1) structure-based pharmacophore modeling, 2) ligand-based pharmacophore modeling, 3) creating shared-feature pharmacophore models, 4) accurate virtual screening and pharmacophore editing in the

*Tutorials in Chemoinformatics*, First Edition. Edited by Alexandre Varnek.
© 2017 John Wiley & Sons Ltd. Published 2017 by John Wiley & Sons Ltd.
Companion website: www.wiley.com/go/varnek/chemoinformatics

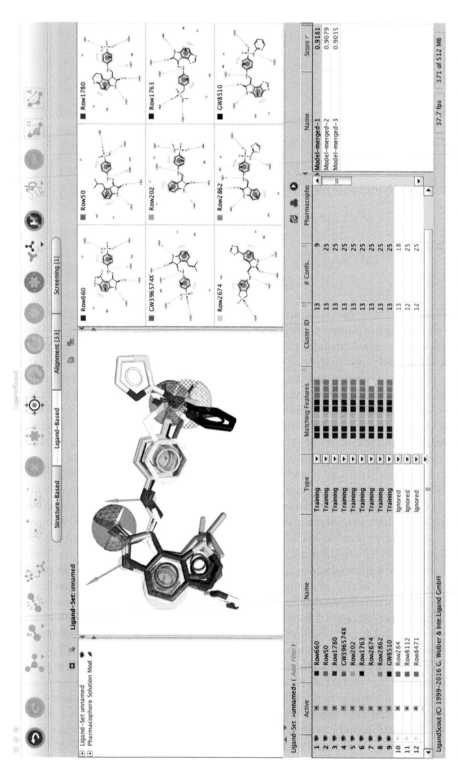

Figure 20.1 The graphical user interface of LigandScout, a molecular design platform that specializes in pharmacophore modeling and virtual screening technology to support hit finding and lead optimization activities in early stage drug discovery research.

active site, 5) hit analysis, and 6) parallel virtual screening. Tutorials presented in this chapter, as well as other advanced workflows for design not covered herein, and data sets are available upon request.[19]

## Theory: 3D Pharmacophores

The tutorials in this chapter involve the creation of 3D-chemical feature pharmacophore models and the use of these models to find biologically active molecules using virtual screening methods. To better understand the academic concepts underlying these models we present in this section the theory behind the concept of a pharmacophore and the feature definitions assigned to a pharmacophore model. The fundamental function of the alignment algorithms and virtual screening modes implemented in LigandScout find foundation in the chemical feature definitions.

Pharmacophore modeling together with virtual screening have become increasingly popular in the last decades and matured to a valuable and efficient basis for a wide variety of computer-aided drug design projects.[20] Due to their simplistic nature, pharmacophores are easy to understand and illustrate, which renders them rather useful as a means to describe and explain ligand-target binding interactions. However, there still exists some confusion about the term pharmacophore which, depending on the background and context, is often attributed with different meanings. Historically, medicinal chemists used (and still use) the term pharmacophore to denote common structural or functional elements that are essential for the activity of a set of compounds toward a particular biological target. However, in 1998, Camille Wermuth, a well-known medicinal chemist, submitted a refined definition of a pharmacophore to the IUPAC that is now officially recognized.[21] It states: "A pharmacophore is the ensemble of steric and electronic features that is necessary to ensure the optimal supra-molecular interactions with a specific biological target structure and to trigger (or to block) its biological response." According to this definition pharmacophores do not represent specific structural motifs of molecules (β-lactams, dihydropyridine) or associations of functional groups (e.g., primary amines, sulfonamides), but are an abstract description of essential steric and electronic properties that are required for an energetically favorable interaction between a ligand and receptor of the macromolecular target.[22] Pharmacophore models can thus be considered the largest common denominator of molecules that show a similar biological profile and are recognized by the same binding site of the target.[23]

## Representation of Pharmacophore Models

To be useful for drug design, pharmacophore models must in some way uniformly represent the physico-chemical properties and location of functional groups involved in ligand-target interactions, as well as the different types of non-covalent bonding and their characteristics in a manner that is easy for humans to comprehend. The most common representation of pharmacophores is a spatial arrangement of so-called chemical (or pharmacophoric) features that describe essential structural elements and/or observed ligand-receptor interactions by means of geometric entities. Although this representation is quite simple, it sufficiently fulfills the above requirements and has found general acceptance among medicinal chemists because they already think of molecules in terms of their pharmacophore space when modifying compounds in lead optimization projects.[24]

However, from the perspective of developing pharmacophore features for computer-aided design purposes, special attention must be made regarding the right level of abstraction of the chemical feature types used in the construction of the pharmacophore models that will be used for virtual screening. Rather general definitions result in models that are universal, at the cost of selectivity. Selectivity, however, is also an important issue for the quality of pharmacophore models and features that are too general may need to be refined to additionally include characteristics of the underlying functional groups instead of merely reflecting universal chemical functionality. Being too restrictive, on the other hand, will increase the number of different feature types at the cost of comparability and the ability to identify novel, structurally unrelated, chemical compounds. Therefore, the design of pharmacophore modeling software faces a trade-off between a generally applicable feature set that is universal and, at the same time, still selective enough to reflect all relevant types of observed ligand-receptor interactions. LigandScout supports the derivation of fourteen feature types with defined geometries and tolerances including hydrogen bond (H-bond) donors, H-bond acceptors, hydrophobic, aromatic interactions, positive and negative ionizable features, coordination to metal ions as well as excluded volumes. The graphical representation of these features is shown in Figure 20.2. The

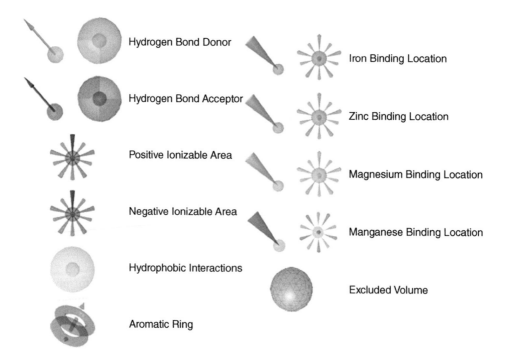

Figure 20.2 Pharmacophore depictions in LigandScout. Hydrogen-bond donors and acceptors can be depicted as vectors or spheres depending on the model requirements.

following sections give a brief overview of the most important types of ligand-receptor interactions and their corresponding geometric representations in pharmacophore models in LigandScout.

## Hydrogen-Bonding Interactions

Hydrogen bonding is an attractive interaction of electropositive hydrogen atoms with an electronegative atom (H-acceptor) like oxygen, fluorine, or nitrogen. The participating hydrogen must be covalently bound to another electronegative atom (H-donor) to create the hydrogen bond. Hydrogen bonding is a relatively strong interaction (~2–167 kJ/mol) and one of the most important for the formation of strong ligand-receptor complexes.[25] To capture the characteristics of hydrogen-bonding interactions (Figure 20.3), they are usually modeled as a position with a certain tolerance for the acceptor (donor) atom and a projected point (also with a certain tolerance) for the position of the corresponding donor (acceptor) atom. Together these two positions form a vector that constrains the direction of the H-bonding axis and also the location of the interacting atom in the target receptor. When the direction constraint is omitted, they become less specific and will match any acceptor/donor atom irrespective of whether essential geometric preconditions for the formation of a hydrogen-bond are actually fulfilled.

Figure 20.3 Hydrogen bonding geometry: The involved N, H, and O atoms are nearly lying on a line. The N-O distance is typically between 2.8 and 3.2 Å. The N-H-O angle is >150° and the C=O-H angle between 100° and 180°.

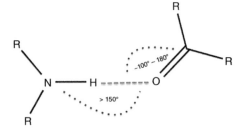

## Hydrophobic Interactions

Hydrophobic (lipophilic) interactions occur when non-polar amino acid side-chains in the protein come into close contact with lipophilic groups of the ligand. Lipophilic groups include, for example, aromatic and aliphatic hydrocarbons, halogen substituents like chlorine and fluorine, and many heterocycles like thiophene, and furan. Since lipophilic areas on the protein and ligand surface are not capable of participating in any polar interactions, attractive forces are negligible for the effect of hydrophobic interactions. They are driven instead by the displacement of water molecules from non-polar areas in the binding pocket to the outside of the protein. This leads to a higher entropy of the system due to a gain in mobility and allows the now unconstrained water molecules to form energetically favorable hydrogen bonds. According to the Gibbs free energy equation, $\Delta G = \Delta H - T\Delta S$, both contributions will lower the change in free energy, $\Delta G$, for the interaction and thus increase the ligand's overall binding affinity. Since hydrophobic interactions are undirected, they can be represented as tolerance spheres (Figure 20.2), which are located in the center of hydrophobic atom chains, branches or groups of the ligand.

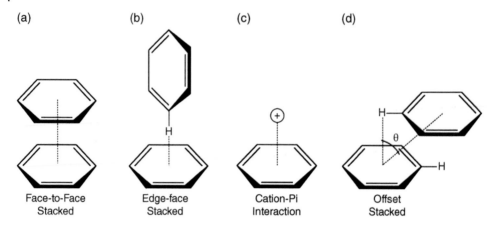

(a)      (b)      (c)      (d)

| Face-to-Face Stacked | Edge-face Stacked | Cation-Pi Interaction | Offset Stacked |

Figure 20.4 Steric configurations of π-π and cation-π interactions.

### Aromatic and Cation-π Interactions

Electron-rich π-systems like aromatic rings are capable of forming strong attractive interactions with other π-systems (π-stacking) and adjacent cationic groups (e.g., metal ions, ammonium cations in protein side-chains).[26] The interaction energies are the same order of magnitude as hydrogen bonds and thus play an important role in various aspects of molecular biology (e.g., stabilization of DNA and protein structures, enzymatic catalysis, and molecular recognition). Since cation-π and π-π interactions require a certain relative geometric configuration of the interacting counterparts (Figure 20.4), they belong to the class of directed interactions. In pharmacophore models, aromatic features are therefore at least represented by tolerance spheres located in the center of the aromatic ring-system (Figure 20.2). To account for the directional aspects of aromatic interactions, they are often attributed with additional information about the spatial orientation of the aromatic ring-system in the form of a ring-plane normal or two points that define this vector.

### Ionic Interactions

Ionic interactions are strong attractive interactions (energies > 400 kJ/mol) that occur between oppositely charged groups of the ligand and the protein environment. Positive or negative ionizable areas can be single atoms (e.g., metal cations, ammonium ions) or groups of atoms (e.g., carboxylic acids, guanidines, aromatic heterocycles) that are likely to be protonated or deprotonated at physiological pH. Ionic interactions are of electrostatic nature and thus undirected which allows the corresponding pharmacophoric features to be represented by simple tolerance spheres.

### Metal Complexation

Some proteins contain metal ions as co-factors. A prominent example involves metalloproteases that contain $Zn^{2+}$ ions that are coordinated to the protein via three amino acids as shown in Figure 20.5.[27] In such proteins, a coordination complex of the metal ion with suitable electron donating atoms or functional groups of the ligand is often the

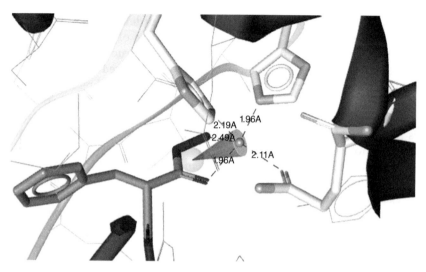

**Figure 20.5** Thermolysin in complex with the hydroxamic acid inhibitor BAN (PDB ID: 5TLN). The Zn$^{2+}$ ion is penta-coordinated with the amino acids Glu166, His142, and His146 of thermolysin, and the hydroxyl- and carbonyl-oxygen of the hydroxamic acid moiety. Source: Matthews, 1988.[28] The Figure was created with LigandScout 4.1 and the structures found in PDB ID:5TLN.

most important contribution to the overall binding affinity and essential for the ligand's mode of action. Functional groups and structural elements that exert a strong affinity for metal ions are, for example, thiols R-SH, hydroxamates R-CONHOH, or sulfur and nitrogen containing heterocycles. In pharmacophore models, metal binding interactions are represented by tolerance spheres located on single atoms or in the center of groups that are capable of interacting with the metal ions. To additionally define or constrain the location of the coordinated metal ion or accommodate a particular coordination geometry, a vector representation is incorporated as well (Figure 20.2).

### Ligand Shape Constraints

The chemical features in a pharmacophore model represent necessary, but not all of the, characteristics active molecules must posses to achieve specific, high affinity binding to a given target receptor. A molecule may be retrieved by a pharmacophore model in a virtual screening exercise because it is capable of matching a set of features that is entirely consistent with the pharmacophore model, but when tested it fails to show activity. A plausible reason for this that can be explained by modeling (rather than biological pathways) is that some part of the molecule experiences a steric clash with the receptor side-chains if it were to bind in the mode described by the pharmacophore model. A common way to avoid this situation is to add exclusion volumes to the model. Excluded volumes are represented by variably sized spheres (Figure 20.2) corresponding to variably assigned tolerances, which indicate regions of "forbidden" space, thereby defining a restricted area that a hit structure should not occupy when it is aligned to the pharmacophore. This means that molecules retrieved in a virtual screening exercise must not only fit the interaction features defined by the model but they must also fit within the region defined by the excluded volumes as well. Obviously, the correct placement of excluded volumes is important in order to avoid the possibility of missing active compounds. The most reliable source of information for a proper placement of excluded volumes is the crystallographic structure of the ligand bound receptor. Such receptor-based excluded volumes are centered on appropriate atoms of the binding-site surface with sizes dictated by the corresponding atomic van der Waals radii. A clash of the aligned molecule with one of the excluded volume spheres directly corresponds to a steric overlap with an atom of the receptor surface and indicates a presumably poor fit of the molecule. When the three-dimensional receptor structure is not available (which is often the case), and ligand-based pharmacophore derivation methods are used, then the placement of exclusion volumes is less straightforward. In this case LigandScout will automatically distribute the location and size of the exclusion volumes spheres based on the union of the molecular shapes of a set of aligned known actives. The user can edit these excluded volumes manually to adjust the selectivity of the model as needed.

## Pharmacophore Modeling

Pharmacophore models can be created by a variety of methods including manual construction, automated perception from the structure of one or more ligands, and receptor-based deduction from a crystallographic structure. The particular method or workflow that is best suited for a given modeling problem depends on a number of factors like the goals of the study, nature and quality of available data, computational resources, and the aim and further use of the created pharmacophore model. The following sections will give an overview of the methodological details and applicability of the most commonly used approaches for creating pharmacophore models.

## Manual Pharmacophore Construction

The simplest way (in terms of algorithmic complexity) to create a pharmacophore model is manual construction based on information about the structure and/or special characteristics of a series of known active ligands. A manually constructed pharmacophore can be quite advantageous, particularly if it is derived from the x-ray structure of a ligand in its binding conformation or from a ligand with low conformational flexibility. In either case, the locations of the pharmacophoric features are essentially pinned down, so that conformational flexibility, one of the biggest uncertainties, is eliminated. However, there is still the question of which features should be incorporated into the model, which is not always easy to infer without additional information such as the structure of a ligand-receptor complex. With the advent of powerful computer-aided methods for pharmacophore modeling, the importance of a manual pharmacophore construction from scratch has largely diminished. Nevertheless, manual pharmacophore generation is still possible using LigandScout. Capabilities, such as, creating new features, modifying a feature type, removing and/or disabling a feature, and changing feature tolerances, will influence the selectivity and the performance of a pharmacophore model at the users discretion.

## Structure-Based Pharmacophore Models

The availability of information about the three-dimensional structure (e.g., from NMR experiments or X-ray crystallography) of a ligand/receptor complex is a tremendous advantage when it comes to the development of high quality pharmacophore models. Knowledge of the 3D structure of the bound ligand and the surrounding receptor surface allows for the analysis of essential interactions and the correct placement of corresponding pharmacophoric features. Furthermore, detailed information about regions that are restricted can be incorporated into the final pharmacophore model in the form of excluded volumes and thus constrain the shape of ligands that are retrieved from virtual screening exercises.

A fundamental step in the development of a receptor-based (often also called structure-based) pharmacophore model is the analysis of the binding site and its associated ligand to identify potential interaction points. A number of methods can in principle be used to identify such regions.[29] *LigandScout*, for example, takes a direct approach and derives a pharmacophore model from a single ligand/receptor complex[2,3] as follows: 1) After the user loaded the protein-ligand complex, LigandScout analyzes and corrects hybridization states of unsaturated bonds and aromatic rings. 2) Following this step, both the ligand and binding pocket amino acids are analyzed for the presence of atoms and groups that can take part in hydrogen-bonding, hydrophobic, aromatic, ionic, and metal binding interactions. If there are complimentary interaction partners between the ligand and binding site functionalities, LigandScout will automatically add a feature to the model. Pharmacophoric feature detection can be customized with respect to interaction specific geometric characteristics like allowed distances and angle ranges. Whether a feature is incorporated into the final pharmacophore model depends on its location relative to a complementary feature in the binding-site. For example, a hydrogen-bond acceptor feature located on an acceptor atom of the ligand is only included if there is an opposing hydrogen-donor feature on the receptor side within a certain distance and angle range. 3) After all complementary feature pairs of the complex have been analyzed and the corresponding ligand-side features have been put into the derived pharmacophore model, exclusion volume spheres are added to mimic the shape of the binding pocket. Figure 20.6 illustrates a structure-based pharmacophore model for a cyclin dependent kinase 2 (CDK2) inhibitor from the PDB ID: 1KE8 that was created automatically using *LigandScout*.

## Ligand-Based Pharmacophore Models

When information about the three-dimensional structure of the receptor is limited or not available, but a sufficient number of actives are available, then ligand-based methods provide an alternative way to leverage the available information and develop pharmacophore models that can be used for virtual screening or lead optimization workflows. An important precondition for ligand-based methods to work and deliver good models is that the ligands used for model generation should bind to the same receptor at the same location in the same way. Otherwise, the resulting pharmacophore models will not represent the correct mode of action and are essentially useless.

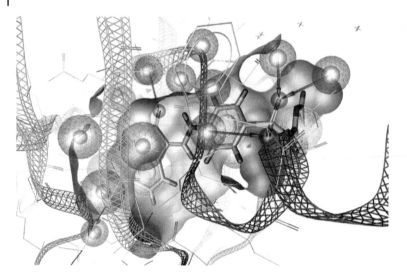

**Figure 20.6** Receptor-based or structure-based pharmacophore generated by LigandScout for the CDK2/inhibitor complex with PDB ID: 1KE8. Gray spheres represent exclusion volumes that model the shape of the receptor surface. Yellow spheres represent hydrophobic interactions, green and red arrows represent hydrogen-bond donor and acceptor features, respectively. Vectors define the direction of the hydrogen bonds.

The derivation of pharmacophore models from a set of ligands involves many different algorithms.[30,31] For example, after the import and preparation of the input structures (SMI, SDF, or MOL files), one will need to generate a sufficiently large and diverse set of low energy ligand conformations. This is done because the bioactive conformations of the input ligands may not be known but it can be assumed that one conformation in each set of generated conformers is at least a good approximation thereof. This multi-conformational set of active ligands is designated the training set by the user and can be computed in an automated manner using the conformer generator iCon in LigandScout from the menu pull-down options or from a shortcut button above the table of ligands. The next step is at the heart of the overall procedure and aims to identify a chemical feature pattern[32] that is common to all training-set ligands and can be superimposed with at least one conformation of each ligand. Since often more than one such pharmacophoric pattern can be found due to multiple conformations of the ligands, LigandScout will produce multiple pharmacophore model solutions and rank them with a fitness function. From these ranked solutions the user can select the best model to fit the project needs. If the best model is not clear, then the user could opt to take the pharmacophore model with the best fit score or test each of the models against a set of active and inactive molecules in a virtual screening validation and ROC curve generation procedure.[31] Methods for pharmacophore validation can be divided into three categories:[30]

1) Statistical significance analysis and randomization tests.
2) Enrichment-based methods involve assuring the ability to recover active molecules from a test database in which a small number of known actives have been hidden among randomly selected compounds. Database mining and the utilization of receiver operating characteristic (ROC) curves[32] fall into this category.
3) Biological testing of matching molecules.

If pharmacophore validation results indicate a generally unsatisfying quality of the generated models, they may be refined manually (e.g., deletion/addition of features, modifying tolerances or features (vector to sphere or interpolation) applicable if only small changes are required) or the whole modeling procedure must be repeated with different set-up (e.g., changes to the composition of the training- and/or test-sets, refinement of ligand conformers, and editing pharmacophore model generation parameters) until acceptable results are obtained. The high number of influential variables and the disregard of the receptor structure makes pure ligand-based modeling relatively error prone and leaves much room for interpretation. The algorithmic power of the employed software, a high expertise of the user in terms of knowledge about the biological target, and a thorough validation of the obtained results are therefore critical for the successful application of this modeling approach.

## 3D Pharmacophore-Based Virtual Screening

Because of their abstract nature and simplicity, 3D pharmacophore models represent efficient filters for the virtual screening of large compound libraries.[33,34] The computational complexity of the hit identification process in virtual screening is greatly reduced by the sparse pharmacophoric representation of ligand–target interactions which results in rapid overall search times. Futhermore, since pharmacophore-based queries are based on pharmacophore feature alignments rather than compound scaffold alignments, they are able to find hit molecules with diverse scaffolds when compared to the original ligands used for the generating the pharmacophore models.[35] This is of special interest for researchers who need to find novel molecules that are patentable or lead candidates with better ADMET properties, and/or higher activity, and/or selectivity toward the target.[36,37]

### 3D Pharmacophore Creation

The first step in a typical pharmacophore-based virtual screening campaign is to create a query pharmacophore model that specifies the type and geometric constraints of the chemical features that have to be matched by the database molecules in a virtual screening experiment. Ligand-based and structure-based pharmacophore models can be created (see Tables 20.1, 20.3 & Fig. 20.9, 20.11) and used separately or in combination via parallel virtual screening workflows in LigandScout if desired. The strategy adopted depends on the goals of the project.

### Annotated Database Creation

One of the biggest challenges in pharmacophore modeling and virtual screening workflows is designing data sets that can be used for pharmacophore model validation.[36] In particular the design of a database of active molecules and inactive molecules that can be used for virtual screening in order to understand how well the pharmacophore models can distinguish between true active and inactive molecules is important before screening a library of commercially available untested compounds. The idea is to edit the pharmacophore model to maximize the retrieval of true active compounds and minimize the retrieval of inactive molecules. Data for creating such data sets may come from in house sources of experimentally tested compounds or they could be extracted from public sources such as ChEMBL or PubMed.[38,39] LigandScout Expert/Knime provides a Knime node[40] for extraction of compounds from ChEMBL. In addition, databases of untested compounds from in house or commercial sources must also be computed for virtual screening campaigns using LigandScout in order to find new hit molecules.[36] An important aspect that needs to be considered when screening molecule libraries against 3D pharmacophores is conformational flexibility. Most major software applications including LigandScout deal with this problem by creating dedicated screening databases that store pre-computed conformations for each of the molecules. Another approach is to tweak the conformation of the molecules on the fly in the pharmacophore fitting process.[41] The advantage of the latter approach is the lower storage requirements. However, it also has the major disadvantage that the screening process is considerably slower and a dramatic reduction of the conformational search space while aligning bears the danger of falling into a local minimum.[29,42] Nowadays enough hard disk storage is available and screening databases with pre-generated conformations are clearly preferred. These databases are usually generated once or can be easily updated without having to recompute the entire library using LigandScout tools and reused whenever needed.

## Virtual Screening-Database Searching

In LigandScout the 3D pharmacophore search of a library of compounds is implemented as a multistep filtering process. First, the user must open and designate the pharmacophore models and databases in the virtual screening perspective of LigandScout. The first step is a fast pre-filtering that aims to quickly identify and eliminate all molecules that cannot be fitted to the query pharmacophore model in 3D space. Only molecules that pass this pre-filtering step need to be processed in the final accurate, but computationally expensive, 3D alignment step. In the alignment step, LigandScout will examine closely the conformations of the remaining molecules that might match the query to see if they are able to match the spatial arrangement of the query features. Special care must be taken in this step because an ultimate decision has to be made whether to reject a database compound or to put it in the hit list and, thus, directly influencing the quality of the obtained screening results. To correctly identify a match of a molecule to the query pharmacophore within the defined feature tolerances an overlay in 3D space is required. This overlay is also necessary to check and score additional constraints imposed by vector features like hydrogen-bond acceptors/donors, plane features like aromatic rings, and exclusion/inclusion volume spheres. Commercial software for pharmacophore modeling that incorporate state-of-the-art screening functionality all perform some sort of geometric alignment in the 3D pharmacophore matching step, which is usually done by minimizing the root-mean-square deviation (RMSD) between associated feature pairs.[43]

## Hit-List Analysis

Hits retrieved from the 3D database search are a good and recommended starting point for the validation and refinement of the query pharmacophore model. There are several useful measures to characterize the obtained hit list like *Sensitivity, Specificity, Yield of Actives,* and many others, which are described elsewhere.[44,45] A modern tool for the assessment of screening results against datasets of active and inactive compounds are *Receiver Operating Characteristic* curves (ROC).[32,46] A ROC curve displays the rate for retrieving true positives (actives) plotted against the rate of retrieving false positives

(Figure 20.7). The Y-coordinate of the ROC curve represents the true-positive rate (rate of retrieving actives), whereas the X-coordinate denotes the appropriate false-positive rate (rate of retrieving inactives or decoys). An ideal curve would rise vertically along the Y-axis until it reaches the maximum true positive rate, which is one (1), and then continues horizontally to the right, which means that the hit list contains all active compounds in the database and that not one of the hits is a false positive. The ROC curve of a random database search is represented by the median.

On the basis of a hit-list analysis with the above measures and tools, the pharmacophore model is often refined to achieve more satisfying results. Adaption of feature definitions, modification of feature tolerances, addition or removal of features and exclusion volumes are some of the adjustments that can help to tune a pharmacophore model. Another possibility is to modify the database by readjusting the number of pre-generated conformations to address molecular flexibility more adequately. Because pharmacophore modeling and database screening are very complex tasks, several iterations of screening, analysis, and refinement are usually necessary to achieve good results. After the model has been suitably refined based on ROC curve analysis to the desired true active over false positive retrieval rate, the model is ready for virtual screening of set of compounds that have not been tested on the target. The hits retrieved will be ranked using a pharmacophore fit score. Higher pharmacophore fit scores represent a better fit to the model.

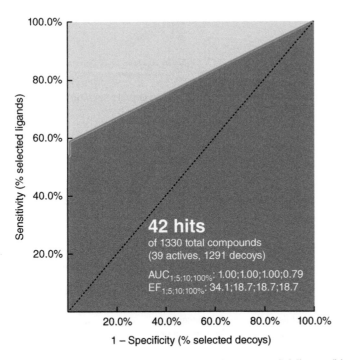

**42 hits**
of 1330 total compounds
(39 actives, 1291 decoys)

$AUC_{1;5;10;100\%}$: 1.00;1.00;1.00;0.79
$EF_{1;5;10;100\%}$: 34.1;18.7;18.7;18.7

**Figure 20.7** Example of a receiver operating characteristic (ROC) curve (blue) automatically generated by LigandScout when the user specifies databases of true active and inactive molecules for pharmacophore model validation. The results depicted are based on a structure-based model derived from PDB ID: 1ke7 screened against true active and inactive molecule data sets derived from ChEMBL. A curve close to the median (dotted line) would indicate that the model is no better than choosing molecules at random (Table 20.1).

# Tutorial: Creating 3D-Pharmacophore Models Using LigandScout

The next two sections detail stepwise tutorials for pharmacophore modeling with LigandScout. Figure 20.8 outlines a recommended and general pharmacophore modeling and virtual screening workflow.

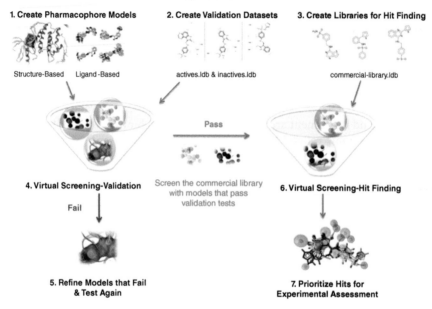

**1. Create Pharmacophore Models**

Structure-Based    Ligand -Based

**2. Create Validation Datasets**

actives.ldb & inactives.ldb

**3. Create Libraries for Hit Finding**

commercial-library.ldb

Pass

**4. Virtual Screening-Validation**

Fail

Screen the commercial library with models that pass validation tests

**6. Virtual Screening-Hit Finding**

**5. Refine Models that Fail & Test Again**

**7. Prioritize Hits for Experimental Assessment**

Figure 20.8  A general and recommended pharmacophore modeling and virtual screening workflow using LigandScout.

## Creating Structure-Based Pharmacophores From a Ligand-Protein Complex

Table 20.1  Description of actions for structure-based pharmacophore creation using LigandScout.[1]

| Experience level: basic | | | Time needed: 5 minutes |
|---|---|---|---|
| **Views** | **Sequence** | **User Controls** | **Advanced Controls (opt.)** |
| • Macromolecule view <br> • Active site view | • Download PDB file using the PDB ID. <br> • Choose the correct ligand <br> • Click on yellow box and zoom into the active site view <br> • Check and correct ligand structure <br> • Create a structure-based pharmacophore | • Ligand box <br> • Create pharmacophore (button or menu) | • Change bond type <br> • Change atom type <br> • Create new bond <br> • Move to core <br> • Move to environment <br> • Compound switching |

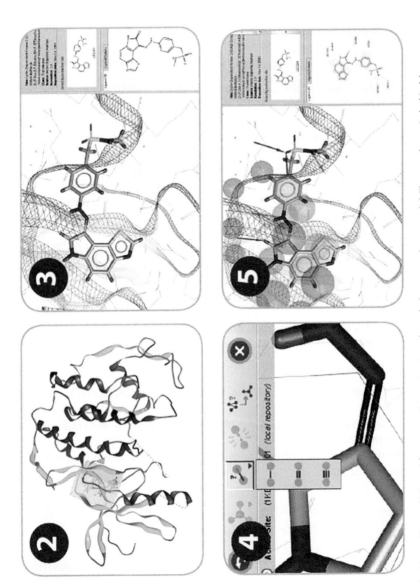

Figure 20.9 Snapshots of the areas of the LigandScout GUI used for generating a structure-based pharmacophore model. The numbers circled in blue correspond to the steps in the description. Source: LigandScout.[1]

## Description: Create a Structure-Based Pharmacophore Model

(Numbering below refers to numbers circled in blue in Figure 20.9).

1) Type the PDB ID: "1ke6" in the upper right area of the screen and press the button "Download 1ke6." The protein will be downloaded from the protein data bank (PDB)[47,48] and displayed. You must be connected to the Internet to access the PDB. If you are not connected to the Internet you can use File → Open from the LigandScout pull down menu. In this case your PDB file should be stored locally on your computer.

2) Click on the yellow box within the protein to zoom into the active site (this is the "macromolecule view").

3) Once inside the active site you will see the ligand. Since structures in the PDB contain incomplete information, you should always check whether the ligand is correctly depicted, for example, whether bond orders are correct.

4) If you find an error, you can edit the bond order by selecting a bond (by clicking on it either in the 2D or 3D view using the left mouse button) and use the bond order icon pull-down to select the correct bond order. Similarly you can select the bond you wish to edit using your left mouse button and use the keyboard numbers such as, "1" to create a single bond, "2" to create a double bond, or "3" to create a triple bond.

5) Once the ligand is chemically correct, create a pharmacophore by pressing Ctrl-F9 on Windows/Linux machines or Cmd-F9 on Apple machines.

**Where to go from here:**

• Move the pharmacophore model to the virtual screening perspective for virtual screening.

• Move the pharmacophore model to the alignment perspective for creating shared pharmacophore models.

• Edit the pharmacophore to make it more or less restrictive based on your project requirements (Table 20.2).

## Create a Shared Feature Pharmacophore Model From Multiple Ligand-Protein Complexes

Table 20.2 Description of actions for creating a shared feature pharmacophore model using LigandScout.[1]

**Experience level:** intermediate                    **Time needed:** 15 minutes

| Views | Sequence | User Controls | Advanced Controls |
|---|---|---|---|
| • Macromolecule view<br>• Active site view<br>• Alignment view | • Download PDB File "1ke6"<br>• Check & correct ligand<br>• Create pharmacophore<br>• Add ligand and pharmacophore to alignment view | • Ligand box<br>• Create pharmacophore (button or menu)<br>• Move molecule to the alignment perspective | • Set reference element<br>• Center all structures<br>• Alignment context menu |

Table 20.2 (Continued)

| Views | Sequence | User Controls | Advanced Controls |
|---|---|---|---|
| | • Repeat these steps for "1ke7" and "1ke8"<br>• Create a "shared feature pharmacophore" from the three pharmacophores<br>• Align the three ligands to the new pharmacophore | • Move pharmacophore model to alignment<br>• Generate shared feature pharmacophore | |

Figure 20.10 Snapshots of the areas of the LigandScout GUI used for generating a shared featured pharmacophore model. The numbers circled in blue correspond to the steps in the description. Source: LigandScout.[1]

## Description: Create a Shared Feature Pharmacophore and Align it to Ligands

(Numbering below corresponds to the numbers circled in blue in Figure 20.10).

First follow the steps described in Table 20.1 & Fig. 20.9 to generate a structure based pharmacophore model for PDB ID: 1KE6.

1) Use the data exchange widget and select "Copy current Ligand to Alignment Perspective" and the molecule "(1ke6) LS2201" will be added to the alignment view. Use the data exchange widget—select "Copy current Pharmacophore(s) to Alignment Perspective" to move pharmacophore model to the alignment perspective. Repeat the steps above using the PDB files with PDB ID: "1ke7," and "1ke8." After generating pharmacophore models for the three different PDB entries you should have six elements (three ligands and three pharmacophore models) in the alignment perspective. Click on the "Alignment" tab to move to the alignment perspective. There you will see your ligands and pharmacophore models in a list.

2) Select 1ke6 pharmacophore and use the first button above the table to set it as the reference (Set Reference). The color of the text will become red. Select the three pharmacophores 1ke6 (red text), 1ke7 and 1ke8 using Control key (Command on Apple) and left mouse button. You will see them appear in the 3D window.
3) Press the align button (icon with two squares) to align the pharmacophore models. In the 3D window you will see that the three models are now aligned.
4) Press the icon next to the align molecules icon to "Generate shared feature pharmacophore" button. A new pharmacophore model called "Shared [LS2201, LS3201, LS4299]" is appended to the list. Deselect all items in the list by clicking the empty space below the table with your left mouse. Now select the shared feature pharmacophore model you just created and mark it as the reference element using the "Set reference" button (the text will turn red). Use the control key (command on Apple) to select the ligands, 1ke6, 1ke7, and 1ke8 identified by the blue icon to the left of each entry and the reference shared pharmacophore model (now red). Hide the three pharmacophores derived from 1ke6, 1ke7, and 1ke8 by clicking on the eye symbols on the right next to these alignment entries. Now press the "Align selected elements" (icon with two squares) button to align the three ligands to the shared feature pharmacophore model.
5) You will see the three ligands from the x-ray structures aligned to the shared feature pharmacophore derived from the three PDB complexes. Move your shared feature model to the screening perspective for virtual screening (Table 20.3).

## Create Ligand-Based Pharmacophore Models

Table 20.3  Description of actions for creating a ligand-based pharmacophore model using LigandScout.[1]

| **Experience level:** intermediate | | | **Time needed:** 15 minutes |
|---|---|---|---|
| **Views** | **Sequence** | **User Controls** | **Advanced Controls (opt.)** |
| • Ligand-based Modeling | • Import ligands<br>• Generate conformations<br>• Cluster ligand-set<br>• Generate ligand-based pharmacophore model | • Generate conformations<br>• 3D Clustering<br>• Generate ligand-based pharmacophore model (button or menu) | • Use different settings for conformational model generation<br>• Use different settings for clustering<br>• Use different settings for ligand-based pharmacophore model generation |

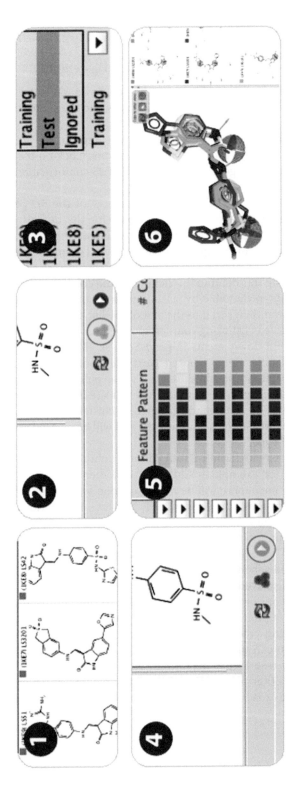

Figure 20.11 Snapshots of the areas of the LigandScout GUI used for generating a ligand-based pharmacophore model. The numbers circled in blue correspond to the steps in the description. Source: LigandScout.[1]

## Description: Ligand-Based Pharmacophore Model Creation

(Numbers below correspond to the numbers in blue in Figure 20.11).

1) Open the SMILES file ("cdk2.smi") by using the pull-down menu the "File → Open." The ligands will appear in a table below the 3D view and are set to "training set" by default.

2) a) Generate conformations (3D structures) by clicking on the button (two blue arrows). LigandScout will prompt you with options. Click on the button "Apply FAST Settings." This will generate a maximum of 25 conformations per ligand in your training set.

b) Cluster the ligand set according to the geometry of the 3D pharmacophore features by clicking on the "cluster" button. Use the default settings when prompted with settings choices and start the clustering process by pressing the "OK" button. LigandScout will create a new column in the table called "Cluster ID."

c) Sort the table using this column by clicking on the column header "Cluster ID."

3) Select all compounds in the table by selecting the first molecule in the list using your left mouse button then scroll to the end of the list and press the shift key and select the last molecule in list using your left mouse button. All molecules should be selected. Use the pull-down menu "Ligand-Set" → "Flag selected molecules as Ignored" to mark all molecules as ignored. Next select only the molecules with cluster id "1" using your left mouse and shift key. Use the pull-down menu "Ligand-Set" → "Flag selected molecules as Training-Set" to indicate that you want to compute a pharmacophore based on the ligands grouped in cluster 1. Only the ligands with cluster ID "1" should be marked as training the rest will be marked as ignored under the column header "Type."

4) Click on the button "Create Ligand-Based Pharmacophore." Give your pharmacophore model a name, for example,—"Cluster-1" and then press "Ok" to accept the default parameters. Since there will be more than one solution to this problem due to multiple conformations, LigandScout will generate 10 solutions (ligand-based pharmacophore models) with fit scores.

5) In the library view you will see a new column appear called "Feature pattern." These patterns of colored squares correspond to the pharmacophore model features. Clicking on a colored square indicates which feature it corresponds to in the pharmacophore model displayed in the 3D-view.

6) The list of pharmacophore solutions will appear to the right of the ligand table. By selecting different pharmacophore models you will see them appear in the 3D view thereby allowing you to view how the ligands fit to each pharmacophore model solution. You can use the data exchange controls on the upper right corner of the 3D view (see Figure 20.10, number 1) to move the currently shown pharmacophore to other perspectives (e.g., the screening perspective) or save it using the "File -> Save as File" menu and selecting the appropriate file type (PML or PMZ).

**Where to go from here:**

- Follow the same procedure to generate a pharmacophore model for each cluster.
- Align structure-based pharmacophore models with ligand-based models.
- Virtual screening in LigandScout using several ligand-based pharmacophore models simultaneously.

## Tutorial: Pharmacophore-Based Virtual Screening Using LigandScout

This section will provide a step-by-step procedure for virtual screening in the structure-based perspective, hit list analysis, and virtual screening in the virtual screening perspective. Using virtual screening methods, the user aims to retrieve the maximum enrichment of active compounds in a hit list by developing selective pharmacophore models. Therefore, before performing a virtual screening experiment on a commercial library of untested molecules, we recommend that the pharmacophore models be tested using virtual screening experiments on a set of known active compounds and a set of inactive compounds for the target of interest (see Table 20.4 & Fig. 20.12). If there are no known inactive compounds the user can create a data set of decoys using various methods described in the literature.[49] Such workflows enable the user to understand how well the pharmacophore models created are able to discriminate between actives and decoys (Table 20.4). A good pharmacophore model will be able to identify a significant portion of known active compounds and as few decoys as possible (Figure 20.7 & 20.8).

## Virtual Screening, Model Editing, and Viewing Hits in the Target Active Site

Table 20.4  Description of actions for virtual screening using LigandScout.[1]

| Experience level: advanced | | | Time needed: 15 minutes |
|---|---|---|---|
| **Views** | **Sequence** | **User Controls** | **Advanced Controls (opt.)** |
| • Macromolecule view<br>• Active site view | • Create pharmacophore for virtual screening<br>• Initiate virtual screening<br>• Refine pharmacophore model<br>• Repeat virtual screening | • Ligand box<br>• Create pharmacophore (button or menu)<br>• Initiate virtual screening | • Virtual screening with omitted features |

## Description: Virtual Screening and Pharmacophore Model Editing

(Numbers below correspond to the numbers circled in blue in Figure 20.12).

In the structure-based perspective, create a structure-based pharmacophore as outlined in Table 20.1, Fig. 20.9 but instead of PDB ID: "1ke6" use PDB ID: "1ke7."

1) Create a pharmacophore based on the 1ke7 complex by pressing the create pharmacophore icon in the tool bar (Figure 20.12). See Table 20.1, Fig. 20.9 if this is not clear.
2) Using the pull-down menu go to "Pharmacophore" → "Screen Pharmacophore Against External Library."

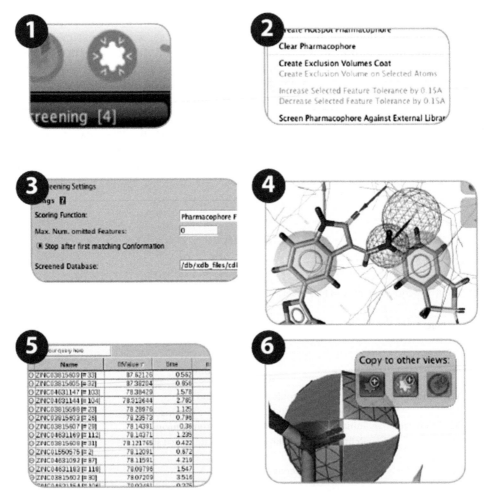

**Figure 20.12** Snapshots of the areas of the LigandScout GUI used for virtual screening and model editing. The numbers circled in blue correspond to the steps in the description. Source: LigandScout.[1]

3) Designate the screening settings in the dialog box. Designate the location and name of the multi-conformational database with known active compounds (file name: "cdk2-ligands.ldb") in the box next to "Screened Database."

4) In the 3D viewer edit the features of the pharmacophore model to test whether or not the edited model can retrieve more hits. Try designating a feature as optional (pull-down menu "Pharmacophore"→"Mark feature as optional") and repeat the screening. You can also designate features as disabled and they will not be considered during screening but should you change your mind you change them back to enabled. The feature tolerances can be modified using the pull-down menu as well (Table 20.4).

5) View the screening results after modifying the pharmacophore model. Select hits in the hit list with your left mouse button and view them in the 1ke7 active site aligned to the pharmacophore model.

6) Use the widget to move the model or hits to other perspectives.

**Where to go from here:**

- Virtual screening of a library of actives and a library of inactives and create a ROC curve
- Docking of hits in the active site
- Analysis of hit list, compute standard properties, and filter based on desired properties.

## Analyzing Screening Results with Respect to the Binding Site

Table 20.5 Description of actions for analyzing virtual screening results using LigandScout.[1]

| Experience level: intermediate | Time needed: 15 minutes | | Prerequisites: Virtual Screening & Pharmacophore Modeling Editing Tutorial |
|---|---|---|---|
| **Views** | **Sequence** | **User Controls** | **Advanced Controls (opt.)** |
| • Library view (Table) <br> • Hierarchy view | • Create or import screening results <br> • Sort virtual screening hit list <br> • Inspect and compare hits <br> • Discover controls of the library view | • Library view (Table) <br> • Viewer controls | • Calculation of Gaussian Shape Similarity Score |

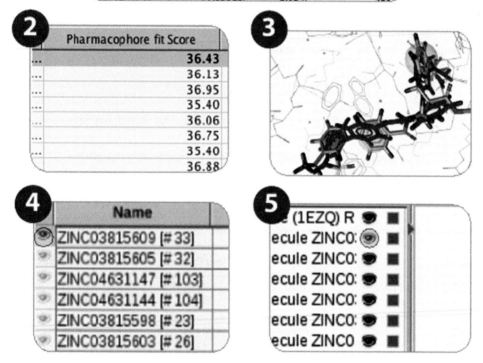

| Name | fitValue ▽ | time | numConfs |
|---|---|---|---|
| 03815609 [#33] | 87.62126 | 0.562 | 400 |
| 03815605 [#32] | 87.38204 | 0.656 | 400 |
| 04631147 [#103] | 78.38429 | 1.578 | 400 |
| 04631144 [#104] | 78.313644 | 2.765 | 400 |
| 03815598 [#23] | 78.28976 | 1.125 | 400 |
| 03815603 [#26] | 78.23573 | 0.796 | 196 |
| 03815607 [#29] | 78.14391 | 0.36 | 174 |
| 04631169 [#112] | 78.14371 | 1.235 | 219 |
| 03815608 [#31] | 78.121765 | 0.422 | 361 |
| 01550575 [#2] | 78.12091 | 0.672 | 400 |
| 04631092 [#87] | 78.11591 | 4.219 | 400 |
| 04631183 [#118] | 78.09796 | 1.547 | 400 |
| 03815602 [#30] | 78.07209 | 3.516 | 400 |
| 04631154 [#106] | 78.03461 | 0.375 | 400 |
| 04631177 [#115] | 78.02224 | 2.297 | 400 |
| 02047781 [#9] | 78.00544 | 0.735 | 400 |
| 03941407 [#43] | 78.00391 | 0.344 | 400 |
| 2007314 [#7] | 77.95323 | 3.954 | 400 |

**Pharmacophore fit Score**

| | |
|---|---|
| ... | 36.43 |
| ... | 36.13 |
| ... | 36.95 |
| ... | 35.40 |
| ... | 36.06 |
| ... | 36.75 |
| ... | 35.40 |
| | 36.88 |

**Name**

| | |
|---|---|
| ZINC03815609 [#33] | |
| ZINC03815605 [#32] | |
| ZINC04631147 [#103] | |
| ZINC04631144 [#104] | |
| ZINC03815598 [#23] | |
| ZINC03815603 [#26] | |

(1EZQ) R
ecule ZINC0:
ecule ZINC0:
ecule ZINC0:
ecule ZINC0:
ecule ZINC0:
ecule ZINC0:
ecule ZINC0

**Figure 20.13** Snapshots of the areas of the LigandScout GUI used for analyzing hit lists from virtual screening. The numbers circled in blue correspond to the steps in the description. Source: LigandScout.[1]

## Description: Analyzing Hits in the Active Site Using LigandScout

(Numbers below correspond to the numbers circled in blue in Figure 20.13).

Generate a structure based pharmacophore using the PDB ID: 1ke7 and perform virtual screening of the CDK2-ligands.ldb database as described in Table 20.4, Fig 20.12.

1) View the hit list in the library view.
2) Sort the hit list by the "Pharmacophore-Fit Score" by left clicking on the "Pharmacophore-Fit Score" column header (higher fit scores indicate a better match to the model). You may invert the sorting order by an additional click.
3) Select a compound of interest by single click on the corresponding table row. The hit ligand will be depicted in the active site view together with the original ligand from the x-ray structure and the pharmacophore model. Multiple compounds can selected when holding shift or control (command on Apple) while clicking.
4) Bring compounds into view or out of view by clicking on the eye symbol to the left of the molecule in the table and to the right of the molecules in the hierarchy view.
5) The visibility of the core molecule (i.e., the original ligand) and other items in the active site view is toggled with the eye symbol located in the hierarchy view. Select a custom color for the core molecule using the square icon next to the eye icon in the hierarchy view. Calculation of the Gaussian Shape Similarity Score offers further functionality for scoring. Use the pull down menu "Library" → "Calculate Gaussian Shape Similarity Score" to compute a Gaussian Shape similarity score. A new property column "Gaussian Shape Similarity Score" will be added in the library view. Similarly you can compute standard properties using the pull-down menu "Library" → "Compute Standard Properties." Several new columns will appear in the table. You can filter the table based on information (text or value) in any column (Table 20.6).

### Where to go from here:

- Pharmacophore modeling: Creating shared and merged feature pharmacophores.
- Customize pharmacophore creation preferences.
- Filter you hit lists and export them to Excel for additional analysis.

## Parallel Virtual Screening of Multiple Databases Using LigandScout

Table 20.6 Description of actions for virtual screening of multiple databases simultaneously using LigandScout.[1]

| Experience level: intermediate | | Time needed: 5 minutes | |
|---|---|---|---|
| **Views** | **Sequence** | **User Controls** | **Advanced Controls (opt.)** |
| • Screening view | • Add pharmacophore model to screening view <br> • Load virtual database <br> • Perform virtual screening | • Load Screening Database (Button) <br> • Perform Screening (Button) | • Get all matching conformations <br> • Get best matching conformation |

# Virtual Screening in the Screening Perspective of LigandScout

In contrast to the virtual screening procedure described in the Virtual Screening, Model Editing and Viewing Hits in the Target Active Site section, which featured a workflow for pharmacophore model refinement and hit viewing within an active site, the Screening Perspective in LigandScout provides additional options for managing virtual screening workflows. Those include such activities as screening multiple pharmacophore models against multiple libraries simultaneously and/or using the Boolean operator to create advanced screening workflows for fast library filtering using a set of pharmacophore models or model validation with automatic generation of ROC curves.

## Description: Virtual Screening Using LigandScout

(Numbers correspond to the numbers circled in blue in Figure 20.14).

In the structure-based perspective, create a structure-based pharmacophore as outlined in Table 20.1, Fig. 20.9 but instead of PDB ID: "1ke6" use "1ke7."

1) Use the widget to copy the pharmacophore model for 1ke7 to the virtual screening perspective.
2) Click on the "Load Screening Database" button to load the screening database "cdk2-ligands.ldb."
3) Next to the database name will be an empty circle. Click on this empty circle to create a green box. By marking this library with a green box you have indicated to

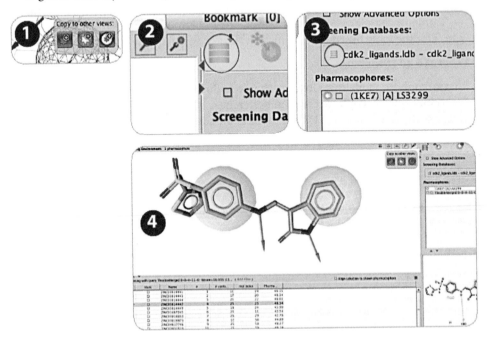

Figure 20.14 Snapshots of the areas of the LigandScout GUI used for virtual screening. The numbers circled in blue correspond to the steps in the description. Source: LigandScout.[1]

LigandScout that this is a database of potentially active compounds that you wish to screen. If there is an empty circle next to the loaded database, it will not be screened. Select the pharmacophore model from the "Pharmacophores" list to make sure it is shown in the 3D view (1ke7 pharmacophore model should be there from step 1). Click on the "Perform Screening" button to initiate the virtual screening.

4) View the results in the table at the bottom of the screen. As you select hits from the table of you will see them in the 3D-view aligned to the pharmacophore model you used for screening.

**Where to go from here:**

- Analyze screening results using filtering functions
- Advanced screening: ROC curves and model combination
- Inject hits into the active site for viewing or for docking experiments.

## Conclusions

The workflows presented herein are certainly not exhaustive by any means when considering the existing advanced functionalities of the molecular design platform LigandScout. However, they do cover the most fundamental aspects of pharmacophore modeling and virtual screening that serve as important starting points for computer-aided molecular design approaches for hit finding in early lead discovery research. The theory covered herein should also provide a deeper understanding of the representations and definitions of pharmacophore models and processes behind these in silico approaches that are used widely across various life science industries, including pharmaceutical drug discovery and academic research.

## Acknowledgments

The authors are grateful to P Adaktylos, F Bendix, M Biely, A Dornhofer, R Kosara, M Langer, A Fuchs, J Hirschl, T Kainrad, L Moshuber, G Wolber, S Bachleitner, Y Aristei, E Bonelli, M Böhler, S Distinto, O Funk, T Huynh Buu, J Kirchmair, E M Krovat, C Laggner, P Markt, D Schuster, G Spitzer, O Wieder, R Martini and T Steindl for their valuable contributions to the development of LigandScout and the Inte:Ligand Scientific Advisory Board for their esteemed scientific guidance.

## References

1 LigandScout 4.09. Inte:Ligand GmbH, http://www.inteligand.com/ligandscout
2 G. Wolber, T. Langer, *J. Chem. Inf. Model.* 2005, **45**, 160–169.
3 G. Wolber, R. Kosara, *Pharmacophores from Macromolecular Complexes with LigandScout* in Pharmacophores and Pharmacophore Searches, Vol. **32**, T. Langer, R. D. Hoffmann (Eds.), Wiley-VCH, Weinheim, 2006, 131–150.
4 G. Wolber, T. Seidel, F. Bendix, T. Langer, *Drug Discovery Today* 2008, **13**, 1–2, 23–29.

5 P.G. Polishchuk, G.V. Samoylenko, T.M. Khristova, O.L. Krysko, T.A. Kabanova, V.M. Kabanov, A.Y. Kornylov, O. Klimchuk, T. Langer, S.A. Andronati, V.E. Kuz'min, A.A. Krysko, A. Varnek, *J Med. Chem.* 2015, **58**, 7681–7694.

6 P. Koch, M. Gehringer, S.A. Laufer, *J. Med. Chem.* 2015, **58**, 72–95.

7 A. Vuorinen, R. Engeli, A. Meyer, F. Bachmann, U. J. Griesser, D. Schuster, A. Odermatt, *J. Med. Chem.* 2014, **57**, 5995–6007.

8 L. De Luca, S. Ferro, F.M. Damiano, C.T. Supuran, D. Vullo, A. Chimirri, R. Gitto, *Eur J Med Chem.* 2014, **71**, 105–111.

9 K. Duwensee, S. Schwaiger, I. Tancevski, K. Eller, M. van Eck, P. Markt, T. Langer, U. Stanzl, A. Ritsch, J.R. Patsch, D. Schuster, H. Stuppner, D. Bernhard, P. Eller, *Atherosclerosis.* 2011, **219**, 109–115.

10 J. Rollinger, D. Schuster, B. Danzl, S. Schwaiger, P. Markt, M. Scmidtke, J. Gertsch, S. Raduner, W.G. Wolber, T. Langer, H. Stuppner, *Planta Medica* 2009, **75**, 195–204.

11 Y. Wei, Y. Ma, Q. Zhao, Z. Ren, Y. Li, T. Hou, H. Peng, *Mol Cancer Ther* 2012, **11**, 1693–1702.

12 F. Rechfeld, P. Gruber, J. Kirchmair, M. Boehler, N. Hauser, G. Hechenberger, D. Garczarczyk, G.B. Lapa, M.N. Preobrazhenskaya, P. Goekjian, T. Langer, R.J. Hoffmann, *J. Med.Chem.* 2014, **57**, 3235–3246.

13 L. De Luca, M.L. Barreca, S. Ferro, F. Christ, N. Iraci, R. Gitto, A.M. Monforte, Z. Debyser, A. Chimirri, *ChemMedChem* 2009, **4**, 1311–1316.

14 V. Corradi, M. Mancini, F. Manetti, S. Petta, M.A. Santucci, M. Botta, *Biorg. Med. Chem. Lett.* 2010, **20**, 6133–6137.

15 S.D. Bryant, T. Langer, *Data Mining Using Ligand Profiling and Target Fishing* in Data Mining in Drug Discovery, R.D. Hoffman, A. Gohier, P. Pospisil, Eds, Wiley-VCH, Weinheim 2013, pp. 257–270.

16 T. Langer, S.D. Bryant, *Computational Methods for Drug Profiling and Polypharmacology* in In Silico Drug Discovery and Design, Future Science Ltd, London, 2013, 178–188.

17 Y. Krautscheid, C.J. Senning, S.B. Sartori, N. Singewald, D. Schuster, H. Stuppner, *J. Chem. Inf. Model.* 2014, **54**, 1747–1757.

18 A.S. Karaboga, J.M. Planesas, F. Petronin, J. Teixido, M. Souchet, V.I. Perez-Nueno, *J. Chem. Inf. Model.* 2013, **24**, 1043–1056.

19 Tutorial materials, datasets and evaluation versions of LigandScout are available by request (support@inteligand.com).

20 S. Yang, *Drug Discovery Today* 2010, **15**, 444–450.

21 C.G. Wermuth, C.R. Ganellin, P. Lindberg, L.A. Mitscher, *Pure Appl. Chem.* 1998, **70**, 1129–1143.

22 T. Langer, S.D. Bryant, *In Silico Screening: Hit Finding from Database Mining* in The Practice of Medicinal Chemistry, 3$^{rd}$ Ed., C.G. Wermuth (Ed.), Elsevier, 2008, 210–227.

23 C.G. Wermuth, Pharmacophores: Historical Perspective and Viewpoint from a Medicinal Chemist in Pharmacophores and Pharmacophore Searches, Vol. **32**, T. Langer, R. D. Hoffmann (Eds.), Wiley-VCH, Weinheim, 2006, 3–13.

24 T. Langer, *Future Med Chem.* 2011, **3**(8), 901–904.

25 M.A. Williams, J.E. Ladbury, Hydrogen Bonds in Protein-Ligand Complexes in Protein-Ligand Interactions: From Molecular Recognition to Drug Design, H.-J. Böhm, G. Schneider (Eds.), Wiley-VCH, Weinheim, 2005, 137–161.

26 J.C. Ma, D.A. Dougherty, *Chem. Rev.* 1997, **97**, 1303–1324.

27 H.J. Böhm, G. Klebe, H. Kubinyi, Metalloprotease-Hemmer in Wirkstoffdesign, Spektrum Akademischer Verlag, Heidelberg-Berlin-Oxford, 1996, 505–520.

28 B.W. Matthews, *Acc. Chem. Res.* 1988, **21**, 333–340.

29 A.R. Leach, V.J. Gillet, R.A. Lewis, R. Taylor, *J. Med. Chem.* 2010, **53**, 539–558.

30 K. Poptodorov, T. Luu, R.D. Hoffmann, Pharmacophore Model Generation Software Tools in Pharmacophores and Pharmacophore Searches, Vol. **32**, T. Langer, R.D. Hoffmann (Eds.), Wiley-VCH, Weinheim, 2006, 17–47.

31 N. Triballeau, H.O. Bertrand, F. Achner, Are You Sure You Have a Good Model? In Pharmacophores and Pharmacophore Searches, Vol. **32**, T. Langer, R. D. Hoffmann (Eds.), Wiley-VCH, Weinheim, 2006, 325–364.

32 N. Triballeau, F. Acher, I. Brabet, P. J. P in, H.O. Bertrand, *J. Med. Chem.* 2005, **48**, 2534–2547.

33 T. Seidel, G. Ibis, F. Bendix, G. Wolber, *Drug Discovery Today: Technologies* 2010, **7**, 4, 221–228.

34 G. Wolber, T. Langer, *Drug Discovery Today: Technologies* 2004, **1**, 3, 203–207.

35 G. Schneider et al., *Angew. Chem. Int. Ed.* 1999, **38**, 19, 2894–2896.

36 T. Langer, R. Hoffmann, S. Bryant, B. Lesur. *Curr. Opin. Pharmacol.* 2009, **9**, 589–593.

37 T. Langer, E. M. Krovat, *Curr. Opin. Drug Discov. Dev.* 2003, **6**, 370–676.

38 ChEMBL. https://www.ebi.ac.uk/chembl/

39 PubChem. https://pubchem.ncbi.nlm.nih.gov/

40 M.R. Berthold, N. Cebron, F. Dill, T.R. Gabriel, T. Kötter, T. Meinl, P. Ohl, K. Thiel, B. Wiswedel, *The Konstanz Information Miner*, in Studies in Classification, Data Analysis and Knowledge Organization, Springer, 2007, pp. 26–31.

41 T. Hurst, *Chem. Inf. Comput. Sci.* 1994, **34**, 190–196.

42 G. Wolber et al., *J. Comput. Aided Mol. Des.* 2006, **20**, 12, 773–788.

43 C. Lemmen, T. Lengauer, *J. Comput.-Aided Mol. Des.* 2000, **14**, 215–232.

44 O.F. Güner, Pharmacophore perception, development, and use in drug design, 2000, Int. Univ. Line

45 J. Kirchmair et al., *J. Chem. Inf. Model.* 2007, **47**, 2182–2196.

46 M.S. Pepe, The Statistical Evaluation of Medical Tests for Classification and Prediction, 2003. Oxford University Press.

47 RCSB Protein Data Bank (PDB). http://www.rcsb.org/

48 H.M. Berman, J. Westbrook, Z. Feng, G. Gilliland, T.N. Bhat, H. Weissig, I.N. Shindyalov, P.E. Bourne, *Nucleic Acids Research* 2000, **28**: 235–242.

49 N. Huang, B.K. Shoichet, J. Irwin, *J. Med. Chem.* 2006, **49**(23), 6789–6801.

Part 7

The Protein 3D-Structures in Virtual Screening

21

# The Protein 3D-Structures in Virtual Screening

*Inna Slynko and Esther Kellenberger*

## Introduction

With the progress in protein crystallography, structure-based computer-aided methods were developed in the early 1970s to design low molecular weight ligands that fit a protein binding site.[1] Nowadays structured-based methods play an important role in drug discovery, in particular allowing the automated database searches for rapid and efficient identification of hits.[2] The two main structure-based methods for virtual screening are pharmacophore and docking. Pharmacophore has been defined by CG Wermuth as "the ensemble of steric and electronic features that is necessary to ensure the optimal supra-molecular interactions with a specific biological target and to trigger (or block) its biological response."[3] A pharmacophore can be generated from the structures of ligands, even when the three-dimensional structure of the target protein is not available. However, as we will see in this chapter, structural information on protein-ligand binding mode helps to prioritize important features in the pharmacophoric hypothesis. Besides, computing tools have been designed especially for the preparation of a pharmacophore from intermolecular interactions detected between a ligand and its target protein in the crystal structure of the complex.[4] Docking methods directly make use of the protein structure to constrain the position and the conformation of a bound ligand, and then estimate the strength of the binding.

In this chapter, you will learn how to interpret structural information for understanding binding between therapeutically relevant protein and "drug-like" ligand. In the first exercise, you will analyze the content of a structure file downloaded from the Protein Data Bank (PDB), evaluate the precision of the structure described in the file, and prepare the molecules for docking. You will also characterize the complex between the protein and the "drug-like" ligand present in the file in order to define a reference binding mode that will guide you in the design of a pharmacophore and help you to post-process docking poses. Last, you will detect binding pockets at the protein surface and observe how cavities are represented for computing purposes. In the second exercise, you will prepare a pharmacophore based on the 3D-aligned structures of three active molecules, and then use this pharmacophore for the retrospective virtual screening of a

*Tutorials in Chemoinformatics*, First Edition. Edited by Alexandre Varnek.
© 2017 John Wiley & Sons Ltd. Published 2017 by John Wiley & Sons Ltd.
Companion website: www.wiley.com/go/varnek/chemoinformatics

chemical library. In the third exercise, you will compare several conditions for the re-docking of the ligand back into its binding site, and select the most efficient settings for the retrospective virtual screening of the chemical library previously tested by the pharmacophore approach.

## Description of the Example Case

### Thrombin and Blood Coagulation

In this tutorial, we will explain the full process of hit finding by structure-based computer-aided drug design (SBCADD) on the example of human thrombin. Thrombin is a protein also named *coagulation factor II* because of its role in the coagulation of blood. In mammals, clot formation is the result of a cascade of enzymatic reactions, which involve the sequential activation of a number of specific serine proteases by limited proteolysis (a protease catalyzes the cleavage of specific peptide bonds in target proteins; in serine protease, serine serves as the nucleophilic amino acid at the active site). Thrombin is present at the end of the cascade, converting soluble fibrinogen into insoluble strands of fibrin, activating several other coagulation factors, and promoting platelet activation.[5]

### Active Thrombin and Inactive Prothrombin

Thrombin represents the active form of the enzyme, produced from the inactive *prothrombin*. *Prothrombin* is synthesized in the liver as a single chain glycoprotein that is post-translationally modified in a vitamin K-dependent reaction before secretion in the blood. Human *prothrombin* is a monomer of 579 residues (numbered 44 to 622 in UniProt P00734 sequence).[6] The enzymatic cleavage of two sites on *prothrombin* by activated Factor Xa removes the activation peptide and releases active thrombin, which is made of two chains, thrombin light chain and thrombin heavy chain (respectively numbered 328 to 363 and 364 to 622 in UniProt P00734 sequence). The catalytic amino acids in thrombin are all on the heavy chain (His406, Asp462, and Ser568 in UniProt P00734 sequence). It is worth knowing that residue numbering in Uniprot and PDB often differs. Thus, in the PDB structure of thrombin studied in this tutorial, the catalytic amino acids correspond to His57, Asp102, and Ser195. From here on, we use the numbering of the residues like it is specified in the PDB structure.

### Thrombin as a Drug Target

Anticoagulants stop or delay blood from clotting and therefore are potential medication for thrombotic disorders such as thrombosis, pulmonary embolism, myocardial infarctions, and strokes.[7] Such substances are naturally occurring in blood-sucking insects and leeches. For example, hirudin, a 65 amino acids peptide present in the salivary glands of medicinal leeches, is the most potent natural inhibitor of thrombin.[8] Lepirudin, a variant of hirudin, is an approved injectable biotech drug for the treatment of heparin-induced thrombocytopenia. Its production was, however, ceased in 2012 because more effective drugs with less severe side effects emerged. For example, synthetic dabigatran etexilate is an oral prodrug converted in the plasma and liver into dabigatran, a competitive and reversible direct thrombin inhibitor.

Thrombin has been actively and extensively studied by x-ray crystallography. The molecular aspects of thrombin function are nicely summarized in the "molecule of the month" highlights of the RCSB PDB.[9] The PDB indeed contains several hundred of entries describing human thrombin, including numerous complexes between the catalytic domain of the protein and inhibitors.[10]

In the tutorial, we will work on the 1OYT PDB file, which describes a hydrated ternary complex between human thrombin, the C-terminal terminus of hirudin, and a synthetic fluorinated inhibitor. The hirudin peptide interacts with the anion-binding exosite of thrombin, leaving the enzyme catalytic centre fully accessible for the fluorinated inhibitor.[11] The 1OYT PDB file reports an accurate and well-defined experimental three-dimensional model built from the high-quality x-ray data, with almost no steric clashes or geometric problems, and, therefore, provides the structural clues for the understanding of thrombin inhibition, and the material for hit finding by SBCADD methods.

The 1OYT PDB file, nevertheless, includes most of the ambiguities commonly observed in the description of molecules in a crystallographic structure. In particular, neither the hydrogen atoms nor the nature of covalent bonds (simple, double, triple) are described, so that multiple valences are possible for the carbon, nitrogen, and oxygen atoms of the fluorinated inhibitor. Importantly, the absence of hydrogen atoms also raises the question of the ionization state of titratable groups at physiological pH (7.4 in blood), as well as that of the tautomeric state of the imidazole ring in neutral histidine, and the position of nitrogen and oxygen in the amide group in asparagine and glutamine. Last, a few amino acids in the file are also ambiguously defined because of missing non-hydrogen atoms, or of a possible alternate set of coordinates.

## Modeling Suite

In this tutorial, we used modeling suite MOE (Chemical Computing Group, Montréal). It is a comprehensive software system which is suitable for a wide range of *in silico* drug design purposes. MOE package combines different modules used for the visualization, simulation, or application development. Other widely used modeling suites include, for example, Schrödinger, Tripos (Certara), or Accelrys. The materials for this tutorial include files in MOE format, as well as files in CSV and SDF formats, which are readable by most modelling suites.

The overview of MOE program, its applications, manuals, and tutorials can be accessed through a comprehensive help page, which can be launched in a browser from the "**MOE | Help**" menu.

## Overall Description of the Input Data Available on the Editor Website

The course material is stored into a compressed archive named *CI-TUTORIAL7.tar.gz*. You need to use the tar command under Unix-like operating systems to open and extract the file.

| tar -zxf CI-TUTORIAL7.tar.gz | The archive from the file called CI-tutorial.tar.gz (-f) is uncompressed with gzip command (-z), extracted to disk in the current directory (-x) |
|---|---|

The path to the input files, as based at the *CI-TUTORIAL7* directory, is detailed in Table 21.1. For the sake of pedagogy, output files are given too.

## Exercise 1: Protein Analysis and Preparation

### Aim

The goal of this exercise is to analyze the structure of thrombin described in the *1OYT* PDB file and to elucidate important aspects (overall and local accuracy of the structure, ambiguities, and errors in the structure), which should be kept in mind during structure preparation, in order to answer the question: "Is the structure suitable for drug design purposes or not?"

### Step 1: Identification of Molecules Described in the 1OYT PDB File

Find the entry 1OYT on the RCSB PDB web site (http://www.rcsb.org) and analyse the additional information available under tabs on the top of the web page, for example, look for the "Summary" and "Sequence" tabs, as well as for the publication accompanying the structure, and answer the following questions:

- What does the structure contain (e.g., protein, ligand, peptide, co-factor, ions)?
- Which organism does the protein belong to?
- Is this protein wild-type or mutant, mature, or precursor?
- Is it a structure of a full-length protein or not? Are there gaps in the structure?

We will now use MOE to investigate the 1OYT PDB file.

Download the PDB file and open it in MOE. This can be done in a few different ways—whether through "File" menu, Ctrl-O keys combination, or by simply clicking "Open" button at the Right Hand Side Button Bar (RHS) and typing code "1oyt" in RCSB PDB tab available on the left side of the "Open" window. Note, that the secondary structure elements are displayed automatically after the PDB complex is loaded in MOE (see Figure 21.1).

Access the MOE sequence editor (**MOE | SEQ**) to display the amino acid sequence in the SEQ window and analyse the content of the PDB file (see Figure 21.1) in order to find out:

- Does the structure contain a drug-like ligand?
- Where is the ligand binding site and which residues are involved in interactions with the drug-like compound?
- Is the ligand binding pocket made of parts of secondary structure elements and/or of loop regions?
- Which molecules are relevant or not for the drug design purposes; what to keep for the next step of complex preparation and what can be removed?

**Table 21.1** Description of input and output files.

| File description | Path to file |
|---|---|
| **Input for exercise 1** | |
| 1OYT original PDB file | *input/1oyt.pdb* |
| **Input for exercise 2** | |
| 3D-structure of the receptor, as prepared in exercise 1 (moe binary file) | *input/1oyt-ThrombinHC-FSN.moe* |
| 3D-structure of the receptor, as prepared in exercise 1 (mol2 format) | *input/1oyt-ThrombinHC-FSN.mol2* |
| 3D- structure of three thrombin inhibitors (sd format) | *input/active3.sdf* |
| 3D-structure of three thrombin inhibitors (moe binary file) | *input/active3.mdb* |
| 2D-structure of molecules in the test library (sd format) | *input/database100.sdf* |
| 2D-structure of molecules in the test library (moe binary file) | *Input/database100.mdb* |
| Low-energy conformers of molecules in the test library (sd format) | *input/database100-multiconf.sdf* |
| Low-energy conformers of molecules in the test library (moe binary file) | *Input/database100-multiconf.mdb* |
| **Input for exercise 3** | |
| 3D-structure of the receptor, as prepared in exercise 1 (moe binary file) | *input/1oyt-ThrombinHC-FSN.moe* |
| 3D-structure of the receptor, as prepared in exercise 1 (mol2 format) | *input/1oyt-ThrombinHC-FSN.mol2* |
| Pharmacophore (moe text file) | *Input/1oyt_ligand.ph4* |
| 3D-structure of molecules in the test library (sd format) | *input/database100-3D.sdf* |
| 3D-structure of molecules in the test library (moe binary file) | *input/database100-3D.mdb* |
| **Ouput of exercise 1** | |
| 2D-structure and descriptors of molecules in the test library (sd format) | *output1/database100-with_descriptors.sdf* |
| 2D-structure and descriptors of molecules in the test library (moe binary file) | *output1/database100-with_descriptors.mdb* |
| **Ouput of exercise 2** | |
| 3D-alignments of three thrombin inhibitors (moe binary file) | *output2/flexalgn.mdb* |
| Top-ranked 3D-alignment of three thrombin inhibitors (sd format) | *output2/active3-moeALI.sdf* |
| Pharmacophore 1 (designed from crystal structures of thrombin inhibitors, moe text file) | *output2/thrombin.ph4* |
| Pharmacophore 2 (designed from 3D-aligned thrombin inhibitors, moe text file) | *output2/thrombin-moeALI.ph4* |
| Virtual screening 1: Hit molecules (best conformation and score, moe binary file) | *output2/ph4out-thrombin.mdb* |

*(Continued)*

Table 21.1 (Continued)

| File description | Path to file |
| --- | --- |
| Virtual screening 1: Hit molecules (best conformation and score, sd format) | *output2/ph4out-thrombin.sdf* |
| Virtual screening 2: Hit molecules (best conformation and score, moe binary file) | *output2/ph4out-thrombin-moeALI.mdb* |
| Virtual screening 2: Hit molecules (best conformation and score, sd format) | *output2/ph4out-thrombin-moeALI.sdf* |
| **Ouput of exercise 3** | |
| Redocking, no refinements (poses and scores, moe binary file) | *output3/1oyt_redock_London.moe* |
| Redocking, no refinements (poses and scores, sd format) | *output3/1oyt_redock_London.sdf* |
| Redocking, no refinements (scores, comma separated text file) | *output3/1oyt_redock_London.csv* |
| Redocking, refinement (poses and scores, moe binary file) | *output3/1oyt_redock_London_grid.moe* |
| Redocking, refinement (poses and scores, sd format) | *output3/1oyt_redock_London_grid.sdf* |
| Redocking, refinement (scores, comma separated text file) | *output3/1oyt_redock_London_grid.csv* |
| Redocking, refinement and rescoring (poses and scores, moe binary file) | *output3/1oyt_redock_London_grid_London.moe* |
| Redocking, refinement and rescoring (poses and scores, sd format) | *output3/1oyt_redock_London_grid_London.sdf* |
| Redocking, refinement and rescoring (scores, comma separated text file) | *output3/1oyt_redock_London_grid_London.csv* |
| Redocking guided by pharmacophore, refinement and rescoring (poses and scores, moe binary file) | *output3/1oyt_redock_London_grid_London_ph4.moe* |
| Redocking guided by pharmacophore, refinement and rescoring (poses and scores, sd format) | *output3/1oyt_redock_London_grid_London_ph4.sdf* |
| Redocking guided by pharmacophore, refinement and rescoring (scores, comma separated text file) | *output3/1oyt_redock_London_grid_London_ph4.csv* |
| Virtual screening 1 (best poses and scores, moe binary file) | *output3/database_London_grid_London.mdb* |
| Virtual screening 1 (best poses and scores, sd format) | *output3/database_London_grid_London.sdf* |
| Virtual screening 1 (scores, comma separated text file) | *output3/database_London_grid_London.csv* |
| Virtual screening 1 after post-processing (best poses and scores, moe binary file) | *output3/database_London_grid_London_PLIF.mdb* |
| Virtual screening 1 after post-processing (best poses and scores, sd format) | *output3/database_London_grid_London_PLIF.sdf* |
| Virtual screening 1 after post-processing (scores, comma separated text file) | *output3/database_London_grid_London_PLIFF.csv* |
| Virtual screening 2 (best poses and scores, moe binary file) | *output3/database_London_grid_London_ph4mdb* |
| Virtual screening 2 (best poses and scores, sd format) | *output3/database_London_grid_London_ph4.sdf* |
| Virtual screening 2 (scores, comma separated text file) | *output3/database_London_grid_London_ph4.csv* |

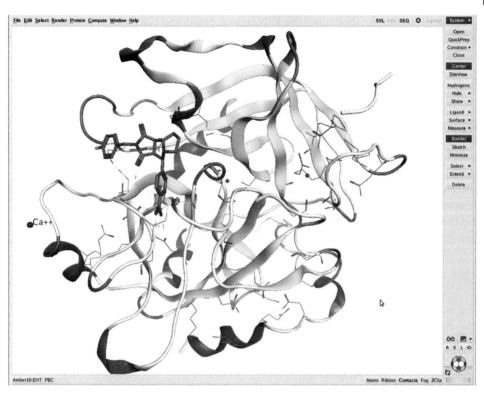

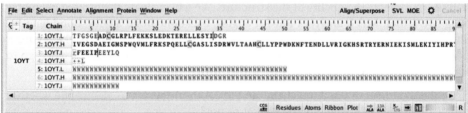

Figure 21.1 Display of 1OYT PDB file in the main GUI of MOE (top) and in the SEQ window (bottom).

## Results

The structure 1OYT contains wild-type human thrombin in the mature state, the peptide inhibitor hirudin-2B (11 residues), and a fluorinated inhibitor (the three-letter HET PDB code is FSN), as well as several metal ions and water molecules. Furthermore, there are gaps in the structure, meaning that the 3D coordinates for some of the residues were not obtained. As we can see from Figure 21.1, the PDB file has 3 chains assigned with letters L, H, and I:

- Chain L containing thrombin and water molecules. The 36 residues of thrombin correspond to the light chain, but only 27 of them were observed experimentally (the electronic density was too poor for the other amino acids, which are displayed in gray color in the SEQ window).

- Chain H containing thrombin, the FSN ligand, Na(I) and Ca(II) ions, and water. The 247 residues of thrombin correspond to the heavy chain, but only 245 of them were observed experimentally.
- Chain I containing hirudin residues 55–65 and water molecules.

Thus, we could see that the 1OYT PDB file contains two thrombin inhibitors—the hirudin peptide and the drug-like molecule FSN. Since this tutorial concentrates on the binding site of a low molecular weight inhibitor and does not cover the topic of protein-protein interactions, chains L and I can be removed from the structure as they do not participate in the interactions with the ligand. Furthermore, we will also ignore salt and water molecules of the chain H, for the sake of simplicity. It is, however, generally worth to pay attention to the solvent molecules, as it is detailed in the next chapter of the textbook.

Select unwanted chains and residues either from the main GUI, or using the SEQ window, and then delete all atoms in the selection. The *delete* command can be accessed from the main GUI (**MOE | Edit | Delete...**) or from the SEQ window (**MOE SEQ | Edit | Delete**).

### Step 2: Protein Quality Analysis of the Thrombin/Inhibitor PDB Complex Using MOE Geometry Utility

The structure was solved using x-ray diffraction method and it has a resolution of 1.67 Å. Structures with resolution of 2.5 Å and less are usually considered applicable for structure-based drug design.[12] Other experimental parameters in the PDB file provide clues on the quality of the structure, yet their interpretation necessitates a deep knowledge in x-ray crystallography.[13]

In the tutorial, we will thus focus on the direct analysis of the coordinates in 1OYT PDB file to evaluate the quality of the structure. We will evaluate the protein geometry using MOE tool, which analyzes Phi-Psi angles, backbone bond angles, bond lengths, dihedrals, and steric clashes. Select the residues within the FSN binding pocket in order to see them highlighted in geometry plots—go to **MOE | Select | Pocket**. Then go to **SEQ | Protein | Geometry** and answer the following questions:

- Are there outliers in the Ramachandran plot of thrombin? If so, determine whether they are located in ligand binding pocket or not?
- Are there other geometric errors in the protein?
- Conclude: rate the quality of PDB structure of the complex between thrombin and inhibitor.
- Decide which errors are critical and should be fixed, and which can be tolerated.

### Results

A high quality of a structure, especially in the area of ligand binding pocket, is a prerequisite for successful application of SBCADD methods. Thus, it should be carefully analyzed and the critical ambiguities should be fixed. The stereochemical analysis confirmed good quality of 1OYT structure of thrombin, showing that all residues are located in the "core" or "allowed" regions of Ramachandran plot. There are no outliers neither within the binding pocket, nor for the whole protein structure (see Figure 21.2).

The values for backbone bond lengths, angles, and dihedrals are compared to averages and standard deviations of a reference PDB database containing high-resolution structures from the May 2007, see **MOE | Help | Panel Index | Protein Geometry** for

(a)

(b)

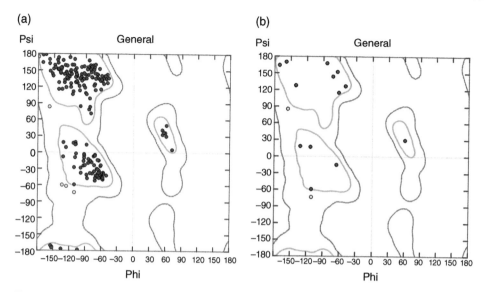

Figure 21.2 Stereochemical analysis of thrombin protein showing phi-psi torsion angles (Ramachandran plot) for the residues in the 1OYT structure: (a) all residues of chain H; (b) binding pocket residues around the FSN ligand. The green areas represent the most favorable "core" combinations of phi-psi angles, the orange lines outline "allowed" areas.

more details. The plots for the bond length and angles are displayed in Figure 21.3a and 21.3b, correspondingly, showing high Z-scores only for N-CA-C angles of two residues (Gly219 and Gly216). There are no problems reported for the protein dihedrals (see corresponding Protein Geometry Tabs).

There is one atom clash identified between the backbone oxygen of residue His91 and the alpha carbon of Ile90 (distance of 2.69 Å, repulsion energy of 0.5511 kcal/mol), nevertheless, it is not located within the ligand binding pocket.

Overall, we have seen that the three-dimensional structure of thrombin 1OYT have good geometric quality and there are very few distortions, which are of limited extent and may be easily fixed by energy minimization.

### Step 3: Preparation of the Protein for Drug Design Applications

The description of a molecular object in a PDB file is generally incomplete and may contain errors.

Use MOE protein preparation tool to correct and complete the thrombin structure.

Go to **MOE | Protein | Structure Preparation** menu, analyze the reported problems, and then press **Correct** button in order to fix the errors and the ambiguities in the protein. Use **Protonate3D** to assign hydrogen atoms for the structure.

### Results

Here we will examine the issues reported by the Structure Preparation application on example of 1OYT PDB file obtained from the previous step. More detailed description of possible errors or warnings can be found at **MOE | Help | Panel Index | Structure Preparation**. As we can see from the report generated in MOE,

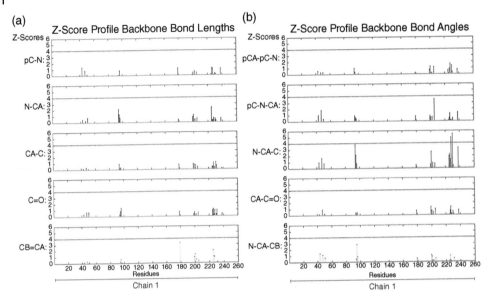

Figure 21.3 The analysis of the quality of backbone bond (b) lengths and (b) angles on a Z-Score scale (standard deviations from expectation) is shown for the binding pocket residues of 1OYT structure. Residues above the red line, which indicates a Z-Score cutoff of 4.0, are considered outliers.

there are totally 25 warnings for chain H of 1OYT structure, concerning alternate positions for 19 residues, wrong states of terminal residues, missing hydrogen atoms and charges, absence of the ligand in the MOE library (see the left summary panel in Figure 21.4).

- *Alternates.* Some atoms have multiple coordinates, each of them corresponding to a partial occupancy. This information is indicated in the *Alternate Location Indicator* column of the PDB file and it raises an issue in MOE. During the correction of alternate locations, MOE sets atom positions to that with the highest occupancy.
- *Termini.* Protein N- or C-terminal residues should be properly charged or capped, since it has an influence on the calculation of an electrostatic potential. If a terminus is preceded or followed by an "empty" residue (i.e., with no coordinates in the PDB file), MOE adds ACE (N-acetyl group) or NME (N-methyl amide), correspondingly, otherwise the terminal residues are charged. Empty residues before N-terminus or after C-terminus are then deleted. In addition, MOE will build loops for chain breaks containing one till seven empty residues. In case this number is equal or greater than seven, the program will delete empty residues and cap chain breaks with ACE/NME groups. Note, that in some cases, adding caps may cause close contacts with surrounding residues.
- *HCount.* If there are atoms with incorrect number of hydrogens (PDB structures usually do not contain hydrogen atoms at all), MOE will add or remove them. Note, that fixing of *HCount* issue only adds hydrogens and does not assign rotamers, protomers, or tautomers, which can be done later using *Protonate3D* functionality.

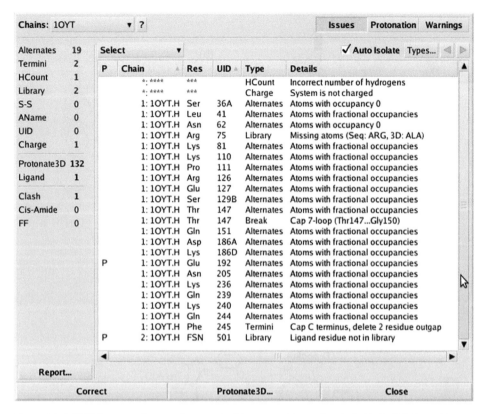

**Figure 21.4** MOE Structure Preparation panel with reported issues for chain H of 1OYT thrombin structure (waters and ions were removed from the structure).

- *Library.* MOE compares the structures of loaded residues to its library (located in $MOE/lib/amber10.mdb) and reports the differences, mismatches, or other problems. For example, there are missing side chain atoms of Arg75 in thrombin structure analyzed here. In addition, the ligand residue (FSN identifier) was not found in the library. While the first issue can be easily fixed—Arg will be mutated to the complete residue (note that it cannot be done for the ligand or solvent molecule), the second one—concerning the ligand structure—should be analyzed by the user and corrected manually if needed. It is important to ensure that the correct atoms types, bonds, and protonation states are assigned for the ligand. Thus, the structure of FSN ligand was compared to the structure from the original publication describing tricyclic thrombin inhibitors (see Figure 21.5 for more details), showing overall correspondence—carbamimidoyl group is charged, aromatic, aliphatic carbon atoms have correct atom types, and so on. Only one difference can be observed— MOE protonated the nitrogen of the tertiary amine (aliphatic amines are basic, with pKa generally higher than 8), which might be a case under the physiological conditions (pH~7.4).
- *AName.* There are two atom pairs with identical atom names (H11 and H22). Correction of this issue will assign new atom names.

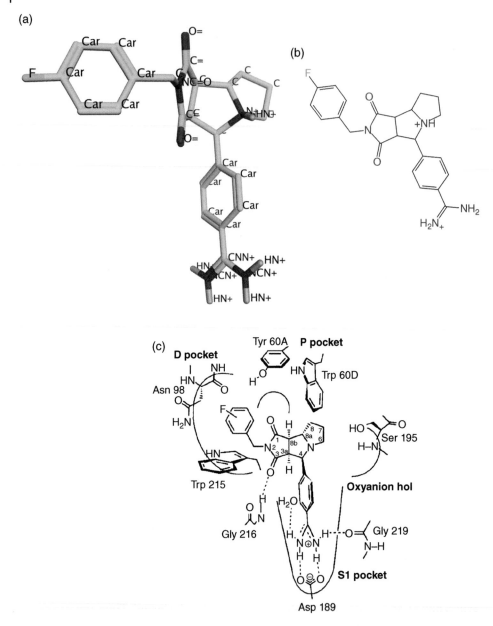

**Figure 21.5** (a) the 3D structure of FSN ligand derived from 1OYT.pdb is depicted in cyan sticks and labelled using atom typing of MM MOE scheme; (b) MOE 2D-depiction of FSN ligand; (c) the binding mode of the tricyclic thrombin inhibitor (FSN). Source: Olsen 2003. Reproduced with permission from John Wiley and Sons.[11]

- *Charge*. Fixing this issue assigns atom charges according to the current force field. A force field, as indicated by its name, allows the definition of the forces exerted on a given atom by the surrounding atoms. By default MOE uses Merck Molecular Force Field— MMFF94x,[14] which was optimized for small organic molecules, but it is also suitable for macromolecules, particularly since we do not perform bulk calculations in this tutorial.

The protonation of the structure using default settings of Protonate3D utility added 2041 hydrogen atoms, that is around half of the total amount of atoms in the structure. Protonate3D automatically assigns protonation states of the residues based upon optimization of hydrogen bond network. For example, we could observe that the side chain of His57 of 1OYT thrombin structure has hydrogen atom on N-delta but not N-epsilon in order to enable the hydrogen bond to the hydroxyl group of Ser195. Possible rotamers, tautomers, and protomers can be analyzed or corrected using **Protonation** tab of the Structure Preparation window—it is recommended to check protonation states of histidines, glutamic acids, aspartic acids, lysines, and cysteines with regard to their local environment, at least within the binding pocket.

Finally, the **Warning** tab reports one clash between NZ atom of Lys107 and NME0 atom of N-methyl amide cap, which was added to the residue Phe245 during fixing of the terminal residues. Since lysine side chain is known to be highly flexible, such close contact can be removed whether through short tethered minimization of the protein (or alternatively only of lysine residue), or by choosing other rotamer of lysine.

Overall, we could see that 1OYT structure describing the complex of thrombin protein and FSN ligand is accurate enough for the drug design purposes—the structure of the ligands shows an agreement with the published one, all residues in the binding site are well resolved, all the issues raised during the structure preparation could be easily fixed.

## Step 4: Description of the Protein-Ligand Binding Mode

### Complementarity of Shape and Physicochemical Properties

Compute the *Molecular Surface* of the binding site. Go to **MOE | Compute | Surfaces and Maps**.

Leave the default settings and press **Apply**. This generates a molecular surface for the receptor atoms around the ligand, and it is colored by default according to molecular properties (atom color). In the next step set the "Active LP" coloring scheme, which shows hydrophobic regions of the surface in green, mildly polar regions (e.g., the hydrogen atoms of benzene) in blue and hydrogen bonding regions in purple. Generate the surface for the ligand in a similar way.

### Results

In general, the shape of FSN ligand complements well with the cavity of thrombin (Figure 21.6). However, the inhibitor does not evenly fit the protein binding site: the benzamidine substructure is deeply buried into the fairly closed pocket, the tricyclic

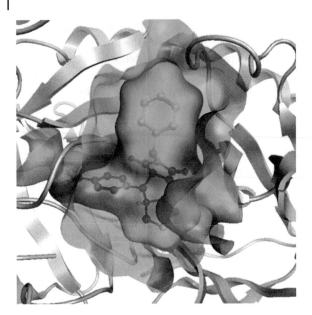

Figure 21.6 Molecular surfaces of protein and ligand colored by ActiveLP method.

scaffold of the inhibitor spans the mildly solvent-exposed part of the cavity, and the fluorophenyl group lies in a shallow hollow at the protein surface.

Figure 21.6 also indicates an overall complementarity of physicochemical properties between the ligand and protein. The single obvious mismatch on lipophilic/hydrophilic potential surfaces concerns one of the two oxygen atoms of the tricyclic scaffold facing hydrophobic residues Tyr60A and Trp60D, whereas this moiety is a potential hydrogen bond acceptor.

### Mapping the Protein Binding Site

Hide the surfaces. Go to **MOE | Compute | Surfaces and Maps** and compute the *Interaction Potential* of the binding site, which can be selected under *Surface* list. The maps for the three probe types are generated after a few seconds, where:

1) N: is a hydrogen bond acceptor probe shown in red color;
2) OH2 is a hydrogen bond donor probe shown in purple color;
3) DRY is a hydrophobic probe shown in green color.

Set the energy levels (in kcal/mol) to the values of −5.5 for the OH2 probe and −2.5 for the others.

### Results

Uncheck boxes for the OH2 and N: probes to view the hydrophobic regions only. Note that one of the oxygen atoms of FSN ligand overlaps with a hydrophobic region, providing one more evidence for the mismatch observed previously. Hide the DRY probe and show OH2 probe to view the hydrophilic regions—we can see that the pyrrolidine of the

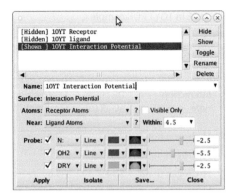

Figure 21.7 The interaction potential map for binding site of 1OYT thrombin structure.

tricyclic ring does not reach the corresponding hydrogen bond donor probe and there is an unfilled cavity in the pocket of the protein (Figure 21.7). In this way, the analysis of the interaction potential map can give the ideas for ligand optimization in order to fill the binding site optimally.

### Intermolecular Non-Covalent Interactions

Hide all surfaces. Use **MOE | Compute | Ligand Interactions** to display non-covalent interactions between FSN ligand and thrombin. From the interaction window, open the **Ligand Interaction Report** to visualize a list of the interactions sorted by type and by strength.

Based on reported interaction energies, identify key interactions between the ligand and the receptor, in particular, hydrogen and ionic bonds, considering that the strength of a bond depends on the nature of interacting atoms/ions as well as on the bond geometry. Typically, the distance between a hydrogen-bond donor (X) and a hydrogen-bond acceptor (A) is comprised between 2.2 Å and 3.2 Å, and the angle between X-H and A is comprised between 120° and 180° (180° is the optimal value).[15] Charged atoms can engage in ionic bonds at longer distance, yet 3.5 Å is a reasonable limit for considering a strong bond. Last, an electrostatic interaction is more efficient in a hydrophobic environment than in a solvent-exposed region.

### Results

The automated analysis of the complex revealed ten non-covalent interactions: six hydrogen bonds and four ionic bonds (Table 21.2, Figure 21.8). The energies of hydrogen bonds range from −9.3 kcal/mol to −1.3 kcal/mol, indicating that both strong and weak interactions were reported. The four strong/ordinary hydrogen bonds (energy around −9 to −4 kcal/mol) include interactions between the carbamimidoyl group of ligand and residues Asp189 and Gly219, and between the succinimide carbonyl and residue Gly216. The two weak hydrogen bonds involve a poor acceptor (Cys220 sulfur atom SG) or a poor donor (carbon atom of the succinimide moiety). All four reported ionic bonds actually represent a single ionic interaction between Asp189 and FSN ligand, since the negative charge on the carboxylate group of Asp189 and the positive charge on the carbamimidoyl group of FSN ligand are delocalized on two oxygens and

Table 21.2 Ligand interactions automatically detected with MOE in 1OYT PDB complex.

| Ligand: FSN | | Receptor: Thrombin heavy chain | | | | |
|---|---|---|---|---|---|---|
| Chemical group | Name of atom (aL) | Residue | Name of atom (aR) | Interaction type | Distance aL-aR (Å) | Energy (kcal/mol) |
| Carbamimidoyl | N3 | Asp189 | OD1 | Hydrogen bond | 2.82 | −9.3 |
| Carbamimidoyl | N1 | Asp189 | OD2 | Hydrogen bond | 2.81 | −6.8 |
| Carbamimidoyl | N1 | GLY219 | O | Hydrogen bond | 2.94 | −4.6 |
| Succinimide | O25 | GLY216 | N | Hydrogen bond | 2.92 | −4.3 |
| Succinimide | C13 | SER214 | O | Hydrogen bond | 3.24 | −1.5 |
| Carbamimidoyl | N1 | CYS220 | SG | Hydrogen bond | 3.99 | −1.3 |
| Carbamimidoyl | N1 | Asp189 | OD2 | Ionic | 2.81 | −5.9 |
| Carbamimidoyl | N3 | Asp189 | OD1 | Ionic | 2.82 | −5.8 |
| Carbamimidoyl | N1 | Asp189 | OD1 | Ionic | 3.59 | −1.6 |
| Carbamimidoyl | N3 | Asp189 | OD2 | Ionic | 3.57 | −1.6 |

two nitrogens, respectively. No interactions were found between the aromatic groups of FSN and the two aromatic side chains of the binding site (Trp60D and Trp215), while the original publication describing the structure of the complex indicates that the fluorophenyl group of the inhibitor undergoes edge-to-face interaction with the indole ring of Trp215 (see Figure 21.5c).[11] The geometric rules in MOE for detection of pi-pi interactions are indeed strict, and although the inter-plane distance between the fluorophenyl and the indole rings is in agreement with pi-pi interaction, the angle between planes is not comprised in the tolerated interval.

In summary, FSN ligand makes several important intermolecular non-covalent interactions to the residues of the thrombin active site:

- a salt bridge between carbamimidoyl group of FSN ligand and carboxylate of Asp189 shows two almost linear hydrogen bonds;
- a hydrogen bond between carbamimidoyl group of FSN ligand and the backbone O of Gly219;
- a hydrogen bond between an oxygen atom of the inhibitor's succinimide and the backbone NH of Gly216;
- an edge-to-face pi-pi stacking between the fluorophenyl of FSN ligand and the indole ring of Trp215.

### Step 5: Detection of Protein Cavities

At this stage, we will further characterize the geometrical and physicochemical properties of the FSN ligand binding pocket in thrombin. With this purpose, we will examine the pocket representation in MOE, because such a representation, which is made of an ensemble of spheres, is also used during a protein-ligand docking in MOE.

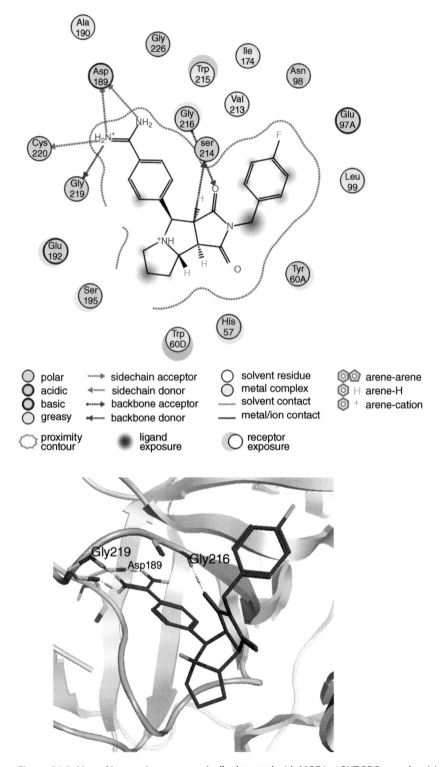

**Figure 21.8** Ligand interactions automatically detected with MOE in 1OYT PDB complex; (a) All interactions summarized in a schematic 2D-view; (b) 3D depiction of amino acids engaged in strong hydrogen and/or ionic bond with FSN ligand.

To identify putative ligand-binding pockets in thrombin, go to **MOE | Compute | Site Finder.** Leave default settings and press **Apply** button. This will calculate possible active sites for the receptor atoms based on its 3D coordinates and summarize them into a list, where:

- the *Size* column indicates the number of spheres comprising a site. Spheres with assigned "hydrophobic" or "hydrophilic" properties are colored in white or red, respectively;
- the *PLB* (Propensity for Ligand Binding) column is a score based on the amino acid composition, it is higher for deeper and more enclosed sites;[16]
- the *Hyd* column indicates the number of hydrophobic contact atoms in the pocket;
- the *Side* column indicates the number of side chain contact atoms in the pocket. Hydrophilic cavity points are colored in red, hydrophobic in white;
- the *Residues* column lists the names of the residues outlining the site.

### Results

Totally, 22 sites were identified for the heavy chain of thrombin using MOE Site Finder, with site 1 being the largest (109 alpha centres) and sites 17 and 20—the smallest (only 6 alpha centres) (Figure 21.9). Indeed, the cavity with the highest ranking (site 1, consisting of 26 residues, PLB score of 3.22) describes the actual binding pocket of FSN ligand from 1OYT complex. Furthermore, the hydrophobic spheres in site 1 overlap consistently with non-polar chemical groups of the FSN ligand (e.g., carbon atoms of pyrrolizidine and benzamidine) and hydrophilic spheres of site 1 match only to the carbamimidoyl group of the inhibitor (Figure 21.9). Noteworthy, the spheres do not cover the whole pocket of the actual inhibitor. In particular, there are no cavity points for the sub-pocket accommodating the fluorophenyl group, or the oxygen atom near Gly216. The placement of spheres with MOE Site Finder depends on the depth and enclosure of the depressions at the protein surface: very small subpocket can be inaccessible to the spherical probes; spherical probes in shallow and solvent-exposed subpockets are not sufficiently buried to be retained in the site description.

As a conclusion, MOE computing approach well describes the binding pocket of the most buried part of FSN ligand, but ignores the subpocket for the binding of the fluorophenyl group, and reveals an additional channel not exploited by this inhibitor.

## Exercise 2: Retrospective Virtual Screening Using the Pharmacophore Approach

### Aim

Retrospective virtual screening is widely used to evaluate computation methods in drug discovery. The test library usually combines active and inactive compounds. Here the test database consists of 10 thrombin inhibitors and 90 "decoys," that is, molecules that are presumed to be inactive on the target protein. Ideally, the search for molecules that superpose well with a three-dimensional pharmacophore would differentiate active and inactive compounds.

(a)

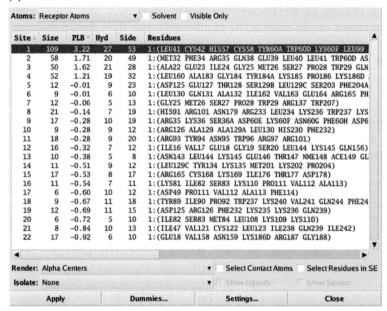

| Site | Size | PLB | Hyd | Side | Residues |
|------|------|------|-----|------|----------|
| 1 | 109 | 3.22 | 27 | 53 | 1:(LEU41 CYS42 HIS57 CYS58 TYR60A TRP60D LYS60F LEU99 |
| 2 | 58 | 1.71 | 20 | 49 | 1:(MET32 PHE34 ARG35 GLN38 GLU39 LEU40 LEU41 TRP60D AS |
| 3 | 50 | 1.62 | 21 | 28 | 1:(ALA22 GLU23 ILE24 GLY25 MET26 SER27 PRO28 TRP29 GLN |
| 4 | 52 | 1.21 | 19 | 32 | 1:(LEU160 ALA183 GLY184 TYR184A LYS185 PRO186 LYS186D |
| 5 | 12 | -0.01 | 9 | 23 | 1:(ASP125 GLU127 THR128 SER129B LEU129C SER203 PHE204A |
| 6 | 9 | -0.01 | 6 | 10 | 1:(LEU130 GLN131 ALA132 ILE162 VAL163 GLU164 ARG165 PH |
| 7 | 12 | -0.06 | 5 | 13 | 1:(GLY25 MET26 SER27 PRO28 TRP29 ARG137 TRP207) |
| 8 | 21 | -0.14 | 7 | 19 | 1:(HIS91 ARG101 ASN179 ARG233 LEU234 LYS236 TRP237 LYS |
| 9 | 17 | -0.28 | 10 | 19 | 1:(ARG35 LYS36 SER36A ASP60E LYS60F ASN60G PHE60H ASP6 |
| 10 | 9 | -0.28 | 9 | 12 | 1:(ARG126 ALA129 ALA129A LEU130 HIS230 PHE232) |
| 11 | 18 | -0.28 | 9 | 20 | 1:(ARG93 TYR94 ASN95 TRP96 ARG97 ARG101) |
| 12 | 16 | -0.32 | 7 | 12 | 1:(ILE16 VAL17 GLU18 GLY19 SER20 LEU144 LYS145 GLN156) |
| 13 | 10 | -0.38 | 5 | 8 | 1:(ASN143 LEU144 LYS145 GLU146 THR147 NME148 ACE149 GL |
| 14 | 11 | -0.51 | 9 | 12 | 1:(LEU129C TYR134 LYS135 MET201 LYS202 PRO204) |
| 15 | 17 | -0.53 | 8 | 17 | 1:(ARG165 CYS168 LYS169 ILE176 THR177 ASP178) |
| 16 | 11 | -0.54 | 7 | 11 | 1:(LYS81 ILE82 SER83 LYS110 PRO111 VAL112 ALA113) |
| 17 | 6 | -0.60 | 10 | 12 | 1:(ASP49 PRO111 VAL112 ALA113 PHE114) |
| 18 | 9 | -0.67 | 11 | 18 | 1:(TYR89 ILE90 PRO92 TRP237 LYS240 VAL241 GLN244 PHE24 |
| 19 | 12 | -0.69 | 11 | 15 | 1:(ASP125 ARG126 PHE232 LYS235 LYS236 GLN239) |
| 20 | 6 | -0.72 | 5 | 10 | 1:(ILE82 SER83 MET84 LEU108 LYS109 LYS110) |
| 21 | 8 | -0.84 | 10 | 13 | 1:(ILE47 VAL121 CYS122 LEU123 ILE238 GLN239 ILE242) |
| 22 | 17 | -0.92 | 6 | 10 | 1:(GLU18 VAL158 ASN159 LYS186D ARG187 GLY188) |

(b)

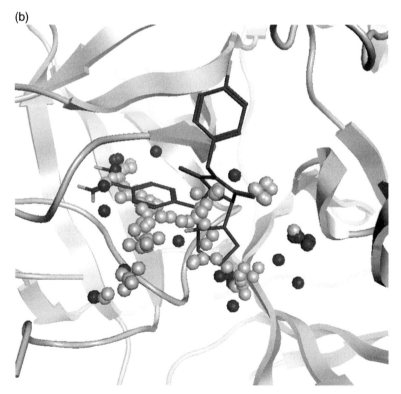

Figure 21.9 (a) Site Finder panel listing the identified cavities in thrombin 1OYT structure and (b) spheres describing binding site for FSN ligand.

The structural requirements for thrombin inhibition will be inferred from 3 of the 10 thrombin inhibitors present in the library. Practically, pharmacophore elucidation involves the 3D-alignment of the three active compounds, and then the selection of essential molecular features for biological activity. Different hypothesis will be tested in the tutorial, and the emphasis will be placed on the importance of 3D-alignment.

### Step 1: Description of the Test Library

The 10 active molecules in the test library were prepared from the SMILE codes retrieved from the RCSB-PDB or the ChEMBL database.[17] Inhibitors were named according to the identifier in the parent database (PDB code followed by the HET code of the ligand or ChEMBL ID). The set of decoys was obtained by random picking in the DUD data set, which was built from the ZINC database.[18] The molecule name of decoys contains the prefix "ZINC." All compounds in the library were subjected to standardization using JChem standardizer (aromatize, clean 2D, convert to Enhanced Stereo, Transform Sulfoxide) (ChemAxon, Budapest, Hungary), then ionized using filter (OpenEye Scientific Software, Santa Fe, NM, USA). The prepared molecules were then stored in the MOE database named **database100.mdb**. For each of the 100 molecules in the library, a low-energy 3D conformation was predicted using Corina v3.40 (Molecular Networks GmbH) and written to **database100-3D.mdb** file. A conformational ensemble was then generated using MOE (default settings for ConfSearch with MMFF94x force field). The conformers were stored in a database named **database100-multiconf.mdb**.

Now we will visualize and briefly analyze the test library to answer the question "Are there chemical features that are obviously specific to the active compounds?" The retrospective screening exercise would indeed be trivial if simple descriptors, such as molecular weight or presence of carbamimidoyl group, could differentiate active and inactive compounds.

Go to **MOE | File | Open…** and open the file **database100.mdb**.

MOE database is a binary file storing data records in cells which are assigned with unique entry numbers (rows) and field names (columns). The molecules are written in the field *mol*, where they are depicted in 2D or 3D projections, which can be viewed and rotated directly inside of the database cell using the middle mouse button. It is possible to modify the size of cells by dragging the mouse. The entries and fields can be easily manipulated using right-click context menu or Edit and Display menu on the top of the viewer (e.g., columns can be sorted by increasing/decreasing order, renamed, moved, colored). More detailed information about MOE database viewer (DBV) can be found at **MOE | Help | Panel Index | Database Viewer.**

Calculate the four following descriptors:

- *Weight* - molecular weight of a compound including implicit hydrogens in atomic mass units,
- *a_don* - number of hydrogen bond donors,
- *a_acc* - number of hydrogen bond acceptors,
- *lip_violation* - the number of violations of Lipinski's Rule of Five, according to which Weight<500 Da, logP<5, number of hydrogen bond donors (OH and NH bonds) <5, number of hydrogen bond acceptors (O and N atoms) <10.[19]

Go to **MOE DBV| Compute | Descriptors | Calculate...** In the window, select the descriptors and then press OK. The computed descriptors are now stored in the database.

## Results

In the MOE database, true thrombin inhibitors can be distinguished from decoys by their names, which are displayed in "*mol*" and "*Name*" fields. In addition, their entries contain word "ACTIVE" in the field "*Activity*."

The chemical library contains organic molecules which mostly fulfill the "Lipinski's Rule of Five, with all but one inhibitor having no more than a single violation. The molecular weights of inhibitors are evenly distributed in the interval [254, 551](Da), the range covered by decoys is slightly narrower [372, 502](Da). Considering the maximal number of hydrogen bond donors and acceptors—6 and 9, respectively—there are no specific trends observed for active compounds. As a conclusion, thrombin inhibitors and decoys cannot be differentiated by simply analyzing the calculated 2D descriptors.

### Step 2.1: Pharmacophore Design, Overview

Three thrombin inhibitors (**1-3**) of the test library were selected for pharmacophore design (Figure 21.10). Note that all three compounds contain a carbamimidoyl group attached to an aromatic ring. In the first exercise of the tutorial, we demonstrated that this chemical functionality binds to the most buried part of the thrombin pocket and it is determinant for FSN inhibition.

The three-dimensional structures of **1, 2,** and **3** were extracted using MOE from the crystal structures of human thrombin complexes (PDB entries 1C5N, 1GHV, and 1C1U, respectively). Atom types were corrected by manual edition. Coordinates were stored in the MOE database named **actives3.mdb**.

The pharmacophore generation requires two steps: first the overlay of the three-dimensional structures of the three inhibitors; second, the selection of common pharmacophoric features.

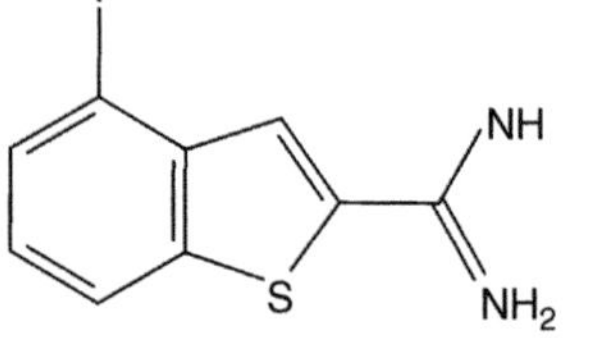

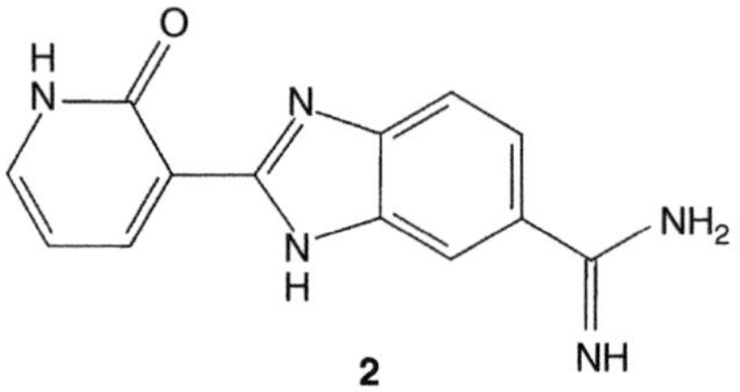

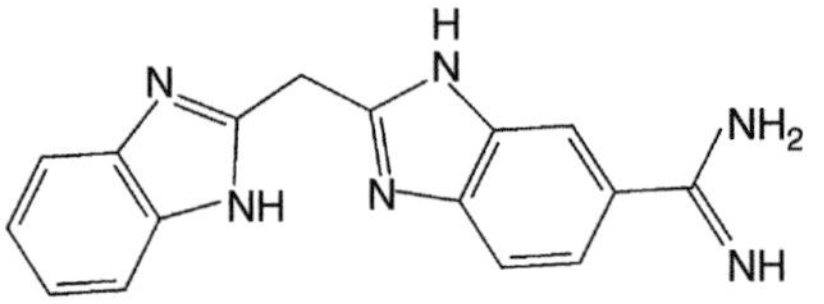

Figure 21.10  Structure of thrombin inhibitors selected for the construction of a pharmacophore.

Open the database containing three active compounds:

---
**MOE | File | Open actives3.mdb**

---

In the MOE database viewer, select the three compounds (hold Shift and click left mouse button on the first and the last compound), then click right button on one of the molecules and select **Send Selected to MOE** in the pop-up menu. To better distinguish the three inhibitors, change the color of carbon atoms of ligands. In right top corner of the MOE main graphical interface, press on **System**, click on the grey box near Ligand and set atom color to **Chain**. Note that inhibitors are already well superimposed, because the protein structures in 1C5N, 1GHV, and 1C1U PDB entries are nicely super-imposed (Figure 21.11a). We will here investigate if MOE alignment tool proposes a relevant superimposition of **1**, **2**, and **3**.

Close the **Database Viewer**. Perform the flexible ligand alignment:

---
**MOE | Compute | Simulation | Flexible Alignment**

---

The calculation lasts for several minutes. If you would like to save time, lower the iteration limit to 20 and the Failure limit to 5 in the "Flexible Alignment" window.

A new database named **flexalgn.mdb** is created, which includes the possible align-ments. Observe the solutions and determinate if there are relevant superimpositions of **1**, **2**, and **3**.

**Results**

Flexible alignment implemented in MOE is a stochastic procedure that randomly searches conformational space of each of input molecules, while generating their rigid body alignment. The scoring of alignments is based on the Gaussian probability density function, which was designed to reproduce x-ray crystallographic alignments.

(a)                                                                (b)

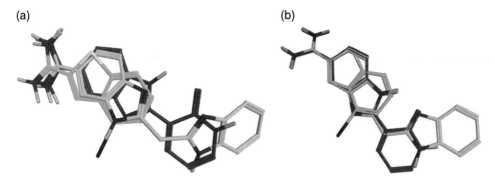

Figure 21.11 (a) Active poses of compounds **1-3**, as observed in crystal structures of complexes with thrombin (after superimposition of the proteins); (b) alignment of compounds **1-3** generated by MOE.

The weights for the terms of the function were trained on 30 sets of aligned ligands from PDB, each set containing at least four diverse ligands (see **MOE | Help | Panel Index | Flexible Alignment** for more details).

In addition to the coordinates of aligned molecules, the output **flexalgn.mdb** database contains following fields:

- **U** – the average strain energy, kcal/mol (sum of the individual force field potential energies divided per number of molecules);
- **F** – the similarity measure of the alignment configuration. **F** is the negative value of the probability density overlap function (lower values for better similarity);
- **S** – the grand alignment score S=U+F (lower values for better alignment);
- **dU, dF, dS** – difference between, correspondingly, **U, F,** and **S** values of the current alignment and the best alignment within the chiral class number;
- **chi** – chiral class number. It can be used to separate an output for the molecules containing unconstrained chiral centers, which are inverted during the search.

The flexible alignment of **1–3** (using default parameters) contains four solutions with U, F, and S scores in the ranges [20.3708, 24.4685], [−103.8838, −88.4845], and [−83.0375, −68.1138] kcal/mol, respectively. Because the search is stochastic, each run returns slightly different solutions. As a consequence, your calculation will not strictly reproduce the number of solutions and score values described here and available in the **flexalgn.mdb** example file.

We now have to select the best alignment to build a pharmacophore. The ranking of solutions according to the increasing S or F score results in the same top-ranked alignment, meaning that, on average, the strain energy is acceptable. If we inspect this 3D-alignment, we can see that it well superimposes the carbamimidoyl groups. From the analysis of the structure of thrombin-inhibitor complex 1OYT we know that carbamimidoyl group plays an important role for the ligand binding. Consequently, the 3D-alignment can be used to make inferences about the nature of thrombin. Considering each inhibitor individually, it appears that although the average strain energy is low, the search procedure tend to flatten the compound **3** (Figure 21.11b).

As a conclusion, the flexible alignment search identified well the pharmacophoric features which are relevant for thrombin binding. However, the optimisation protocol produced an exact 3D-match of these features at the expense of small distortion in molecules. In the next step, we will generate two pharmacophores, the first one based on the crystal structures of thrombin with bound inhibitors **1–3** (Figure 21.11a) and the second one based on the best computed 3D-alignment of **1–3** (Figure 21.11b).

**Step 2.3: Pharmacophore Design, Query Generation**

We will first generate a pharmacophore based on the crystal structures of bound inhibitors **1–3**. The query will be constructed on the basis of pharmacophore consensus for all three compounds in order to select four features placed on the well-aligned and common functional groups of thrombin inhibitors—the carbamimidoyl group and the aromatic ring system.

Close all DBV. Clear the MOE main graphical interface, and then open the database containing the three active compounds:

---

**MOE | File | Close**
**MOE | File | Open   actives3.mdb**

---

In the MOE DBV, select the three compounds (hold Shift and click left mouse button on the first and the last compound), then click right button on one of the molecules and select **Copy Selected to MOE** in the pop-up menu. Close the **Database Viewer** and check that only the superimposed structures of **1–3** are loaded in MOE.

Open the pharmacophore query editor to identify all features which are common to at least two inhibitors, then select two features describing the carbamimidoyl group, and two features describing the two fused rings linked to the carbamimidoyl group.

---

**MOE | Compute | Pharmacophore | Query Editor...**
In the "Pharmacophore Editor" window, give a title to the query (e.g., thrombin) and click on **Consensus...** to open the "Pharmacophore Consensus" window, and select **MOE** as input then click on **Calculate**: pharmacophoric features common to active atoms are displayed into the MOE main GUI and listed in the "Pharmacophore Consensus" windows. Select the four features (see Table 21.3 for details) and click on **Load Selected**. Close the "Pharmacophore Consensus" window. In the "Pharmacophore query editor" check that the radius of each feature is set to the value specified in Table 21.3, then check the box for **Partial Match** search with at least four features to be fulfilled, and save your query as **thrombin.ph4**. Close the window.

---

Repeat all the procedure to generate the second pharmacophore: close all DBV; clear MOE main graphical interface; open **flexalgn.mdb** and copy to MOE the top-ranked alignment; open the pharmacophore query editor and **Consensus...** window; select the four features and set their radiuses as it is specified in the Table 21.3; in the "Pharmacophore Editor" check the box for **Partial Match** search with at least four features to be fulfilled and save the query as **thrombin-moeALI.ph4**.

### Results

Pharmacophore consensus suggests pharmacophoric features from a set of aligned structures. Each feature is specified with its location, radius, and annotation (e.g., cation **Cat**, hydrogen bond donor **Don**, hydrogen bond acceptor **Acc**, Aromatic **Aro**, hydrophobic **Hyd**). In addition, some of the features are assigned with projections (e.g., **Don2**, **Acc2**, **PiN**) and projection vectors, which indicate the positions of the complementary features on the receptor.

The pharmacophore consensus based on the best solution in **flexalgn.mdb** identified a total of 18 features (ID G1 to G18), where 13 of them are common to the inhibitors **1–3** and 5 features are matching only two inhibitors (Table 21.3). In comparison, only 16 features are common to the three inhibitors if the alignment was based on the protein

Table 21.3 Summary of common and selected pharmacophoric features.

| Alignment of 1-3 | Based on protein | | | computed with MOE | | |
|---|---|---|---|---|---|---|
| Input file | actives3.mdb | | | flexalgn.mdb | | |
| Output pharmacophore | thrombin.ph4 | | | thrombin-moeALI.ph4 | | |
| Total of common features | 16 | | | 18 | | |
| Selected features in output file | ID | Radius (Å) | Expression | ID | Radius (Å) | Expression |
| 1 | G13 | 1.6 | CN2\|Cat\|Don | G12 | 1.0 | Cat&Don |
| 2 | G9 | 0.9 | Don2 | G13 | 1.0 | Cat&Don |
| 3 | G7 | 1.0 | Aro\|Hyd | G5 | 1.1 | Aro\|Hyd |
| 4 | G6 | 0.9 | Aro\|hyd | G6 | 1.1 | Aro\|Hyd |

superimposition (from **active3.mdb**). Because of the looseness of this second alignment, the type of the feature describing the carbamimidoyl group is different. Particularly, the two nitrogen atoms of the carbamimidoyl group are both described by cationic and hydrogen bond donor features (**Cat&Don**) in the **thrombin-moeALI.ph4** pharmacophore, while in the **thrombin.ph4** the carbamimidoyl group is described with two different features: the first one is either a bioisostere annotation (**CN2** for NCN+ bioisostere) or a cation or a hydrogen bond donor (**CN2|Cat|Don**), and has a radius comprising the full carbamimidoyl group; the second feature, a projected point for hydrogen bond donor (**Don2**), defines the orientation of the carbamimidoyl group (only one was selected in order to have the same number of features in the two pharmacophores) (Figure 21.12).

As a conclusion, the pharmacophore named **thrombin** is fuzzier than the pharmacophore named **thrombin-moeALI** since it is defined by features with more generic types and of increased tolerance.

## Step 3: Pharmacophore Search

Open the **database100-multiconf.mdb** (conformers have been pre-computed for all molecules) and sequentially perform the pharmacophore search using the two pharmacophores.

---

**MOE | File | Open** database100-multiconf.mdb
**MOE DBV | Compute | Pharmacophore | Search**
   In the Query section of the "Pharmacophoric Search" window, Browse... the pharmacophore query file **thrombin.ph4**. Output results in a database named **ph4out-thrombin.mdb**. Report only Hits: Best Per Molecule. Click on Search to start the calculation.

---

Repeat all the procedure to screen the database using the query file **thrombin-moeALI. ph4** and save the results in a database named **ph4out-thrombin-moeALI.mdb**.

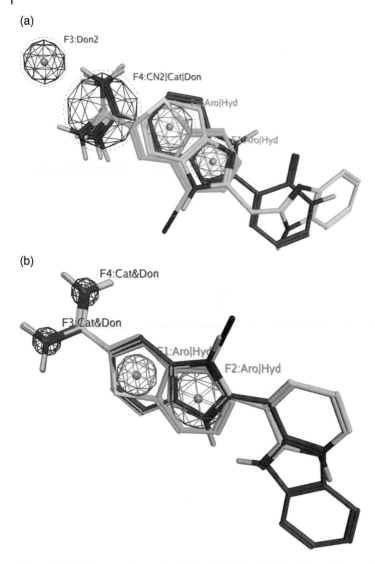

Figure 21.12 Pharmacophores build from inhibitors **1-3** (a) using active pose in crystal structures of complexes with thrombin (after superimposition of the protein); (b) using alignment generated by MOE.

### Results

The molecules that have passed the pharmacophore filter are stored in the output database containing the following fields:

- **mol** – the conformation of the molecule from the input database, which was rotated and translated to match the pharmacophore query;
- **mseq** – an integer sequence number, which starts from 1 and is incremented with each new hit;
- **rmsd** - the root of the mean square distance between the features of the pharmacophore query and corresponding ligand annotation points.

Table 21.4 Summary of pharmacophoric search results.

| Query pharmacophore | thrombin.ph4 | | | thrombin-moeALI.ph4 | | |
|---|---|---|---|---|---|---|
| Input file | database100-multiconf.mdb | | | database100-multiconf.mdb | | |
| Output | ph4out-thrombin.mdb | | | ph4out-thrombin-moeALI.ph4 | | |
| Hits | Number of | rmsd min (Å) | rmsd max (Å) | Number of | rmsd min (Å) | rmsd max (Å) |
| True positive | 10 | 0.2149 | 0.6551 | 9 | 0.1394 | 0.7671 |
| False positive | 81 | 0.1249 | 0.7713 | 17 | 0.2307 | 0.8110 |

The first pharmacophore screening, using fuzzy **thrombin.ph4** query, was able to retrieve all 10 thrombin inhibitors, while the second screening retrieved only nine of them (1cn5_ESI, 1ghv_120, 1gj4_132, 1o5g_CR9, 1oyt_FSN, 1ypg_UIR, 2pks_G44, CHEMBL344790, and CHEMBL358110). Some decoys also match the queries, but to a different extent in the two screenings. Totally, 91 hits are identified using the query **thrombin.ph4**, whereas only 26 molecules match the query **thrombin-moeALI.ph4** (Table 21.4).

Considering that the full database contains a hundred molecules, **thrombin.ph4** is much less efficient than **thrombin-moeALI.ph4** in rejecting decoys. The decoys with the lowest rmsd in the two screenings, ZINC01534345 and ZINC01553076, are displayed in Figure 21.13ab. The chemical moiety of ZINC01553076 matching the pharmacophore features of **thrombin-moeALI.ph4** is similar to the active fragment of the thrombin inhibitors **1–3** used to build the query, whereas the substructure of ZINC01534345 matching the pharmacophore features of **thrombin.ph4** largely differs from a benzamidine.

The **thrombin-moeALI.ph4** query also allows a better match of true inhibitors than **thrombin.ph4** (see the alignment of the true hit, 1ypg_UIR, to both pharmacophore queries in Figures 21.13cd). However, the alignment of 1oyt_FSN ligand (Figure 21.13e) or CHEMBL358110 (Figure 21.13f) to the pharmacophore query **thrombin-moeALI. ph4** is not corresponding to the expected one, where the phenyl ring linked to the carbamimidoyl group is overlapping well with the feature F1 Aro|Hyd. There is one inhibitor missing in the final database, which contains the results of the screening for **thrombin-moeALI.ph4** query, namely 1d6w_00R. This compound is more flexible than the other thrombin inhibitors in the test library (eight rotatable bonds), and it is represented by 279 conformers in the multi-conformational database. Nevertheless, none of these conformers fits pharmacophore query well.

In summary, the two pharmacophore searches were able to identify most of the true thrombin inhibitors. Only the more precise query, built from the 3D-alignment of **1–3**, efficiently rejects decoys and can therefore be further used for virtual screening purposes. In addition, we could observe that the compounds containing conjugated aromatic ring systems attached to the carbamimidoyl group (e.g., 1cn5_ESI, 1ghv_120, 1gj4_132, 1o5g_CR9) are effectively identified by the pharmacophore search—their corresponding hits are well-aligned to all four features and have low rmsd scores. In contrast, the pharmacophore alignments for compounds 1oyt_FSN, CHEMBL358110 are not correct, suggesting that further improvements for the pharmacophore model

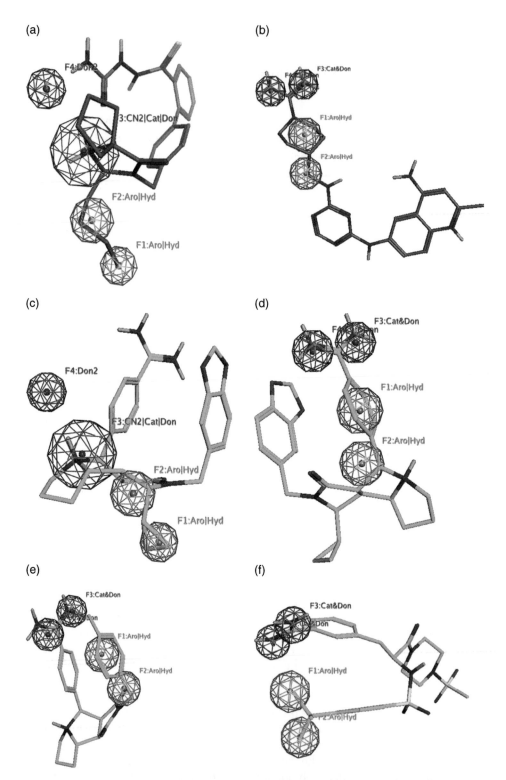

**Figure 21.13** 3D-alignment of hits with query pharmacophore: (a) ZINC01534345 and **thrombin.ph4** (rmsd = 0.1249 Å); (b) ZINC01553076 and **thrombin-moeALI.ph4** (rmsd = 0.2307 Å); (c) 1ypg_UIR and **thrombin.ph4** (rmsd = 0.5207 Å); (d) 1ypg_UIR and **thrombin-moeALI.ph4** (rmsd = 0.5114 Å); (e) 1oyt_FSN and **thrombin-moeALI.ph4** (rmsd = 0.7671 Å); (f) CHEMBL358110 and **thrombin-moeALI.ph4** (rmsd = 0.2092 Å).

can be made. For example, additional features, volumes, or constrains can be included to the pharmacophore model, and the radius or properties of the features can be tuned in order to decrease the number of false positives in the hit list.

## Exercise 3: Retrospective Virtual Screening Using the Docking Approach

### Aim

According to literature, docking has become the prevalent method for structure-based virtual screenings of chemical libraries.[20] Success in screening requires that the docking tool ranks active compounds before inactive compounds, implying that the scoring function can differentiate relevant binding modes from irrelevant binding modes. In this tutorial, we will first test the ability of MOE docking tool to correctly predict the binding mode of FSN ligand to thrombin. We will ask two questions: "Can docking reproduce the binding mode of FSN inhibitor to thrombin as it is observed in the crystal structure 1OYT?" and "Can scoring distinguish correct docking poses and wrong docking poses?" Then, we will test if MOE docking can distinguish thrombin inhibitors from decoys. Last, we will see that it is possible to filter out false positives (i.e., decoys with a higher score) by postprocessing poses based on protein-ligand interaction fingerprints (PLIF).

### Step 1: Description of the Test Library

The test library combines active and inactive compounds, here the data set consists of 10 thrombin inhibitors and 90 "decoys," that is, molecules which are presumed to be inactive on the target protein. For the description and the basic chemoinformatics analysis of the test library, please refer to step 1 of Exercise 2.

### Step 2: Preparation of the Input

Docking in MOE requires the all-atom representation of the protein site. For the detailed preparation of the protein file, please refer to Exercise 1.

Docking in MOE considers the site specified in the settings of the docking protocol. In the tutorial, we will define the docking site around FSN ligand in 1OYT file (residues at less than 4.5 Å).

Open the file containing the target and the ligand structures.

| **MOE | File | Open | 1oyt-ThrombinHC-FSN.moe | OK** |
|---|

### Step 3: Re-Docking of the Crystallographic Ligand

In the re-docking experiment, the FSN ligand, which was extracted from the crystal structure of the 1OYT complex, is docked back into its thrombin binding site prepared from the same crystal structure. In this excercise, we will test different settings of MOE docking tool and evaluate the accuracy of pose geometry and ranking. In particular, we will study the effect of pose refinement, and also the bias in placement if a pharmacophore is used to guide docking.

The re-docking is successful if the pose which is ranked first by the docking score is close to the crystallographic pose, typically, if the root mean square deviation (rmsd) between the coordinates of non-hydrogen atoms of the two poses is lower than 2 Å.

MOE provides several scoring functions. In the tutorial, we will systematically use *London dG*, which is set by default. The *London dG* is an empirical scoring function which estimates the free energy of binding from a given ligand pose. The function consists of the following contributions to the ligand binding: 1) gain or loss of rotational and translational entropy; 2) the energy term describing the loss of ligand flexibility; 3) hydrogen bond energy; 4) metal ligation energy; 5) the difference in desolvation energies for each atom (see **MOE | Help | Panel Index... | Dock**). The terms for hydrogen bond or metal ligation energy are computed as the multiplication products of the corresponding geometric imperfection coefficients (value in a range [0,1]) and the energy of an ideal hydrogen bond or metal ligation.

Taking into account that the performance of most scoring function is ligand- and target-depended, the test of various combinations of settings and scoring methodologies in the re-docking experiment helps to select the best experimental conditions for virtual screening purposes.

Run docking calculation—set the setting like it is explained below and press the Run button.

---

**MOE | Compute | Dock**

| | |
|---|---|
| Output | **1oyt-redock_London.mdb** |
| Receptor | Receptor Atoms |
| Site | Ligand Atoms (the residues close to the ligand atoms will define the docking site) |
| Pharmacophore | None |
| Ligand | Ligand Atoms |
| Placement | Triangle Matcher |
| Rescoring 1 | *London dG* |
| Retain | 10 |
| Refinement | None |
| Rescoring 2 | None |
| Retain | 10 |

---

The docked poses and scores are then written to the "**1oyt_redock_London.mdb**" output database. The docking results will appear in a new DBV window. The final refined poses are ranked by the *London dG* scoring function and written in the **S** field.

Repeat re-docking with the settings as indicated in Table 21.5 (don't forget to modify output database name for each run, example names are given in Table 21.1). The pharmacophoric constraint in run 4 defines ligand placement according to at least two of the three following features: two hydrogen bond donor and cation features (corresponding to FSN carbamimidoyl group in 1OYT complex) and one hydrogen bond acceptor feature (corresponding to FSN carbonyl oxygen involved in polar contact with Gly216 in

**Table 21.5** Experimental settings for the re-docking experiments.

| Run | Pharmacophore | Placement | Rescoring 1 | Retain | Refinement | Rescoring 2 | Retain |
|-----|---------------|-----------|-------------|--------|------------|-------------|--------|
| 1 | None | Triangle Matcher | *London dG* | 10 | none | none | 10 |
| 2 | None | Triangle Matcher | *London dG* | 10 | GridMin | none | 10 |
| 3 | None | Triangle Matcher | *London dG* | 10 | GridMin | *London dG* | 10 |
| 4 | **1oyt-ligand.ph4** | Triangle Matcher | *London dG* | 10 | GridMin | *London dG* | 10 |

1OYT complex). You can visualize the pharmacophore by opening the **1oyt-ligand.ph4** file. Be aware that if the pharmacophore is visible, it is considered as a constraint during docking. Close the pharmacophore before starting calculation runs 1, 2, and 3 (Table 21.5).

---

**MOE | File | open | 1oyt-ligand.ph4 | OK**

---

Use the database browser to compare the docking poses to the crystallographic pose of the ligand which is already present in the MOE window. Open "**1oyt_redock_London.mdb**" database and go to

---

**MOE DBV | File | Browse…**

---

**Results**

Depending on the settings used, the final database contains the following fields: output poses (**mol**); the receptor sequence number for the multiple receptor structures (**rseq**); molecule sequence number (**mseq**); the final score (**S**); the root mean square deviation between the pose and the original ligand, or between the poses before and after refinement (**rmsd** or **rmsd_refine**, Å); energy of conformer (**E_conf**); and scores for successive docking stages: placement, rescoring, and refinement (**E_place**, **E_score1**, **E_score2**, **E_refine**). Lower final S-scores indicate more favourable poses.

Table 21.6 summarizes the results of four re-docking experiments performed for the inhibitor FSN into 1OYT thrombin structure. If no refinements or rescorings are performed (run 1), the first-ranked solution is not in agreement with the conformation of the ligand in the crystal structure 1OYT (rmsd > 2 Å), however the poses ranked 4 and 5 have rmsd values lower than 2 Å. This shows that the *London dG* scoring function was not able to rank a correct pose of FSN inhibitor on the top of the list. In addition, most of the poses show high rmsd values (more than 2 Å). The docking poses can be further refined by means of energy minimization using the conventional molecular mechanics set-up (run 2). This option allows to optimize the protein-ligand geometry, remove clashes, and even to take into account a receptor induced fit effect (see the option "Side Chains: Free" in the refinement configurations). Induced fit effect is not considered in this tutorial, but it is studied in the next chapter.

Table 21.6 Results of the re-docking of FSN ligand back into its thrombin cavity.

| | Number | rmsd$^a$ (Å) | | Number of correct | S score (kcal/mol) | | Score |
|---|---|---|---|---|---|---|---|
| Run | of poses | First pose$^b$ | Last pose$^b$ | poses$^c$ | First pose$^b$ | Last pose$^b$ | accuracy$^d$ |
| 1 | 10 | 7.55 | 6.98 | 2 | −14.59 | −12.82 | no |
| 2 | 10 | 0.86 | 7.57 | 2 | −28.29 | 10.75 | yes |
| 3 | 10 | 7.04 | 7.63 | 2 | −14.20 | −12.56 | no |
| 4 | 8 | 1.21 | 1.05 | 8 | −15.37 | −12.69 | yes |

a)  rmsd of the docked pose to the crystallographic conformation in 1OYT.
b)  First and last pose in the list ranked by S score.
c)  Correct pose reproduces the crystallographic pose with rmsd ≤ 2 Å.
d)  Does the score S discriminate correct and wrong poses?

(a)                                                    (b)

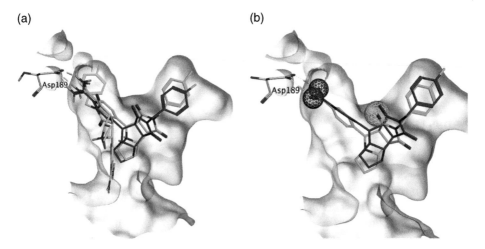

Figure 21.14 The comparison of the experimentally observed binding mode of FSN ligand in the crystal structure 1OYT (ligand is shown in green) and docking poses generated by MOE Dock (ranking is done according to the ascending S-score); (a) the top-ranked pose without any refinement (docking run 1, ligand is shown in cyan sticks, rmsd = 7.55 Å), and the top-ranked pose after the refinement using GridMin (docking run 2, ligand is shown in magenta sticks, rmsd = 0.86 Å); (b) the top-ranked pose of docking run 4 (rmsd = 1.21 Å) which was combined with the pharmacophore query (magenta ligand).

As we can see from the Figure 21.14a, the refinement of the docking poses using the grid minimization improves the protein-ligand interaction pattern—the refined pose is well-overlapped with the crystal structure ligand (rmsd = 0.86 Å) and the bidentate ionic interactions with Asp189 can be observed, in contrast to the top-ranked docking pose which did not undergo the refinement stage (rmsd = 7.55 Å). Noteworthy, the potential energy well discriminates correct and wrong poses (run 2), whereas the rescoring of the refined poses using *London dG* scrambles correct and wrong docking solutions (run 3). If the placement is biased by a pharmacophoric constraint (run 4), only 8 poses are saved in the output database **1oyt_redock_London_grid_London_ph4.mdb.** As we can see from the Table 21.6, run 4 has resulted in fewer poses comparing to previous runs, which can be explained by the current docking settings: 1) the pharmacophoric constraint filters out non-matching poses; 2) the docking solutions should be diverse (the default MOE Dock option "Remove Duplicates" filters out poses with similar ligand-receptor interaction patterns). The top-ranked pose of run 4 is well superimposed with the crystallographic ligand (Figure 21.14b), suggesting that in this case the docking poses are prioritized more efficiently by geometric criteria than by energetic considerations.

### Step 4: Virtual Screening of a Database

In this exercise, we will carry out two docking-based virtual screenings using settings as described in the re-docking section: in the first screening poses are refined using grid minimization and then re-scored using *London dG*, finally, the docking solutions are post-processed based on intermolecular interaction filters (binding modes which differ from that observed between FSN ligand and thrombin in 1OYT are discarded); the second screening uses the same settings as the first, but instead of post-processing of docking poses, the pharmacophore constrains are used on the stage of ligand placement.

*First virtual screening.*
Prepare screening run.

---

**MOE | Compute | Dock...**
Output               **database_London_grid_London.mdb**
Receptor          Receptor
Site                 Ligand Atoms
Pharmacophore None
Ligand            MDBfile                    Browse "**database100_3D.mdb**"
Placement       Triangle Matcher
Rescoring 1     *London dG*
Retain           30
Refinement     GridMin
Rescoring 2     *London dG*
Retain           1

---

The calculation lasts for about 15 minutes on a single CPU. The docked poses and scores are then written to the "**database_London_grid_London.mdb**" output database. Only the poses with the best *London dG* score are retained in the database. Sort molecules either according to ascending *E_refine* or to ascending *London dG* score (S field).

---

**MOE DBV | Right click on E_refine column | Sort | Ascending...**

---

## Results

As it was explained previously in the description of the test library, the screening database contains 10 true inhibitors and 90 decoys from the ZINC database. After the docking, 99 compounds are left (see file **database_London_grid_London.mdb**), and one docking pose is missing, namely for the compound ZINC03320084 containing adamantane moiety. A docking pose is proposed for all the ten thrombin inhibitors, however, they are not always in agreement with crystal structures, for example, see the correct docking solution for 1c5n_ESI ligand and partially wrong docking solution for 1d6w_00R ligand in Figure 21.15.

Nevertheless, the current docking set-up could discriminate active compounds from decoys if the refinement score (*E_refine*) is used to rank molecules—there are five inhibitors (true positives) recovered among the top 10% of the sorted database. The remaining five molecules are false negatives. The 10 thrombin inhibitors are distributed between rank 1 and 98 and have *E_refine* score is in a range of [−33.52; 179.76] kcal/mol. Most of the decoys have positive values of *E_Refine* score. Discrimination of active compounds from decoys is worse if the database is sorted by S score.

### *Post-processing results of the first virtual screening.*

A post-processing step based on Protein-Ligand Interaction Fingerprint (PLIF) allows the filtering of poses based on inter-molecular binding mode. We will select all molecules that establish an ionic interaction with Asp189.

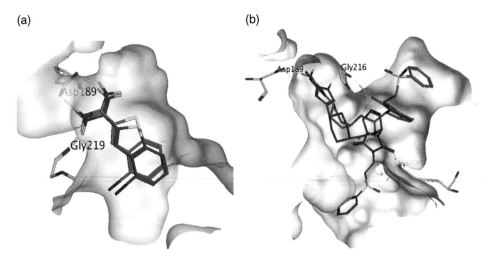

**(a)**            **(b)**

Figure 21.15 Docking poses of (a) 1c5n_ESI and (b) 1d6w_00R ligands into 1OYT protein structure (the docking solutions are shown in magenta sticks) in comparison to their binding modes observed in crystal structures 1C5N and 1D6W, correspondingly (ligands are shown in blue sticks, important protein residues are shown in cyan sticks, protein surface is shown in white).

> **MOE DBV | Compute | PLIF | Generate...**
> **MOE PLIF Setup | Prepare**
> **MOE PLIF Setup | Generate**
>
> **MOE DBV | Compute | PLIF | Analyze... | Select button**
> click *Asp189/ionic interaction*
> click *Apply* and *Close*
> **MOE PLIF | Select in DBV**
> **MOE DBV | Entry | Hide unselected**

### Results

If no selections are made on score (either $S$ or $E\_refine$), the final hit list contains only 25 molecules, among them there are 8 true positives and 17 decoys, which means that most of inactive molecules are filtered out during PLIF post-processing. In this way, the post-processing of docking results using interaction fingerprints allows to select those docking poses which show the interactions of interest. At the same time, two thrombin inhibitors are missing (false negatives), indicating that the required ionic interaction to Asp189 is not fulfilled and their docking poses are wrong. To overcome this issue several poses for each molecule could be saved and then post-processed for the presence of the desired interactions. Alternatively, the docking poses can be filtered directly after the placement stage using the pharmacophore query, like it is demonstrated in the next exercise (Figure 21.16).

(a)                                              (b)

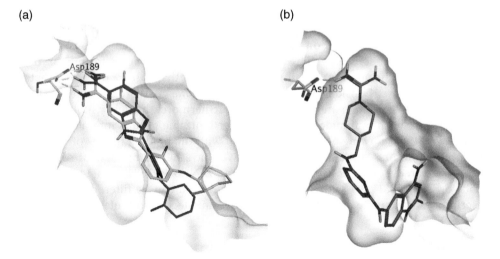

Figure 21.16 The example docking poses after PLIF post-processing: (a) thrombin inhibitor 1o5g_CR9 (magenta ligand) in comparison to its binding mode from the crystal structure 1O5G (cyan ligand); (b) the docking pose of decoy molecule ZINC01553076 (grey ligand). Protein surface is shown in white for the residues of the binding pocket.

***Dock a database of ligands using a pharmacophore query for pose filtering.***
Prepare screening run.

```
MOE | Compute | Dock
Output            database_London_grid_London_ph4.mdb
Receptor          Receptor
Site              Ligand Atoms
Pharmacophore 1oyt_ligand.ph4
Ligand            MDBfile                      Browse "database100_3D.mdb"
Placement         Triangle Matcher
Rescoring 1       London dG
Retain            30
Refinement        GridMin
Rescoring 2       London dG
Retain            1
```

The calculation lasts for less than 15 minutes on a single CPU. The docked poses and scores are then written to the "**database_London_grid_London_ph4.mdb**" output database. Only the poses with the best *London dG* score are retained in the database. Sort molecules either according to ascending *E_refine* or to ascending *London dG* score (S field).

```
MOE DBV | Right Click on E_refine field | Sort | Ascending...
```

**Results**

After the docking with the pharmacophore query 30 compounds are left in the database (see file **database_London_grid_London_ph4.mdb**), which is considerably less than in the first screening. The output database contains poses for nine of the actives (1d6w_00R is missing). Again, *E_refine* is better than *S score* to discriminate actives from decoys. Considering a loose cut-off *E_refine* <0, there are nine hits left in the list and five of them are true positives. Furthermore, the decoy molecules in hit list also have carbamimidoyl group or other ionizable groups interacting with Asp189, and their docking poses are consistent with the binding modes of some of thrombin inhibitors. Some of the false positives have higher number of rotatable bonds compared to true actives in the list (e.g., ZINC03791681 with 14 rotatable bonds or ZINC03817241 with 13 rotatable bonds) and can be filtered out by this descriptor. The example docking poses for active and decoy molecules are shown in Figure 21.17.

**Conclusion**

The summary of the two docking screenings can be found in Table 21.7. In the absence of pharmacophore constraint, pose placement is still an issue for some ligands, especially the most flexible ones. More importantly, scoring does not effectively differentiate

(a) (b)

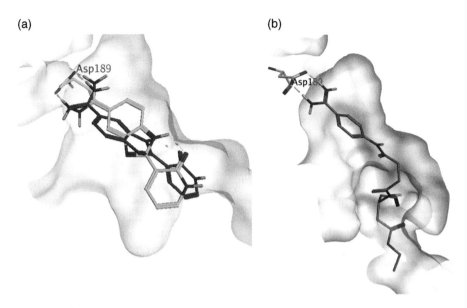

Figure 21.17 The example docking poses of hits identified by docking combined with pharmacophore query: (a) Thrombin inhibitor 1ghv_120 (magenta) in comparison to its binding mode from the crystal structure 1GHV (cyan ligand); (b) The docking pose of decoy molecule ZINC03791681 (gray ligand). Protein surface is shown in white for the residues of the binding pocket of 1OYT structure.

Table 21.7 Summary table of screening results.

| | Docking | | | Number of docked compounds | | | |
| | | | | No score thresholds | | $E\_refine < 0$ | |
| Screening | Pharmacophore | Scoring | Post-processing using PLIF | All | Active | All | Active |
| --- | --- | --- | --- | --- | --- | --- | --- |
| 1 | none | *London dG* | none | 99 | 10 | 15 | 6 |
| 1 | none | *London dG* | Asp189 ionic | 25 | 8 | 9 | 4 |
| 2 | 1oyt-ligand.ph4 | *London dG* | none | 30 | 9 | 9 | 5 |

correct and wrong poses, a fortiori does not accurately distinguish thrombin inhibitors from decoys. Many false positives can, however, be identified from binding mode analysis. Here we have discarded all poses which do not agree with the expected key interaction for thrombin binding. Bias in placement using a pharmacophoric constraint is an efficient alternative to post-processing filters. The two approaches are nevertheless not interchangeable, as indicated by differences in hit lists (typically only three thrombin inhibitors with *E_refine >0* are common to both hit lists).

## General Conclusion

In this chapter, we have seen that a good knowledge of a studied protein, especially of its key residues playing an important role in ligand binding, is determinant for the design of structure-based screening experiment. In the case of human thrombin, the interaction with the carboxylic acid of Asp189 is essential for the efficient anchoring of inhibitors, which usually contain a carbamimidoyl group (or a bioisostere). It allows a very strong interaction between protein and ligand: a bidentate H-bond assisted ionic interaction, well buried into a hydrophobic environment.

In the pharmacophore design, the knowledge about the essential role of the carbamimidoyl group helps to prioritise the pharmacophoric features which are common to the various thrombin inhibitors. In docking, most of the wrong poses can be easily identified by analysing their binding modes. Moreover, constraints in the placement (typically to enforce ligand to establish polar interaction with Asp189) drastically decrease the number of wrong solutions.

The screening of the test database, which consists of 10 thrombin inhibitors and 90 decoys, using a pharmacophore query built from the three thrombin inhibitors and containing their common pharmacophoric features, was able to retrieve 9 out of 10 actual thrombin inhibitors and 17 decoys. The pharmacophore-based approach thus efficiently and rapidly filters the database. The number of hits is still high as compared to the size of the database, and therefore it is tempting to tune the pharmacophore in order to improve the statistics in the retrospective exercise. But a pharmacophore that is too optimized for a test set may have poor performance on a real prospective exercise. Especially, an increased precision in the pharmacophore definition has a negative impact on the structural diversity of hits. A common approach to refine hit selection is docking, which is more time-consuming than pharmacophore search. Docking proposes binding mode for selected ligand, therefore, discards molecules that do not fit a protein binding site.

In the tutorial we have seen that docking is able to predict the correct binding mode of thrombin inhibitors, but it is not necessarily well scored. Score is also an issue for ligand ranking. Overall, docking results are poorly reliable for highly flexible ligands (with more than 10 rotatable bonds). However, if there is extra pharmacophore information, docking of more flexible ligands may still produce reasonable results and many well-scored decoys are filtered out. As a result, the final selection contains a small number of molecules, which are mostly true positives.

Despite the scoring errors, performances of docking to discriminate thrombin inhibitors from decoys are overall good (Figure 21.18). In addition, the receiver operating curves indicate that all the screening conditions described in the tutorial produced an early enrichment in thrombin inhibitors of hit list, thereby suggesting that docking is suitable for prospective hit identification. For the sake of comparison, the screening of the full DUD data set (65 inhibitors, 229 decoys), using the program Dock, similarly retrieves a high number of true actives while discarding many decoys (13.7 enrichment at 1%).[18]

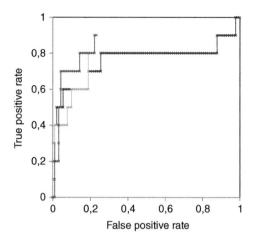

True positive rate (y-axis)
False positive rate (x-axis)

Figure 21.18 Receiver operating characteristic (ROC) curves for two virtual screenings by docking: first screening (blue curve) was done using grid minimization and rescoring and the second screening (red curve) was done in a combination with a pharmacophore query. The dark and light blue curves were obtained for rough and PLIF-filtered data, respectively. The true positive rate is the number of thrombin inhibitors in hit list divided by the total number of thrombin inhibitors in the database (here 10). The false positive rate is the number of decoys in the hit list divided by the total number of decoys in the database (here 90). Red and light blue curves are interrupted because no rankings were produced for PLIF- or pharmacophore-filtered molecules.

## References

**1** Goodford, P.J. (1984). *J. Med. Chem.* 27, 558–564.

**2** Lounnas, V., Ritschel, T., Kelder, J., McGuire, R., Bywater, R.P., and Foloppe, N. (2013). *Comput. Struct. Biotechnol. J.* 5, 1–14.

**3** Wermuth, C.G., Ganellin, C.R., Lindberg, P., and Mitscher, L.A. (1998). *Pure Appl. Chem.* 70.

**4** Wolber, G., and Langer, T. (2005). *J. Chem. Inf. Model.* 45, 160–169.

**5** Degen, S.J., and Davie, E.W. (1987). *Biochemistry (Mosc.)* 26, 6165–6177.

**6** The UniProt Consortium (2015). *Nucleic Acids Res.* 43, D204–D212.

**7** Di Nisio, M., Middeldorp, S., and Büller, H.R. (2005). *N. Engl. J. Med.* 353, 1028–1040.

**8** Rydel, T.J., Ravichandran, K.G., Tulinsky, A., Bode, W., Huber, R., Roitsch, C., and Fenton, J.W. (1990). *Science* 249, 277–280.

**9** Goodsell, D.S. (2002). *RCSB Protein Data Bank.*

**10** Berman, H., Henrick, K., and Nakamura, H. (2003). *Bank. Nat. Struct. Biol.* 10, 980–980.

**11** Olsen, J.A., Banner, D.W., Seiler, P., Obst Sander, U., D'Arcy, A., Stihle, M., Müller, K., and Diederich, F. (2003). *Angew. Chem. Int. Ed.* 42, 2507–2511.

**12** Blow, D.M. (2002). *Acta Crystallogr. D Biol. Crystallogr.* 58, 792–797.

**13** Ilari, A., and Savino, C. (2008). In Bioinformatics, J.M. Keith, ed. (Totowa, NJ: Humana Press), pp. 63–87.

**14** Halgren, T.A. (1996). *J. Comput. Chem.* 17, 490–519.

**15** Jeffrey, G.A. (1997). An introduction to hydrogen bonding (New York: Oxford University Press).

**16** Soga, S., Shirai, H., Kobori, M., and Hirayama, N. (2007). *J. Chem. Inf. Model.* 47, 400–406.

**17** Bento, A.P., Gaulton, A., Hersey, A., Bellis, L.J., Chambers, J., Davies, M., Krüger, F.A., Light, Y., Mak, L., McGlinchey, S., et al. (2014). *Nucleic Acids Res.* 42, D1083–D1090.

**18** Huang, N., Shoichet, B.K., and Irwin, J.J. (2006). *J. Med. Chem.* 49, 6789–6801.

**19** Lipinski, C.A., Lombardo, F., Dominy, B.W., and Feeney, P.J. (1997). *Adv. Drug Deliv. Rev.* 23, 3–25.

**20** Ripphausen, P., Nisius, B., Peltason, L., and Bajorath, J. (2010). *J. Med. Chem.* 53, 8461–8467.

Part 8

Protein-Ligand Docking

22

# Protein-Ligand Docking

*Inna Slynko, Didier Rognan, and Esther Kellenberger*

## Introduction

Molecular docking is a computational method for simulating molecular recognition. Protein-ligand docking focuses on complexes between low molecular weight compounds and their targeted protein binding sites, attempting to predict the conformation and relative orientation of both molecules so as to minimize the free energy of the overall system.

Molecular docking is commonly used in drug design for hit identification and optimization, for that diverse or focused compound libraries are screened.

Many programs have been designed to be suitable for such high throughput applications, therefore they can process one ligand on a second-to-minute time scale on a single CPU. For that reason the underlying methodologies work on a simplified problem, where protein is entirely or mainly treated as rigid, and free energy is approximated with elementary scoring functions (for example, the so-called empirical scores are computed from individual contributions of non-covalent intermolecular interactions).[1]

Docking benchmarks revealed that prediction accuracy depends on the studied system. Typically, failures can be caused by 1) insufficient conformational sampling, especially when the ligand is highly flexible; 2) inappropriate representation of the protein conformation, for example, when a ligand induces conformational changes of a protein—so-called "induced fit effect"; 3) poor ranking of the solutions.[2–4] The preparation of the input file directly impacts calculations, since conformational, tautomeric, ionization, and protonation states of a molecule define its binding capacities.[5,6]

The tutorial emphasizes the key role of modelers to reproduce an experimental binding mode using the docking approach. The first exercise investigates the effect of ligand ionization state and the presence of water in the binding site on the accuracy of pose prediction. This exercise contains a series of *re-docking* experiments, where the ligand, which was extracted from the crystal structure of a ligand/protein complex, is docked back into the protein binding site prepared from the same crystal structure. The second exercise demonstrates the limitations of the flexible ligand/rigid receptor docking approach on an example of a protein which undergoes conformational rearrangement upon ligand

*Tutorials in Chemoinformatics*, First Edition. Edited by Alexandre Varnek.
© 2017 John Wiley & Sons Ltd. Published 2017 by John Wiley & Sons Ltd.
Companion website: www.wiley.com/go/varnek/chemoinformatics

binding. To that end, *re-docking* experiment is compared to *cross-docking* experiment, which uses a protein structure from a complex containing a different ligand.

## Description of the Example Case

In this tutorial we will challenge docking methods in, for example, acetylcholinesterase (AChE). AChE is found mainly at synapses between nerve cells and muscle cells, where it hydrolyzes the neurotransmitter acetylcholine into choline and acetic acid (Figure 22.1a). The enzymatic reaction is very fast (~0.1 ms) and efficiently stops the communication between the two nerves. Choline and acetate are then recycled to build new neurotransmitter molecules for the next signal.

AChE blockade by irreversible inhibitors such as insecticides (e.g., malathion) or poison gas (e.g., sarin) are detrimental to health, because it causes the accumulation of acethylcholine in the synapse, thereby paralyzing the muscle. By contrast, reversible inhibitors can benefit in cognition and/or behavior for people with central nervous system diseases. In particular, the loose nerve signal in people with Alzheimer's disease can be reinforced by taking drugs that target AChE. Tacrine (Cognex®) was the prototypical AChE inhibitor for the treatment of Alzheimer's disease, but has been discontinued due to concerns over safety. Donepezil (Aricept®) is another widely used Alzheimer's disease treatment, which is now sold as a generic by multiple suppliers. Figure 22.1 shows the chemical structure of AChE inhibitors.

AChE three-dimensional structure was studied in *Torpedo californica*—an electric fish where this protein is particularly abundant. The crystal structures of AChE showed that the catalytic triad of amino acids (serine200-histidine440-glutamate327) is located at the bottom of a long channel that allows substrate and competitive ligands to slip down inside the enzyme (Figure 22.2). In the active site, the positive quaternary amine group of acetylcholine is recognized by an anionic subsite, yet its binding involves non-covalent interactions with aromatic residues and not with the negatively charged amino acids.[7–9]

Literature reports many computational studies on AChE, in particular for prediction of inhibitor binding mode and for identification of new inhibitors. Besides, docking benchmarks demonstrate that performance for AChE largely depends on experimental conditions, especially on the input protein structure, the presence of water, and the decoy set used in virtual screening experiment.[10,11] This tutorial investigates how ligand and protein structure as well as the way they were prepared impact the docking results. To that end we will test: (1) two protonation states of tacrine, (2) three inhibitors based on two chemotypes (tacrine and tacrine-pyridone inhibitors with acrinidine scaffold, and donepezil containing benzyl-4-piperidyl moiety), (3) presence or absence of crystal water in AChE active site, and (4) two different conformations of AChE (PDB 1ACJ and 1EVE).

## Methods

This tutorial describes input preparation, calculation, and analysis of docking results for three reversible AChE inhibitors using FlexX technology in LeadIT, version 2.1.8 (BioSolveIt GmbH, Sankt Augustin, Germany). The program LeadIT was chosen

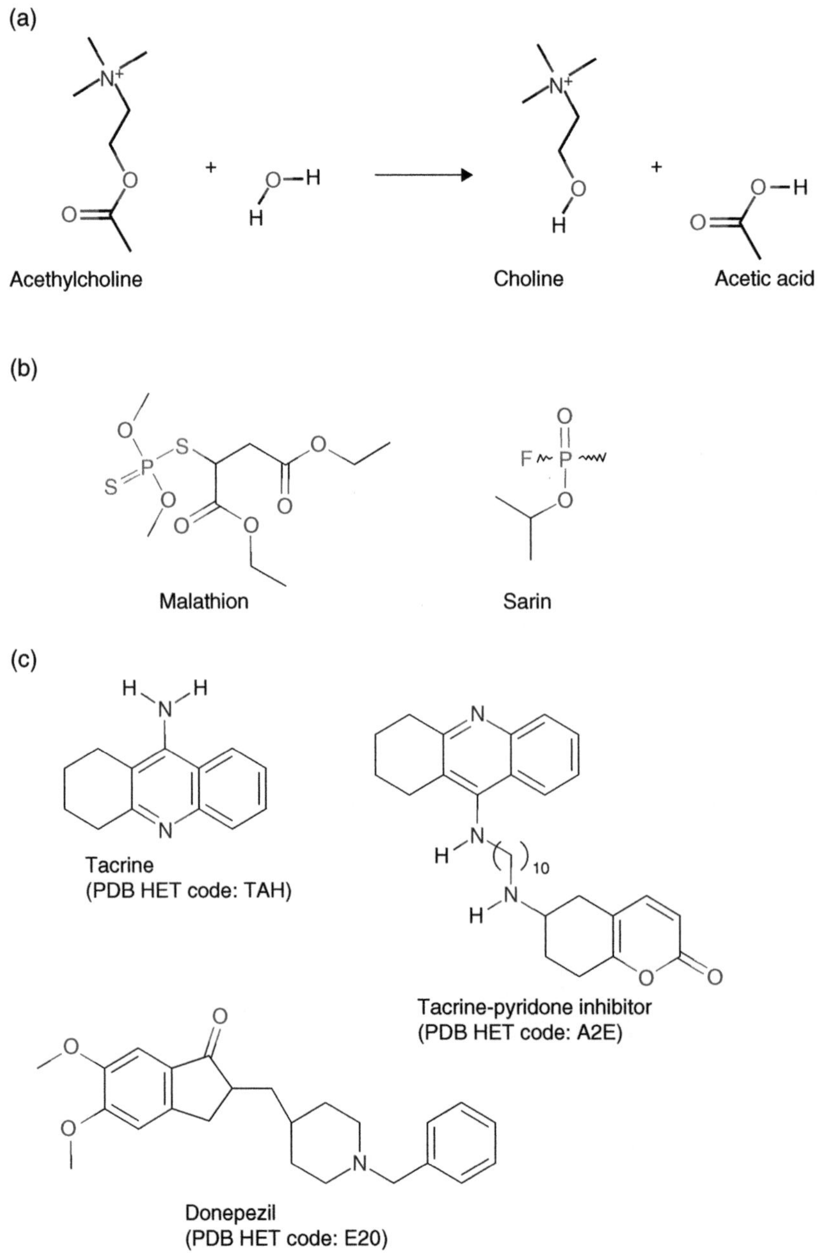

(a)

Acethylcholine                    Choline          Acetic acid

(b)

Malathion                          Sarin

(c)

Tacrine
(PDB HET code: TAH)

Tacrine-pyridone inhibitor
(PDB HET code: A2E)

Donepezil
(PDB HET code: E20)

Figure 22.1 Ligands of AChE. (a) Substrate and products of the reaction catalyzed. (b) Irreversible inhibitors. (c) Reversible inhibitors.

because of its user-friendly graphical interface which provides relevant guidelines to prepare the docking run and to analyze the results. The techniques described here are, however, transferable to any protein-ligand system using other docking programs.[12,13]

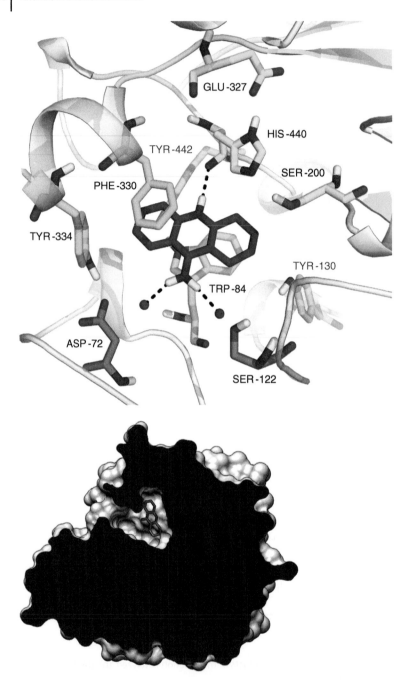

Figure 22.2 Crystal structure of tacrine bound to AChE (PDB 1ACJ). Key residues of the protein active site are represented using CPK-colored sticks (carbon atoms of the aromatic residues, the catalytic triad, and polar residues outlining tacrine binding site are shown in green, yellow, and orange, respectively). (top) Two conserved waters interacting with amine group of tacrine are depicted as red spheres. Hydrogen bonds are shown as black dashed lines. The picture was prepared using PyMOL (http://sourceforge.net/projects/pymol/). (bottom) Clipped solvent-excluded surface of the protein. The picture was prepared using Chimera (http://www.cgl.ucsf.edu/chimera)

## Ligand Preparation

Ligand input for docking is an all-atom description of the molecule in mol, sdf or mol2 format. The protocol for ligand preparation consists of (1) standardization of chemical functional groups, (2) ionization of titratable groups at the considered pH and choice of the tautomeric state, and (3) generation of three-dimensional coordinates. In the tutorial we will use ready-to-dock ligands in mol2 files prepared from PDB complexes as follows:

- Atom types conform to chemical structures in Figure 22.1. Carbon and nitrogen atoms in aromatic rings are typed as aromatic (C.ar and N.ar, respectively). Aliphatic nitrogen atoms are either trigonal pyramidal (N.3 type) or planar, and trigonal if attached to an aromatic ring (N.pl3 type). The sp2 and sp3 oxygen atoms are typed as O.2 and O.3, the aliphatic sp2 and sp3 carbon atoms as C.2 and C.3. Note that the ketone tautomer, which is generally more stable than the enol tautomer is preferred in donepezil.
- Trigonal pyramidal nitrogen atoms are basic and considered as positively charged at pH=7. N.3 type is therefore moved to N.4 in tacrine-pyridone and donepezil. The nitrogen atom in the tetrahydroacridine ring of tacrine is either considered as neutral (N.ar type unchanged) or positively charged (N.ar type moved to N.pl3). The nitrogen atom in the tetrahydroacridine ring of tacrine-pyridone inhibitor is positively charged (N.ar type moved to N.pl3).
- The coordinates of tacrine heteroatoms are obtained from 1ACJ PDB file. The coordinates of tacrine-pyridone inhibitor and donepezil are prepared from 1ZGC and 1EVE PDB files, respectively, after structural alignment of the protein backbone atoms in 1ZGC or 1EVE to the protein in 1ACJ.
- The hydrogen atoms are added to the inhibitors using Sybyl (Tripos, Certara USA, Inc., Princeton, NJ, USA).

## Protein Preparation

Protein input for docking is, generally, the all-atom description of the molecule in mol2 or in another proprietary format. It should be noted that protein structures available in Protein Data Bank (PDB) are always incomplete (missing hydrogen atoms, truncated amino acids, missing residues), and can contain errors (e.g., inconsistent atom name or residue numbering) and ambiguities (alternate position of atoms, undefined ionization state of titratable protein side chains, undefined tautomeric state of amide and imidazole groups). Moreover, the starting PDB file often includes additional molecules, for example, other copies of the protein, covalent ligands from post-translational modifications such as bound hexoses, non-covalent ligands such as nucleic acids, crystallization agents, metal ions, etc. ... and water molecules.[14]

In LeadIT, the protein preparation flowchart automatically fixes the ambiguities found in the uploaded file. Starting with an original PDB file, the steps are the following:

1) *Definition of protein:* selection of protein chain(s), bound co-factor(s) and metallic ion(s).
2) *Definition of binding site:* selection of the residues flanking the binding pocket. In the tutorial we consider that the binding site is made of all residues close enough from a reference ligand. LeadIT also allows a sphere-based selection (the user defines center and radius). LeadIT cannot explore the full protein surface.

3) *Analysis of the protein coordinates:* the program asks the user to set (1) ionization state of GLU, ASP residues—negatively charged or neutral, (2) ionization state of HIS, LYS, and ARG residues—positively charged or neutral, (3) tautomeric state of neutral HIS residues—protonation of N-delta or N-epsilon, and (4) position of polar hydrogen atoms in hydroxyl groups of SER, TYR, and THR residues, in amine groups of LYS residues, and in the side chain amide groups of ASN and GLN residues.

4) *Analysis of crystal water in binding site:* water molecules with two or more hydrogen bonds to the protein are hardly displaceable by a ligand, and can be defined as part of the protein. In LeadIT, there are three types of crystal water molecules, the « *oriented water* » where hydrogen atoms are fully defined, the « *freely rotatable water* » which is a freely spinning molecule and the « *freely rotatable, displaceable water* » which can be replaced by a ligand atom during the docking run.

5) *Definition of metal coordination type (not used in this tutorial):* if a binding site contains a metal ion, docking can be biased toward completion of the metal coordination sphere by ligand polar heteroatom(s).

6) *Addition of hydrogen atoms:* hydrogen atoms are added to fill out the remaining open valences of the receptor. In LeadIT, the last step of the preparation protocol automatically adds hydrogen atoms to the protein before saving the protein representation for docking.

### Docking Parameters

This tutorial mainly focuses on protein preparation and uses the default parameters of the program FlexX implemented in LeadIT. However, the docking popup menu of the graphical interface allows the user (1) to choose the method for base fragment placement during the incremental construction of ligand in protein site, (2) to downscale the score of ligand poses at the rim of the binding pocket (*Access scaling scoring* option, which is switched on by default), (3) to modulate tolerance to steric clashes, and (4) to set the number of docking solutions.

## Description of Input Data Available on the Editor Website

The course material is stored into a compressed archive named *CI-TUTORIAL8.tar.gz*. You need to use the tar command under Unix or Linux operating systems to open and extract the file.

`tar -zxf CI-TUTORIAL8.tar.gz`    The archive from the file called CI-tutorial.tar. gz (–f) is uncompressed with gzip command (–z), extracted to disk into the current directory (–x)

The paths to input files from the *CI-TUTORIAL8* directory, are detailed in Table 22.1. For the sake of pedagogy, output files are given too.

**Table 22.1** Description of input and output files.

| File description | Path to file |
|---|---|
| **Input: Crystal structure of acethylcholinesterase** | |
| Original PDB file | *input/pdb/pdb1acj.ent* |
| All-atom description of the protein prepared from 1acj PDB file | *input/receptor/1acj_ TAHsite65.mol2* |
| | *input/receptor/1acj_TAHsite65WATmol2* |
| "TAH," "A2E," or "E20" in filename indicates which ligand was used for the definition of binding site. "WAT" in filename indicates the presence of water in active site | *input/receptor/1acj_A2Esite65WAT.mol2* |
| | *input/receptor/1acj_E20site65WAT.mol2* |
| All-atom description of the protein, including one water molecule in the active site, prepared from 1eve PDB file after structural alignment of protein in 1eve to protein in 1acj | *input/receptor/1eve_ali_WAT.mol2* |
| **Input: Crystal structure of ligands bound to acethylcholinesterase** | |
| Neutral form of tacrine, prepared from 1acj PDB file | *input/ligand/TAH_1acj.mol2* |
| Positively charged tacrine, prepared from 1acj PDB file | *input/ligand/TAH_1acj+.mol2* |
| Tacrine-pyridone inhibitor, prepared from 1zgc PDB file after structural alignment of protein in 1eve to protein in 1acj | *input/ligand/A2E_1zgc.mol2* |
| Donepezil, prepared from 1eve PDB file after structural alignment of protein in 1eve to protein in 1acj | *input/ligand/E20_1eve.mol2* |
| **Results: output data of Exercise 1 (scores are in csv file, poses are in SDF and MOL2 files)** | |
| re-docking of TAH into TAH binding site in 1acj no water in site | *flexx/1acj_TAHsite65-TAHredock.mol2* |
| | *flexx/1acj_TAHsite65-TAHredock.sdf* |
| | *flexx/1acj_TAHsite65-TAHredock.csv* |
| re-docking of TAH+ into TAH binding site in 1acj no water in site | *flexx/1acj_TAHsite65-TAH+redock.mol2* |
| | *flexx/1acj_TAHsite65-TAH+redock.sdf* |
| | *flexx/1acj_TAHsite65-TAH+redock.csv* |
| re-docking of TAH+ into TAH binding site in 1acj two water molecules in site | *flexx/1acj_TAHsite65WAT-TAH+redock.mol2* |
| | *flexx/1acj_TAHsite65WAT-TAH+redock.sdf* |
| | *flexx/1acj_TAHsite65WAT-TAH+redock.csv* |
| **Results: output data of Exercise 2 (scores are in csv file, poses are in SDF and MOL2 files)** | |
| cross-docking of A2E into extended TAH binding site in 1acj, two water molecules in site | *flexx/1acj_A2Esite65WAT-A2Ecrossdock.mol2* |
| | *flexx/1acj_A2Esite65WAT-A2Ecrossdock.sdf* |
| | *flexx/1acj_A2Esite65WAT-A2Ecrossdock.csv* |
| cross-docking of E20 into extended TAH binding site in 1acj, two water molecules in site | *flexx/1acj_E20site65WAT-E20crossdock.mol2* |
| | *flexx/1acj_E20site65WAT-E20crossdock.sdf* |
| | *flexx/1acj_E20site65WAT-E20crossdock.csv* |
| re-docking of E20 into E20 site in 1eve one water molecule in site | *flexx/1eve_E20site65WAT-E20redock.mol2* |
| | *flexx/1eve_E20site65WAT-E20redock.sdf* |
| | *flexx/1eve_E20site65WAT-E20redock.csv* |

## Exercises

Program can be launched as/softs/biosolveit/leadit-2.1.8-Linux-x64/leadit
This tutorial uses four commands in the menu bar of LeadIT graphical interface:

- To launch the panel for protein preparation, issue `Molecules|Prepare receptor…`, load protein file and follow the instructions.
- To save prepared protein, issue `Molecules|Export receptor…`.
- To launch the panel for ligand preparation, issue `Molecules|Choose Library…`, and load the ligand file.
- To launch the panel for ligand to protein docking, issue `Docking|Define FlexX Docking…` and press *Apply and Dock!* Button.

In the two exercises, each docking lasts for a few seconds to several minutes, depending on the number of rotatable bonds in ligand.

Starting from 1ACJ PDB file, which describes tacrine complexed to AChE, you will prepare protein and execute a series of *re-docking* experiments for the neutral form of tacrine in absence of water, for the positively charged form of tacrine in absence of water, and for the positively charged form of tacrine in presence of water.

The original 1ACJ PDB file contains the crystal structure of AChE in complex with tacrine (HET code: TAH), and 82 water molecules. It describes a single copy of the complex. The binding site is defined as selection of protein residues within 6.5 Å distance (default value) from tacrine molecule. Polar hydrogen atoms were added to AChE binding site to optimize intramolecular interactions, and side chains of acidic residues at the rim of the binding site were neutralized (see Table 22.2).

Table 22.2 Manual modification of residues in tacrine binding site of AChE.

| Amino acid | Protonation state | $\chi_1$ torsion angle (°) | Comment |
|---|---|---|---|
| ASP-72-A | asp1 | 0 | Side chain not in the pocket |
| SER-81-A | ser | 180 | Default |
| ASN-85-A | asn | | Default |
| TYR-121A | tyr | 0 | Default |
| SER-122-A | ser | 330 | H-bond to a crystal water molecule |
| TYR-130-A | tyr | 300 | H-bond to a crystal water molecule |
| GLU-199-A | glu- | | Default |
| SER-200-A | ser | 120 | H-bond to HIS440A N-epsilon nitrogen |
| TYR-334-A | tyr | 150 | H-bond to ASP72A carboxylic acid |
| HIS-440-A | his1 | | H-bond to SER200 hydroxyl group |
| TYR-442-A | tyr | 180 | Polar hydrogen does not point toward the cavity |

Issue **Molecules|Prepare** receptor... to prepare protein, following the steps detailed below (press the green arrow to move to next step). Screen captures are annexed to the tutorial.

| | |
|---|---|
| 1) `Load protein` | Load protein from file |
| | `input/pdb/pdb1acj.ent` |
| 2) `Choose receptor component` | Chain(s) of the receptor: |
| | ↕ `chain A` |
| 3) `Define binding site` | Specify Binding Site using Reference Ligands |
| | Choose a Reference Ligand |
| | ↕ `THA-999-A` |
| | Include amino acids within 6.5 Å radius |
| 4) `Resolve chemical ambiguities` | Residues: |
| | Set protonation state and conformer according to values given in Table 22.2. |
| | Water: |
| | Deselect all molecules (`HOH-604-A`, `HOH-6344-A` and `HOH-643-A`) |
| | Small Molecules: |
| | No actions required (the single small molecular weight compound in input file is defined as reference ligand, thus not included in receptor) |
| 5) `Confirm receptor` | Save Receptor Definition as: |
| | `1acj_TAHsite65` |

Issue `Molecules|Export   receptor`... to save protein file as `input/receptor/1acj_TAHsite65.mol2`.

Docking of Neutral Tacrine, then of Positively Charged Tacrine

Although pKa prediction in LeadIT suggests that tacrine is not basic, the experimental pKa of tacrine in aqueous solution is 9.95.[15] As a consequence, the formal charge of tacrine at physiological pH is most probably +1. Here we will test the effect of the two possible ionization states of tacrine on docking into AChE.

Issue `Molecules|Choose   Library`... to load the ligand file `input/ligand/TAH_1acj.mol2`.

Issue `Docking|Define FlexX Docking`... to start calculation (press *Apply and Dock!* Button).

Repeat operations for the positively charged tacrine (file `input/ligand/TAH_1acj+.mol2`). IMPORTANT: LeadIT automatically *protonates molecules as in aqueous solution* (pH=7). This option must be turned off, otherwise the positively charged form of tacrine will be neutralized.

A summary of these two docking runs is given in Table 22.3. The root mean square deviation (RMSD) to the crystallographic solution, as computed by LeadIT over the coordinates of non-hydrogen atoms, is higher than 2 Å for all poses of the neutral form of tacrine, meaning that no correct docking solution is found. The experimental binding of the positively charged tacrine is well predicted by the docking program

Table 22.3 Statistics for the re-docking of tacrine into AChE.

| Site | Ligand | Number of poses | | Top pose (best score) | | | Best pose (best RMSD) | | |
| | | Total | with RMSD<2Å | Rank | Score | RMSD (Å) | Rank | Score | RMSD (Å) |
| --- | --- | --- | --- | --- | --- | --- | --- | --- | --- |
| 1acj_TAH_site65 | TAH_1acj | 10 | 0 | 1 | −20.18 | 2.93 | 7 | −19.07 | 2.57 |
| 1acj_TAH_site65 | TAH_1acj+ | 10 | 8 | 1 | −21.34 | 0.59 | 10 | −20.05 | 0.58 |
| 1acj_TAH_site65_WAT | TAH_1acj+ | 10 | 10 | 1 | −25.50 | 0.72 | 3 | −24.43 | 0.43 |

(8 out of 10 docking solutions are correct), however, the pose with the lowest RMSD value does not get the highest docking score in the pose ensemble. The additional hydrogen atom of the positively charged nitrogen is engaged in a hydrogen bond with the backbone carbonyl group of the catalytic histidine (His-440-A).

Docking of Positively Charged Tacrine in AChE in Presence of Water

In 1ACJ PDB file, the tacrine binding site of AChE contains seven water molecules, establishing one or two hydrogen bonds with the protein. Two of them make additional hydrogen bonds with the exocyclic amine of tacrine. The presence of these two bridging waters in AChE can therefore facilitate the correct placement of tacrine in the protein active site. In the following docking calculation, hydrogen atoms of water molecules are not fully defined in order to limit bias in ligand placement.

Issue `Molecules|Prepare receptor…` to include the two water molecules in protein, and go to panel "`Resolve chemical ambiguities.`"

| | |
|---|---|
| `4. Resolve chemical ambiguities` | Residues:<br>     No actions required<br>Water:<br>     Select HOH-634-A and HOH-643-A<br>     Type: freely rotatable (blue disk)<br>Small Molecules:<br>     No actions required |
| `5. Confirm receptor` | Save Receptor Definition as:<br>1acj_TAHsite65_WAT |

Issue `Molecules|Export receptor…` to save protein file as `input/receptor/1acj_TAHsite65WAT.mol2`.

Issue `Molecules|Choose Library…` to load the ligand file input/ligand/TAH_1acj+.mol2.

Uncheck option "Protonate as in aqueous solution (recommended)."

Issue `Docking|Define FlexX Docking…` to start calculation (press *Apply and Dock!* Button).

A summary of this docking run is given in Table 22.3. All poses of positively charged tacrine are correct, with a RMSD to the experimental solution, as computed over the coordinates of non-hydrogen atoms, lower than 2 Å. All three intermolecular hydrogen bonds between tacrine and AChE observed in 1ACJ crystal structure, including the two water-mediated, are well predicted by the docking program. Note that consideration of water significantly improves the score of docking pose (~4 to 5 kcal/mol gain).

## Conclusions

The binding mode of tacrine to AChE is difficult to predict, because:

- The protein cavity size largely exceeds the ligand volume. In the current exercise this problem is bypassed, since docking is restrained from exploring cavity out of the AChE subsite occupied by tacrine.

- There is only one directional interaction between ligand and protein—a hydrogen bond between the protonated nitrogen of acridine moiety in tacrine and the backbone carbonyl group of histidine 440 (Figure 22.3).
- Two water molecules bridge the interactions between tacrine exocyclic amine and AChE (Figure 22.3).

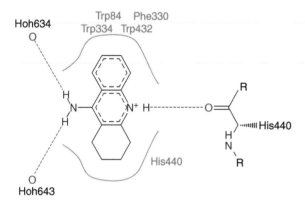

Figure 22.3 Experimental binding mode between tacrine and AChE (1ACJ PDB file). The picture was prepared using PoseView (BioSolveIt GmbH, Sankt Augustin, Germany).

Docking of tacrine into AChE fails when ligand hydrogen-bonding pattern is erroneous (neutral form), because shape complementarity cannot prevail over incorrect polar interactions. Proper anchoring of tacrine at the bottom of AChE pocket is observed, however, not systematically, when the correct ligand hydrogen-bonding pattern (positively charged form) is used. In addition, water molecules restrict the available space during ligand placement and provide two additional hydrogen bond donors at key positions in AChE binding site.

### Cross-Docking of Tacrine-Pyridone and Donepezil Into AChE

Starting from 1ACJ PDB file, which describes tacrine in complex with AChE, you will prepare the protein to execute docking of tacrine-pyridone and donepezil. The docking solutions are then compared to the structures of tacrine-pyridone (1ZGC PDB file) and donepezil (1EVE PDB file) co-crystallyzed with AChE.

### Preparation of AChE From 1ACJ PDB File

The original 1ACJ PDB file contains the crystal structure of AChE in complex with tacrine (HET code: TAH), 82 molecules of water. It describes a single copy of the complex. In general, the protocol for protein preparation, which was described in the section "Preparation of AChE From 1ACJ PDB File" - page 362, suits also the *cross-docking* experiments.

However, special attention should be paid to the AChE binding site, since the pocket defined around tacrine in 1ACJ PDB file is indeed too small to accommodate larger ligands.

Issue Molecules|Prepare receptor... to prepare protein, and follow the protocol detailed in the section "Preparation of AChE From 1ACJ PDB File" - page 362 (be careful to keep the two bridging water molecules). If you want to skip steps 2 and 4, launch `input/receptor/1acj_WAT.mol2` instead of `input/pdb/pdb1acj.ent`. Then go to the panel "`Define Binding Site`" and prepare AChE binding site as follows:

- for docking of tacrine-pyridine inhibitor

| | |
|---|---|
| `3. Define binding site` | Specify Binding Site using Reference Ligands |
| | 1) Choose a Reference Ligand |
| | from file `input/ligand/A2E_1zgc.mol2` |
| | 2) Include amino acids within 6.5 Å radius |
| `5. Confirm receptor` | Save Receptor Definition as: |
| | `1acj_A2Esite65WAT` |

- for docking of donepezil:

| | |
|---|---|
| `3. Define binding site` | Specify Binding Site using Reference Ligands |
| | 1) Choose a Reference Ligand |
| | from file `input/ligand/E20_1eve.mol2` |
| | 2) Include amino acids within 6.5 Å radius |
| `5. Confirm receptor` | Save Receptor Definition as: |
| | `1acj_E20site65WAT` |

Note that binding site definitions in `1acj_E20site65WAT` and `1acj_A2Esite65WAT` are the same. However, the appropriate reference ligands required for RMSD calculation between crystallographic and docked poses are different.

### Cross-Docking of Tacrine-Pyridone Inhibitor and Donepezil in AChE in Presence of Water

Issue `Molecules | Choose Library...` to load ligand file `input/ligand/A2E_1zgc.mol2`. IMPORTANT: LeadIT automatically *protonates molecules as in aqueous solution* (pH=7). This option must be turned off, otherwise the positively charged form of tacrine-pyridine inhibitor will be neutralized.

Issue `Docking | Define FlexX Docking...` to start calculation (press *Apply and Dock!* Button).

Repeat operations for donepezil (file `input/ligand/E20_1eve.mol2`).

A summary of these docking runs is given in Table 22.4. The RMSD to the X-ray solution, as computed over the coordinates of non-hydrogen atoms, is less than 2 Å for 3 poses of tacrine-pyridine inhibitor (A2E). Docking program was able to identify pose of tacrine-pyridine inhibitor which fits well into 1ACJ AChE binding site. Nevertheless, it differs from the bioactive (crystallographic) one, especially at the flexible alkyl chain part of the molecule. This is not surprising, since the presence of multiple rotatable bonds in the compound increases the number of degrees of freedom to be explored. Furthermore, the docking favored poses, where pyridone moiety of the inhibitor A2E forms hydrogen bonds to Ser286 and Phe288, is not in agreement with the experimentally determined binding mode (see Figure 22.4). Binding of tacrine-pyridine inhibitor causes conformational change of AChE protein, which is known as induced fit effect, and it requires other rotamer of TRP-279-A (as observed in 1ZGV PDB file).

Table 22.4 Statistics for docking of tacrine-pyridone inhibitor and donepezil into AChE structures 1ACJ and 1EVE.

| | | Number of poses | | Top pose (best score) | | | Best pose (best RMSD) | | |
| Site | ligand | Total | with RMSD<2Å | Rank | score | RMSD (Å) | Rank | Score | RMSD (Å) |
| --- | --- | --- | --- | --- | --- | --- | --- | --- | --- |
| 1acj_A2E_site65WAT | A2E_1zgv | 10 | 3 | 1 | −33.03 | 2.02 | 7 | −30.03 | 1.52 |
| 1acj_E20_site65WAT | E20_1eve+ | 10 | 0 | 1 | −13.75 | 9.37 | 5 | −12.97 | 3.98 |
| 1eve_E20_site65_WAT | E20_1eve+ | 10 | 1 | 1 | −19.86 | 2.66 | 4 | −19.16 | 1.43 |

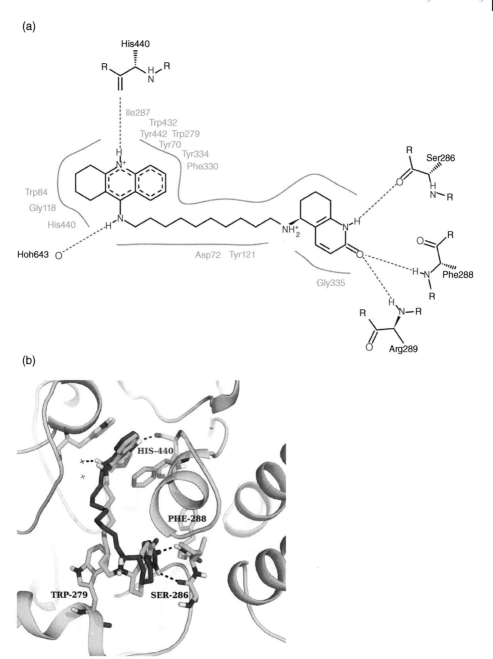

**Figure 22.4** (top) LeadIT pose view for the lowest RMSD pose of tacrine-pyridone inhibitor docked into 1ACJ structure; (bottom) comparison of crystallographic pose of tacrine-pyridone inhibitor (ligand from 1ZGV PDB file is shown in cyan) and the lowest-RMSD FlexX solution (cross-docking was done into 1ACJ structure of AChE, protein ribbon as well as the important residues of the binding pocket are shown in pale green while ligand is depicted in magenta).

The RMSD to the X-ray solution is higher than 3.98 Å for all poses of donepezil (E20), meaning that predicted and experimental binding modes of donepezil to AChE are very different.

Accurate docking prediction for donepezil into AChE structure, which was prepared from 1ACJ file, is out of reach because of inappropriate rotamer of a phenylalanine (Phe-330-A), whose side chain actually fills the space required for the ligand piperidine group.

### Re-Docking of Donepezil in AChE in Presence of Water

Difficulty of predicting donepezil binding mode is further addressed by a *re-docking* experiment. For that purpose, protein was prepared from 1EVE PDB file, where glyco-sylated AChE is complexed with donepezil (HET code: E20) in presence of 396 water molecules. The binding site was defined as a set of protein residues within the distance of 6.5 Å from donepezil ligand atoms. Positions of polar hydrogen atoms were set as it is described in Table 22.2.

Issue Molecules | Prepare receptor... to load the prepared protein from input/ receptor/1eve_ali_WAT.mol2 file. Go to the panel "Define Binding Site" and prepare AChE binding site by setting Reference Ligand from file input/ligand/ E20_1eve.mol2 and include amino acids within 6.5 Å radius. Then go to the panel "Confirm receptor" and save receptor definition as AChE binding site as 1eve_ E20site65WAT.

Issue Molecules | Choose Library... to load ligand file input/ligand/E20_1eve. mol2.

Issue Docking | Define FlexX Docking... to start calculation (press *Apply and Dock!* Button).

Docking results are summarized in Table 22.4 and indicate that *re-docking* of donepezil back into AChE binding site, which was extracted from 1EVE PDB entry, yields to wrong solutions (only one solution has RMSD value less than 2 Å, but it is not ranked as the best). Closer look into 1EVE PDB structure shows that donepezil fits to AChE binding site optimally in lock-and-key fashion. Despite this favorable geometric match, docking fails, most probably due to the fact that donepezil binding to AChE is driven by interaction with aromatic residues. No hydrogen or ionic bonds are observed between donepezil and AChE in the crystal structure of the complex (1EVE PDB file), although donepezil is positively charged and AChE binding site contains negatively charged residues (Figure 22.5).

## Conclusions

AChE is a flexible protein, which undergoes structural rearrangement upon ligand binding. *Cross-docking* experiments using 1ACJ PDB file fail for *tacrine-pyridone inhibitor and donepezil* because of inappropriate conformations of Phe300 and Trp279—two aromatic residues involved in allosteric control of the enzyme.[16] Proper docking of tacrine-pyridone inhibitor is hindered by using wrong AChE conformation. Docking of donepezil is inaccurate even when appropriate AChE structure is used, due to inca-pacity of docking algorithm to prioritize the true binding mode, which is characterized predominantly by aromatic interactions.

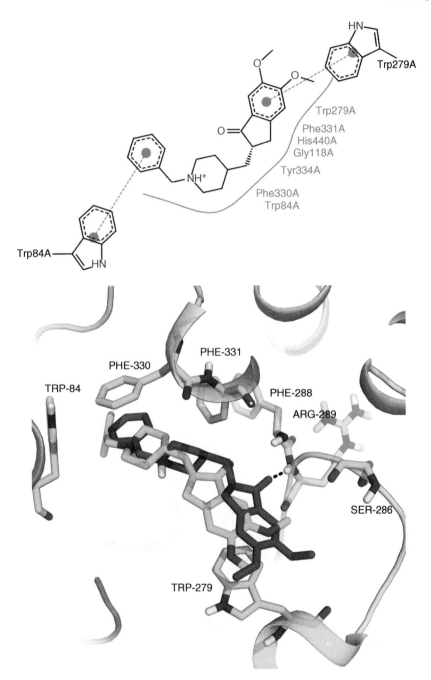

Figure 22.5 (top) experimental binding mode between donepezil (E20) and AChE (1EVE PDB file). (bottom) comparison of crystallographic pose of donepezil (1EVE PDB file, protein is depicted in pale green, ligand is shown in cyan) and the top-scored FleXX solution derived after re-docking experiment (ligand is depicted in magenta).

## General Conclusions

We have demonstrated that protein-ligand docking cannot be used as a black box, and requires a good knowledge of the studied system. Particular attention has to be paid to ligand and protein preparation. Modeler has also to be aware of biases that can be introduced by docking algorithm and scoring functions.

Docking of donepezil to AChE is especially challenging due to its specific binding mode, where hydrogen or ionic bonds between ligand and protein are absent, but at the same time charged and polar atoms are present in both structures. Tacrine and tacrine-pyridone inhibitor are easier to dock in AChE, provided they are correctly ionized. The accurate placement of their acridine substructure is driven by a hydrogen bond to the catalytic histidine of AChE.

Finally, ligand recognition by AChE has an induced fit mechanism. As a consequence the rigid docking approach is not successful when using inappropriate protein structure.

## Annex: Screen Captures of LeadIT Graphical Interface

1) Load protein.

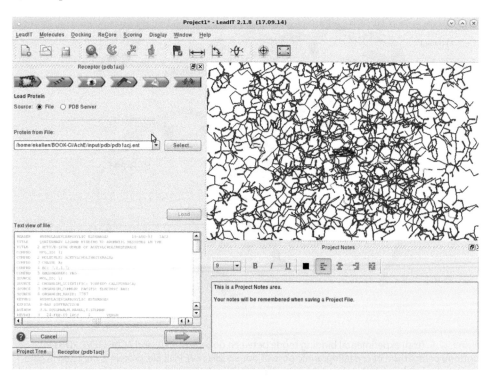

## 2) Choose receptor component.

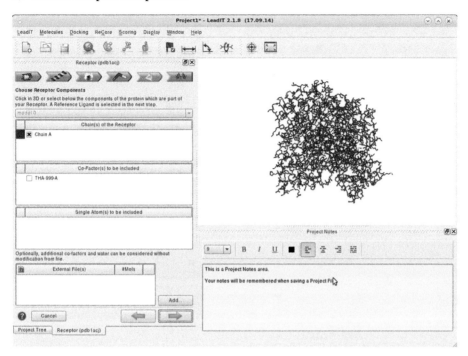

## 3) Define binding site

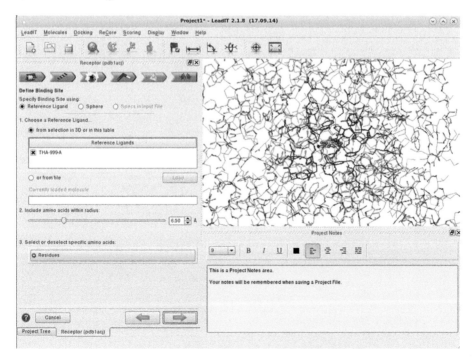

4) Resolve chemical ambiguities.

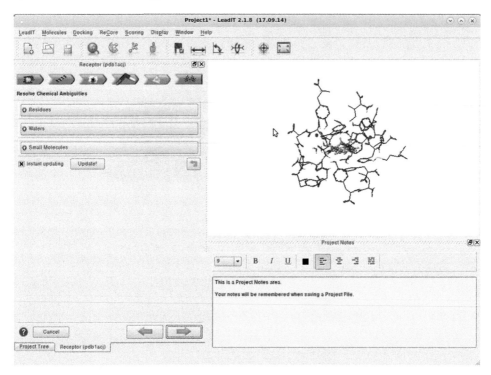

5) Confirm receptor.

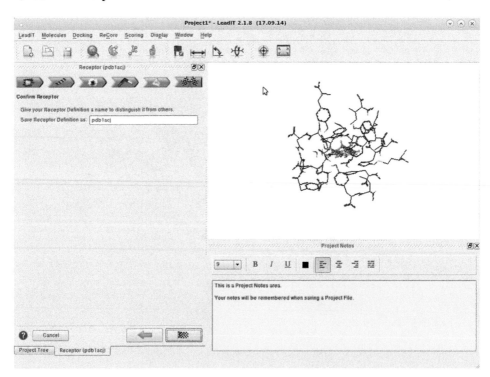

# References

1 Yuriev, E., Holien, J., and Ramsland, P.A. (2015). *J. Mol. Recognit.* doi:10.1002/jmr.2471.
2 Huang, S.Y., Grinter, S.Z., and Zou, X. (2010). *Phys. Chem. Chem. Phys.* 12, 12899.
3 Leach, A.R., Shoichet, B.K., and Peishoff, C.E. (2006). *J. Med. Chem.* 49, 5851–5855.
4 Lexa, K.W., and Carlson, H.A. (2012). *Q. Rev. Biophys.* 45, 301–343.
5 Ibrahim, T.M., Bauer, M.R., and Boeckler, F.M. (2015). *J. Cheminformatics* 7, 21.
6 Kellenberger, E., and Rognan, D. (2013). *In Data Mining in Drug Discovery*, R.D. Hoffmann, A. Gohier, and P. Pospisil, eds. (Weinheim, Germany: Wiley-VCH Verlag GmbH & Co. KGaA), pp. 1–24.
7 Harel, M., Schalk, I., Ehret-Sabatier, L., Bouet, F., Goeldner, M., Hirth, C., Axelsen, P.H., Silman, I., and Sussman, J.L. (1993). *Proc. Natl. Acad. Sci. U. S. A.* 90, 9031–9035.
8 Kryger, G., Silman, I., and Sussman, J.L. (1999). *Struct. Lond. Engl.* 1993 7, 297–307.
9 Sussman, J.L., Harel, M., Frolow, F., Oefner, C., Goldman, A., Toker, L., and Silman, I. (1991). *Science* 253, 872–879.
10 Huang, N., Shoichet, B.K., and Irwin, J.J. (2006). *J. Med. Chem.* 49, 6789–6801.
11 Therrien, E., Weill, N., Tomberg, A., Corbeil, C.R., Lee, D., and Moitessier, N. (2014). *J. Chem. Inf. Model.* 54, 3198–3210.
12 Moitessier, N., Englebienne, P., Lee, D., Lawandi, J., and Corbeil, C.R. (2009). *Br. J. Pharmacol.* 153, S7–S26.
13 Villoutreix, B.O., Renault, N., Lagorce, D., Sperandio, O., Montes, M., and Miteva, M.A. (2007). *Curr. Protein Pept. Sci.* 8, 381–411.
14 Kellenberger, E., and Dejaegere, A. (2011). Chapter 10. In *RSC Biomolecular Sciences*, A. Podjarny, A.P. Dejaegere, and B. Kieffer, eds. (Cambridge: *Royal Society of Chemistry*), pp. 300–346.
15 Perrin, D.D. (1972). *Dissociation constants of organic bases in aqueous solution - Supplement* 1972. London: Butterworths.
16 Rosenberry, T.L., Johnson, J.L., Cusack, B., Thomas, J.L., Emani, S., and Venkatasubban, K.S. (2005). *Chem. Biol. Interact.* 157–158, 181–189.

Part 9

Pharmacophorical Profiling Using Shape Analysis

23

# Pharmacophorical Profiling Using Shape Analysis

*Jérémy Desaphy, Guillaume Bret, Inna Slynko,*
*Didier Rognan, and Esther Kellenberger*

## Introduction

Over the previous decades, drug discovery efforts have focused on the design of selective drugs, assuming that targeting a key protein in a single biological process causes the beneficial therapeutic effect. Growing experimental evidences have recently shifted the single-target to multi-target paradigm, thereby boosting the development of computational approaches to identify all possible ligands for all possible targets. The new research field, called chemogenomics or *in silico* polypharmacology, has firstly proposed efficient ligand-centric methods for structure-activity data mining. Hence, empirical models developed from experimental binding properties of active low molecular weight compounds may suggest biological activities of a new compound. Protein-centric methods have then complemented the toolbox, thus allowing the consideration of orphan targets (i.e., proteins with no known ligands). However, their usage necessitates that the protein three-dimensional (3D) molecular structure has been determined experimentally or modelled by computing approaches.[1–5]

In this tutorial, we will work on the issue of ligand profiling and answer the question "Can we find secondary targets of a ligand whose primary target is known?" To that end, we will test two methods based on 3D-shape comparison:

- *A ligand-centric method: 3D similarity between ligands*

A conformational ensemble representing the ligand constitutes the reference, which is compared to the low energy structures of high affinity binders collected for a selection of therapeutically relevant proteins.

- *A protein-centric method: 3D similarity between protein binding sites*

The ligand-binding site in its specific protein constitutes the reference, which is compared to each entry of a dataset made of ligand-binding sites in therapeutically relevant proteins. The sites are defined from the crystal structure of ligand/protein complexes.

*Tutorials in Chemoinformatics*, First Edition. Edited by Alexandre Varnek.
© 2017 John Wiley & Sons Ltd. Published 2017 by John Wiley & Sons Ltd.
Companion website: www.wiley.com/go/varnek/chemoinformatics

The basic idea behind the ligand-centric method is that ligands with similar shapes are prone to bind to the same protein.[6] The protein-centric method assumes that two similar binding sites can accommodate the same ligand.[7] The general screening strategy is summarized in Figure 23.1.

## Description of the Example Case

### Aim and Context

In this tutorial we will challenge shape comparison methods to identify known and putative secondary targets of haloperidol (Figure 23.2). Haloperidol is a phenyl-piperidinyl-butyrophenone that is used primarily to treat schizophrenia and other psychoses. It is also used in schizoaffective disorder, delusional disorders, ballism, and tourette syndrome and occasionally as adjunctive therapy in mental retardation and the chorea of huntington disease. It is a potent antiemetic and is used in the treatment of intractable hiccups. Haloperidol have common adverse effects (>1% incidence) because it is not highly selective to its primary targets, which are dopamine receptors (D2, D1A and D3 subtypes, http://www.drugbank.ca/drugs/DB00502). For example, the interaction of the drug with the receptors of acethylcholine causes constipation. Hypotension consequent to adrenergic receptor blockade is another example of haloperidol promiscuous binding.

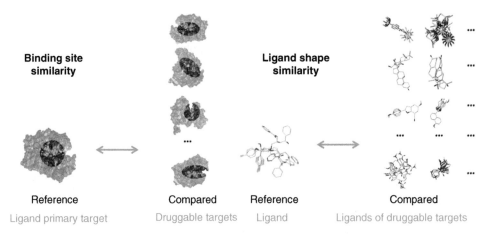

| Binding site similarity | | Ligand shape similarity | |
|---|---|---|---|
| Reference | Compared | Reference | Compared |
| Ligand primary target | Druggable targets | Ligand | Ligands of druggable targets |

Figure 23.1 Protein-centric (left) and ligand-centric (right) strategies for ligand profiling.

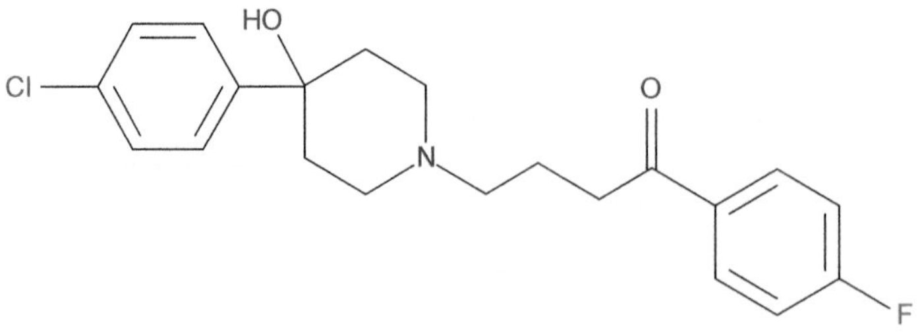

Figure 23.2 Chemical structure of haloperidol.

Haloperidol will be profiled against fifteen proteins that are present in the Protein Databank and represent a diverse set of therapeutical targets (Table 23.1).[8] Of note, the dopamine D3 receptor belongs to the class A of G-protein coupled receptors. The data set contains three other class A G-protein coupled receptors, one class B G-protein coupled receptor, two nuclear receptors, a chaperone, and eight enzymes. For each target, ten different ligands were collected from the sc-PDB (ligand co-crystallized with the protein) and from ChEMBL (drug targeting the protein, or ligand with $IC_{50} < 50$ nM).[9,10]

The three-dimensional structures of proteins were downloaded from sc-PDB (protein. mol2 files). The 3D–structures of ligands were either directly retrieved from sc-PDB (ligand.mol2 files) or built from 2D-structure extracted in ChEMBL using CORINA v3.40 (Molecular Network, Erlangen, Germany) after an ionization step using FILTER v4.40 (OpenEye Scientific Software, Santa Fe, NM, USA). All ligands were submitted to conformational sampling using OMEGA2 v2.4.6 (OpenEye Scientific Software, Santa Fe, NM, USA). Default parameters set the maximal number of conformers to 200.

The reference in the ligand-centric method is haloperidol 3D-structure, prepared as follows: the 2D structure was downloaded from ChEMBL, ionized using FILTER v4.40 (OpenEye Scientific Software, Santa Fe, NM, USA) and folded using CORINA v3.40 (Molecular Network, Erlangen, Germany). A total of 590 conformers were obtained using omega2 v2.4.6 with default settings except that the number of conformers was unlimited (OpenEye Scientific Software, Santa Fe, NM, USA). The reference in the protein-centric method is the crystal structure of the human dopamine D3 receptor as described in complex with eticlopride (PDB ID: 3pbl). The protein mol2 file was prepared as described in[10] and contains only protein residues, including hydrogen atoms.

## Methods

Ligand shape analysis will be performed using ROCS.[6] Site comparison will be performed using VolSite and Shaper.[11] A brief overview of the methods is given below.

ROCS is a fast ligand shape comparison application. It represents the molecular volume as atom-centered Gaussian functions. It maximizes the shared volume between the query molecule and the compared molecule by optimizing the overlap of their Gaussian functions (rigid body motions). Best solutions are ranked depending on the match of functions, using a *Tanimoto* score as defined in Equation 23.1:

$$\frac{O_{r,c}}{I_r + I_c - O_{r,c}} \tag{23.1}$$

where $I_r$ and $I_c$ are the self-volume overlaps of the reference and the compared molecules, respectively, and $O_{r,c}$ represents the overlap between the two molecules. The *ShapeTanimoto* score is computed considering volumes, the *ColorTanimoto* score

Table 23.1 Description of ligandable targets and their ligands.

| Functional class[a] | Target Name[a] | Organism | Database ID[b] Protein | Ligand |
|---|---|---|---|---|
| Class A G-protein coupled receptor: adrenoceptor | β1-adrenoceptor | turkey human | 2ycw 2rh1 | 2rh1,2ycw, atenolol, oxprenolol, penbutolol, protokylol, timolol, CHEMBL1788270, CHEMBL276659, CHEMBL51667 |
| Class A G-protein coupled receptor: acethylcholine receptors (muscarinic) | M2 receptor | human | 3uon | 3uon, CHEMBL10272, CHEMBL135645, CHEMBL1779036, CHEMBL194837, CHEMBL495531, procyclidine, scopolamine, tolterodine, trospium |
| Class A G-protein coupled receptor: Chemokine receptor | CCR5 | human | 4mbs | 4mbs, CHEMBL1178786, CHEMBL182940, CHEMBL207004, CHEMBL2178576, CHEMBL322251, CHEMBL322693, CHEMBL392659, CHEMBL481068, CHEMBL540366 |
| Class B G-protein coupled receptor: corticotropin-releasing factor receptor | CRF1 | human | 4k5y | 4k5y, CHEMBL115142, CHEMBL1819077, CHEMBL1939593, CHEMBL2087552, CHEMBL482950, CHEMBL484158, CHEMBL497653, CHEMBL525716, CHEMBL573978 |
| 3-Ketosteroid receptor | androgen receptor | mouse human | 2qpy 3b5r | 1e3g, 1gs4, 1t7r, 1z95, 2ax6, 2axa, 2hvc, 2pnu, 2qpy, 3b5r |
| Estrogen receptor | ER beta | human rat | 2fsz 2j7x | 1hj1, 1l2j, 1nde, 1qkn, 1u3r, 1u3s, 1x76, 1x78, 2fsz, 2j7x |
| heat shock protein | HSP90alpha | human | 3owd 4efu | 1yet, 2bz5, 2qf6, 2qg0, 2qg2, 2vcj, 2xhr, 3ekr, 3owd, 4efu |

| Enzyme class[a] | Enzyme | Organism | Identifier[b] | PDB identifiers |
|---|---|---|---|---|
| Carboxylic ester hydrolase | acetylcholinesterase | electric ray | 1zgc 3i6m | 1e66, 1eve (aricept), 1qon, 1zgc, 2gyw, 2ha6, 2xi4, 3i6m, 4arb, 4b7z |
| Protein-serine/threonine kinase | cyclin-dependent kinase 2 (CDK2) | human | 1gij 1w0x | 1di8, 1dm2, 1e9h, 1gij, 1ke8, 1oit, 1p2a, 1w0x, 2bts, 2c5x |
| Protein-serine/threonine kinase | aurora kinase | human | 2np8 2x81 | 2np8, 2x81, 3d14, 3dj5, 3lau, 3myg, 3o50, 3p9j, 3r21, 3unz |
| Aspartic endopeptidase | beta secretase | human | 2fdp 4djv | 2b8v, 2f3f, 2fdp, 2oah, 2q15, 2qu3, 2vij, 3exo, 3pi5, 4djv |
| Aspartic endopeptidase | Renin | human | 2g1o 3vye | 2bks, 2bkt, 2g1n, 2g1o, 2g1r, 2g1s, 2g1y, 2g20, 2g24, 3vye |
| Serine endopeptidase | thrombin | human | 3rlw 3sv2 | 3da9, 3p17, 3qwc, 3rlw, 3rml, 3shc, 3sv2, 3u98, 3utu, 4bah |
| Methyltransferase | thymidylate synthase | pneumocystis carinii | 1ci7 3uwl | 1axw, 1ci7, 1f28, 1f4g, 1itq, 2aaz, 2fto, 3uwl, 4fog, 4lrr |
| Carbon-oxygen lyase | carbonic anhydrase | human | 1a42 3mhm | 1a42, 2nnv, 3bet, 3dbu, 3f4x, 3ffp, 3k2f, 3m67, 3mhm, 3n0n |

a) As defined in ENZYME for enzymes (enzyme.expasy.org/), IUPAR-DB for non-enzymatic receptors (www.iuphar-db.org/).
b) Identifier in Protein Databank (www.rcsb.org/pdb), drug name or identifier in ChEMBL (www.ebi.ac.uk/chembl/).

considering pharmacophoric features (H-bond donor and acceptor, anion, cation, hydrophobe, and rings). The *TanimotoCombo* score is the sum of *ShapeTanimoto* and the *ColorTanimoto* scores.

ROCS is developed and distributed by OpenEye Scientific Software (www.eyes open.com).

### VolSite and Shaper

The program VolSite detects cavities in a protein as illustrated in Figure 23.3: *step1*, the protein or its binding site is placed into a cubic grid; *step2*, grid is pruned according to the protein atomic coordinates; *step3*, grid is further pruned in order to discard non-buried points (and optionally the points sitting too far from any ligand atom). Remaining grid points are colored according to the pharmacophoric properties of nearest protein atoms (hydrophobic, aromatic, H-bond acceptor, negative ionizable, H-bond acceptor/donor, H-bond donor, positive ionizable, null); *step4*, each ensemble of contiguous cells defines one cavity, adjacent cavities are merged. By default, the grid is centered on the ligand, the grid edge is 20 Å, and the grid resolution is 1.5 Å.

The cavity grid points generated for binding sites by VolSite are input for the 3D-alignment program Shaper. Shaper superimposes a query site to a compared site by maximizing the geometric overlap of corresponding pruned colored grids. Each grid point is represented with a Gaussian function. The compared cavity is 3D-aligned to a reference cavity by optimizing the overlap of their Gaussian functions (rigid body motions). Best solutions are ranked depending on the match of functions of overlaid grid points having the same pharmacophoric property, using a *refTversky* score as defined in Equation 23.2:

$$\frac{O_{r,c}}{0.95I_r + 0.05I_c} \tag{23.2}$$

where $O_{r,c}$ represents the overlap between the two cavities, $I_r$ the reference grid and $I_c$ the compared grid.

VolSite and Shaper are available upon request to Didier Rognan (rognan@unistra.fr). They are freely available for academic purposes. Shaper requires licensing for OEChem TK. OEChem TK is developed and distributed by OpenEye Scientific Software, Santa Fe, NM, USA (www.eyesopen.com).

### Other Programs for Shape Comparison

A non-exhaustive list of other available programs for shape comparison is given in Table 23.2.

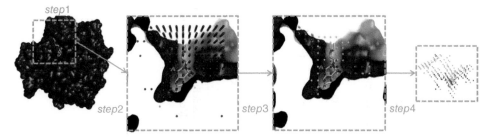

**Figure 23.3** Principle of cavity detection in VolSite.

Table 23.2 Programs for shape comparison.

| Program | Developer and access | |
|---|---|---|
| **Protein centric** | | |
| CavBase | Cambrige Crystallographic Data Center, Cambridge, UK (Hendlich et al., 2003)[12] | Module in Relibase+ |
| SiteEngine | University of The-Aviv, Israel[13]<br>The program is available from http://bioinfo3d.cs.tau.ac.il/SiteEngine/ | Free for non-commercial research and/or teaching purposes |
| SiteAlign | University of Strasboug, France[14]<br>The program is available from http://bioinfo-pharma.u-strasbg.fr | Free for non-commercial research and/or teaching purposes |
| CPASS | Web server at the University of Nebraska-Licoln, USA[15]<br>http://cpass.unl.edu/ | Free access for academic users |
| SOIPPA | University of California San Diego, CA, USA<br>The algorithm has been implemented in software SMAP, which can be download from http://www.mybiosoftware.com/ | SMAP-WS web service for protein site comparison is no more available [16,17] |
| **Ligand centric** | | |
| Forge, Blaze | Cresset, Cambridgeshire, United Kingdom http://www.cresset-group.com | Commercial program |
| HPCC | Distributed by Harmonic Pharma, Villers les Nancy, France, www.harmonicpharma.com/<br>Developed by LORIA, Nancy, France[18] | Commercial program |
| SQW | Merck Research Laboratories<br>See benchmarking in the references[19] | Non-commercial program |
| ElectroShape | Distributed by InhibOx Ltd, Oxford, UK http://www.inhibox.com/<br>developed by the University of Oxford, UK[20] | ElectroShape Polypharmacology server is available from http://ub.cbm.uam.es/chemogenomics/ |
| ShaEP | Developed and distributed by Åbo Akademi University, Finland[21] | Free<br>The program is available from http://users.abo.fi/mivainio/shaep |

## Description of Input Data Available on the Editor Website

The course material is stored into a compressed archive named *CI-TUTORIAL9.tar.gz.* You need to use the tar command under Unix like operating systems to open and extract the file.

| | |
|---|---|
| tar -zxf CI-TUTORIAL9.tar.gz | The archive from the file called CI-tutorial.tar.gz (-f) is uncopressed with gzip command (-z), extracted to disk in the current directory (-x). |

The path to input files, as based at the *CI-TUTORIAL9* directory, are detailed in Table 23.3. Scripts are also available to loop commands over the 15 targets. For the sake of pedagogy, output files are given too.

Table 23.3 Description of input and output files. See Table 23.1 for the list of Name and ID.

| File description | Path to file |
| --- | --- |
| **Input: Reference 3D protein structure** | |
| Crystal structure of dopamine receptor | *REF/D3receptor-3pdbl_protein.mol2* |
| Binding cavity in dopamine receptor | *REF/D3receptor-3pdbl_cavity6.mol2* |
| Ligand co-crystallized with dopamine receptor | *REF/D3receptor-3pdbl_ligand.mol2* |
| **Input: Reference 3D ligand structure** | |
| Lowest energy conformer of haloperidol | *REF/haloperidol.mol2* |
| 590 conformers of haloperidol | *REF/haloperidol_multiconf.mol2* |
| **Input: Compared 3D protein structures** | |
| 27 crystal structures of 15 proteins | *TARGET/${Name}-${ ID}_protein.mol2* |
| 27 binding cavities in 15 proteins | *CAVITY/${Name}-${ ID}_cavity6.mol2* |
| **Input: Compared 3D ligands** | |
| diastereoisomers of ligands | *LIGAND/${Name}-${ ID}_ligand.mol2* |
| Conformers of ligands (max. 200 conformers per ligand) | *LIGAND/${Name}-${ ID}_ ligandmulticonf.mol2* |
| **Execution: scripts** | |
| Define environment variable before running calculations | *EXE/environment.bash* |
| Compute cavities | *EXE/volsite.bash* |
| Screen by binding site similarity | *EXE/shaper.bash* |
| Align proteins based on the alignment of cavities | *EXE/align.bash* |
| Screen by ligand shape (single conformer for the query) | *EXE/rocs.bash* |
| Screen by ligand shape (multiple conformers for the query) | *EXE/rocsmulticonf.bash* |
| **Output: ligand shape analysis** | |
| Log file | *ROCS/${Name}-${ ID}_1.log* |
| Parameter | *ROCS/${Name}-${ ID}_1.param* |
| Job summary | *ROCS${Name}-${ ID}_1. status* |
| Overlay of reference and compared ligands | *ROCS/${Name}-${ ID}_1.mol2* |
| Ranked list of ligands | *ROCS/rocs_res.csv* |
| **Output: binding site comparison** | |
| Ranked list of target proteins | *SHAPER/Shaper_res.csv* |
| Compared cavity 3D-aligned to reference cavity | *SHAPER/Shaper_${Name}-${ ID}.pdb* |
| Compared protein (and ligand) 3D-aligned to the reference (using the transformation for cavity alignment) | *ALIGN/rot_${Name}-${ ID}_protein. mol2* <br> *ALIGN/rot_${Name}-${ ID}_ligand.mol2* |

# Exercises

The computational procedure given below is intended for **Linux operating systems. You will launch programs and run calculations using command lines in a terminal.** Before starting the exercises, you have to define the following environment variables:

- WORKDIR: the working directory, absolute path to the CI-TUTORIAL9 directory.
- OE_DIR and OE_LICENCE: path to openeye installation directory and to oe_license. txt file, respectively.
- ICHEM_LIC and ICHEM_LIB: path to ichem.lic license file and to ichem libraries, respectively.

We suggest you edit the *environment.bash* file, then to load the file into the current shell:

| | |
|---|---|
| `source environment.bash` | In linux terminal, define environment variables |

In the two exercises, basic calculations last for a few minutes using a single CPU. The advanced calculation (ROCS comparison using multiconformers for the query) is longer, about one hour.

Haloperidol is profiled by direct comparison of computed 3D-structures. Haloperidol is the reference, ten ligands are considered for each of the fifteen protein targets.

### How to Run ROCS to Compare Two Ligands

| | |
|---|---|
| `cd $WORKDIR` | In linux terminal, go to working directory. |
| `rocs -help simple` | Get the instructions to execute the program. |
| `rocs -query haloperidol.mol2`<br>`      -dbase LIGAND/CCR5-4mbs_ligand\`<br>`                      multiconf.mol2`<br>`      -prefix   CCR5-4mbs`<br>`      -oformat mol2`<br>`      -maxhits 1` | Compare two ligands.<br>Here the reference is the haloperidol, and the compared ligand is the ligand in complex with CCR5 (in the PDB file: 4mbs). The reference ligand is represented with the lowest energy conformer, and the compared ligand with a conformation ensemble. |

The run produces five files:

- *CCR5-4mbs_1.rpt* contains the alignment scores of the best conformer (as ranked by TanimotoCombo score)
- *CCR5-4mbs_1.status* gives a summary of the job
- *CCR5-4mbs_hits_1.mol2* contains the 3D-structure the reference liagnd and the 3D-structure of the compared ligand (best conformer) 3D-aligned onto the reference ligand
- *CCR5-4mbs.log* is the logfile
- *CCR5-4mbs.parm* details the job parameters.

### How to Perform Profiling Using ROCS

To screen the full data set, you have to iterate the command line over the 150 entries (10 ligands per target, 15 different targets). An example of script, *rocs.bash*, shows how to loop over the ligand filenames and build the summary result file *rocs_res.csv* from the 150.rpt files that have been generated.

Display *rocs_res.csv* and analyze results. The compared ligands are ranked by decreasing TanimotoCombo scores, which evaluate the 3D-similarity between two ligands depending on their capability to establish non-covalent bonds with protein. A TanimotoCombo score equal to two indicates that the two ligands share exactly the same shape and pharmacophore, whereas a TanimotoCombo score equal to zero denotes the absence of common features for protein binding.

| | |
|---|---|
| cd $WORKDIR | In linux terminal, go to working directory. |
| ./EXE/rocs.bash | Execute *rocs.bash* script. |
| cat rocs_res.csv | Display *rocs_res.csv*. |

The scores of two proteins have exceeded 1.0: β1-adrenoreceptor (with compared ligand protokylol) and acethylcholinesterase (with compared ligand aricept). Both are true targets of haloperidol. The 3D-structure of the two ligands 3D-aligned to the reference revealed common pharmacophoric features: conserved positively charged nitrogen in the center of molecule, directly surrounded by two hydrophobic groups, and terminal aromatic rings (Figure 23.4).

### Going Further...if You Don't Know the Active Structure of the Reference Ligand

Note that the screening was based on a single conformer for the reference and an ensemble of conformers for each of the compared ligands. You can compare these results to the results obtained using multiple conformers for the reference (*haloperidol_multiconf.mol2* file, -mcquery true option in ROCS), and conclude that the top ranked proteins are the same, yet an overall increase of scores is observed.

### Binding Site Comparison

Haloperidol is profiled by comparing its primary target, namely the D3 dopamine receptor, to 15 protein targets.

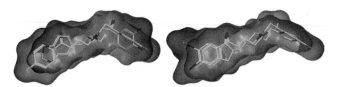

Figure 23.4 Best hits of the ligand-centric screening. On the left side, the ligand of β1-adrenoreceptor, protokylol (*adrenoreceptor-protokylol_hits_1.mol2*), is 3D-aligned with the reference (*haloperidol.mol2*, in cyan). On the right side, the ligand of acethylcholinesterase, aricept (*acethylcholine_1eve_hits_1.mol2*), is 3D-aligned with the reference (*haloperidol.mol2*, in cyan). Molecular shapes are outlined using Connolly surfaces. The figure was created using MOE v2014 (Chemical Computing Group, Montreal, Canada).

### How to Run Volsite to Generate Cavities for a Protein?

| | |
|---|---|
| `cd $WORKDIR` | In linux terminal, go to working directory. |
| `IChem` | Get the instructions to execute the program. |
| `IChem volsite`<br>`REF/D3receptor-3pbl_protein.mol2`<br>`REF/D3receptor-3pbl_ligand.mol2` | Compute the cavity around the bound ligand in the protein. Here exemplified on the reference binding site. |

Note that VolSite requires input files in mol2 format. By default, it ignores hydrogen atoms, water molecules, cofactors, or nucleic acids in the protein file. The ligand file is given to restrict the cavity definition around it.

### What are VolSite Output Files?

The run produces six files:

- The five *CAVITY_N1_X.mol2* files, where X can be *4, 6, 8, 12,* or *ALL*, contain cavity grid points. They differ by the number of points conserved after the last pruning step, when a point is removed if too far from the ligand atoms, if any, in the cavity ($>4\text{\AA}$, $>6\text{\AA}$, $>8\text{\AA}$, $>12\text{\AA}$, and no limits, respectively). The files can be visualized like standard molecular files.
- VolSite_Stat.csv gives statistics and druggability scores of cavities.

  To prepare cavities for the 27 structures of the fifteen targets, you have to iterate the command line (see an example in script *volsite.bash*). In practice, all cavities have been pre-computed. The course material only provides the cavity defined using the $6\text{\AA}$ radius.

### How to Run Shaper to Compare Two Protein Binding Sites?

| | |
|---|---|
| `cd $WORKDIR` | In linux terminal, go to working directory. |
| `Shaper` | Get the instructions to execute the program. |
| `Shaper -r REF/D3receptor-3pbl_cavity6.mol2`<br>`      -c CAVITY/CCR5-4mbs_cavity6.mol2`<br>`      -o Shaper_CCR5-4mbs.mol2`<br>`      -rn D3receptor`<br>`      -cn CCR5-4mbs` | Compare two cavities. Here exemplified on dopamine D3 receptor and CCR5. |

### What are VolSite Output Files?

The run produces two files:

- Shaper_CCR5-4mbs_cavity6.mol2 is the cavity grid points of the compared protein superimposed to the reference.
- Shape_res.csv reports statistics and alignment scores.

### How to Perform Profiling Using Shaper?

To screen the full data set, you have to iterate the command line over the 27 cavities (27 structures for the 15 targets). An example of script, *shaper.bash*, shows how to loop over the cavity filenames and build the summary result file *Shaper_res.csv* from the 27 *Shape_res.csv* files that have been generated.

Display *Shaper_res.csv* and analyze results. The compared cavities are ranked by decreasing ColorRefTversky scores, which evaluate the 3D-similarity between two cavities depending on their capability to establish non-covalent bonds with ligands. A ColorRefTversky above 0.45 is highly significant, since this value represent three times the standard deviation added to the mean score of the normal distribution obtained for the comparisons of the reference binding site to the 9 427 entries of sc-PDB, release 2013.

The refTversky of the top-ranked protein, β1-adrenoreceptor (2ycw), is equal to 0.47. The literature supports the cross binding of the ligands of the dopamine receptors to adrenergic receptors. The refTversky score of the next protein in the list, acetylcholinesterase (3i6m, 1zgc), is equal 0.43, just under the scoring threshold. Haloperidol is known to bind receptors of acetylcholine. Here data suggest that an enzyme whose substrate is acetylcholine can also be a target of haloperidol.

### Going Further…Observe Aligned Binding Sites

| | |
|---|---|
| `cd $WORKDIR` | In linux terminal, go to working directory. |
| `mkdir ALIGN`<br>`cd ALIGN` | Create a directory, go to the ALIGN directory. |
| `RotaMole`<br>`    -r SHAPER/Shaper-adrenoreceptor\`<br>`            -2ycw_cavity6.pdb`<br>`    -c CAVITY/adenoreceptor\`<br>`            -2ycw_cavity6.mol2`<br>`    -a TARGET/adenoreceptor\`<br>`            -2ycw_protein.mol2`<br>`    LIGAND/adenoreceptor-`<br>`            2ycw_ligand.mol2` | Create in the current directory aligned files:<br><br>`rot_adenoreceptor-2ycw_pro-`<br>`tein.mol2`<br>`rot_adenoreceptor-2ycw_`<br>`ligand.mol2` |

The 3D-structure of the β1-adrenoreceptor and acethylcholinesterase 3D-aligned to the dopamine receptor D3 reference revealed common features, in particular the presence of an aspartic acid facing a tyrosine, and four other conserved aromatic residues (phenylalanine or tryptophane). These residues defined a binding site that well accommodates the pharmacophore defined from haloperidol, protokylol, and aricept (see exercice "Ligand shape analysis" – page 387) (Figure 23.5).

## Conclusions

We have demonstrated that 3D computing methods are suitable to the target profiling of haloperidol. Similarity between protein binding sites and similarity between ligands yielded the same results and identified two potential secondary targets that are likely to be true positives, as supported by the experimental evidences. Noteworthy, in the ligand-centric approach only one of the ten ligands selected for the hit targets matched the shape of the query (haloperidol). The protein-centric method showed dependency to the definition of cavity in a protein, with only one of the two copies of β1-adrenoreceptor found similar to the query cavity (dopamine D3 receptor). Hence, both the ligand-centric and the protein-centric approaches bias the screening results toward the used reference, and are therefore complementary to achieve reliable predictions.

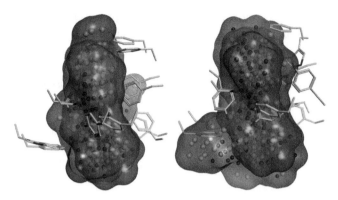

Figure 23.5 Best hits of the protein-centric screening. On the left side, β1-adrenoreceptor (*rot_adenoreceptor-2ycw_protein.mol2* and *Shaper_adenoreceptor-2ycw_cavity6.mol2*, in grey), is 3D-aligned with the reference (*D3receptor-3pdb_protein.mol2* and *D3receptor-3pdb_cavity6.mol2*, in cyan). On the right side, the ligand of acethylcholinesterase (*rot_acethylcholine_1zgc_protein.mol2* and *Shaper_acethylcholine_1zgc_cavity6.mol2*, in grey) is 3D-aligned with the reference (*D3receptor-3pdb_protein.mol2* and *D3receptor-3pdb_cavity6.mol2*, in cyan). Cavities are represented with color points and delimited by surface. Key side chains in protein site are shown using capped sticks. The figure was created using MOE v2014 (Chemical Computing Group, Montreal, Canada).

## References

1 Koutsoukas, A., Simms, B., Kirchmair, J., Bond, P.J., Whitmore, A.V., Zimmer, S., Young, M.P., Jenkins, J.L., Glick, M., Glen, R.C., et al. (2011). *J. Proteomics 74*, 2554–2574.

2 Meslamani, J., Bhajun, R., Martz, F., and Rognan, D. (2013). *J. Chem. Inf. Model. 53*, 2322–2333.

3 Peters, J.U. (2013). *J. Med. Chem. 56*, 8955–8971.

4 Reddy, A.S., and Zhang, S. (2013). *Expert Rev. Clin. Pharmacol. 6*, 41–47.

5 Rognan, D. (2007). *Br J Pharmacol 152*, 38–52.

6 Hawkins, P.C.D., Skillman, A.G., and Nicholls, A. (2006). *J. Med. Chem. 50*, 74–82.

7 Vulpetti, A., Kalliokoski, T., and Milletti, F. (2012). *Future Med. Chem. 4*, 1971–1979.

8 Rose, P.W., Prli, A., Bi, C., Bluhm, W.F., Christie, C.H., Dutta, S., Green, R.K., Goodsell, D.S., Westbrook, J.D., Woo, J., et al. (2015). *Nucleic Acids Res. 43*, D345–D356.

9 Davies, M., Nowotka, M., Papadatos, G., Dedman, N., Gaulton, A., Atkinson, F., Bellis, L., and Overington, J.P. (2015). *Nucleic Acids Res. 43*, W612–W620.

10 Desaphy, J., Bret, G., Rognan, D., and Kellenberger, E. (2015). sc-PDB: *a 3D-database of ligandable binding sites--10 years on. Nucleic Acids Res. 43*, D399–D404.

11 Desaphy, J., Azdimousa, K., Kellenberger, E., and Rognan, D. (2012). *J. Chem. Inf. Model. 52*, 2287–2299.

12 Hendlich, M., Bergner, A., Günther, J., and Klebe, G. (2003). *J. Mol. Biol. 326*, 607–620.

13 Shulman-Peleg, A., Nussinov, R., and Wolfson, H.J. (2005). *Nucleic Acids Res. 33*, W337–W341.

14 Schalon, C., Surgand, J.-S., Kellenberger, E., and Rognan, D. (2008). *Proteins 71*, 1755–1778.

15 Powers, R., Copeland, J.C., Stark, J.L., Caprez, A., Guru, A., and Swanson, D. (2011). Searching the protein structure database for ligand-binding site similarities using CPASS *v.2. BMC Res. Notes 4*, 17.

**16** Ren, J., Xie, L., Li, W.W., and Bourne, P.E. (2010). *Nucleic Acids Res. 38*, W441–W444.

**17** Xie, L., and Bourne, P.E. (2008). *Proc. Natl. Acad. Sci. U. S. A. 105*, 5441–5446.

**18** Karaboga, A.S., Petronin, F., Marchetti, G., Souchet, M., and Maigret, B. (2013). *J. Mol. Graph. Model. 41*, 20–30.

**19** McGaughey, G.B., Sheridan, R.P., Bayly, C.I., Culberson, J.C., Kreatsoulas, C., Lindsley, S., Maiorov, V., Truchon, J.-F., and Cornell, W.D. (2007). *J. Chem. Inf. Model. 47*, 1504–1519.

**20** Armstrong, M.S., Morris, G.M., Finn, P.W., Sharma, R., Moretti, L., Cooper, R.I., and Richards, W.G. (2010). *J. Comput. Aided Mol. Des. 24*, 789–801.

**21** Vainio, M.J., Puranen, J.S., and Johnson, M.S. (2009). *J. Chem. Inf. Model. 49*, 492–502.

Part 10

Algorithmic Chemoinformatics

# 24

## Algorithmic Chemoinformatics

*Martin Vogt, Antonio de la Vega de Leon, and Jürgen Bajorath*

## Introduction

This chapter focuses on the implementation of basic chemoinformatic tasks and fundamental algorithms in a high-level programming language. The methods described and implemented herein are typically already integrated in chemoinformatics software tools like the Molecular Operating Environment (MOE),[1] Pipeline Pilot,[2] and KNIME[3] or in chemoinformatics toolkits like CDK,[4] Daylight,[5] ChemAxon,[6] Openbabel,[7] OpenEye,[8] or RDKit.[9] The purpose of the tutorials is not to provide a replacement for these implementations. Instead, the descriptions and code presented herein are designed to enable a detailed understanding of the methods and gain insights into the process how theoretical descriptions of methods and algorithms can be implemented "from scratch" in a general purpose programming language.

This chapter is separated into four sections. The first section focuses on writing a virtual screening program in Python. A similarity search program using simple data fusion techniques will be realized that is based on fingerprint representations of data sets of molecules. This part does not use specific chemoinformatics or statistics toolkits and will be implemented in "pure" Python without the aid of additional packages, except for the initial generation of fingerprints.

Sections 'Canonical SMILES: The Canon Algorithm', 'Substructure Searching: The Ullmann Algorithm', and 'Atom Environment Fingerprints' are oriented toward the implementation of fundamental chemoinformatic algorithms and rely heavily on chemoinformatics toolkits for the handling and manipulation of molecules. The first two of these describe the implementations of two fundamental and essential tasks of algorithmic chemoinformatics:

1) The generation of unique representation of molecules by creating canonical SMILES,
2) Substructure searching, that is, the identification of query substructures in a molecule.

The final section describes the generation of fingerprints based upon atom environments.

In this chapter, all code will be implemented in Python 2.7 (http://www.python.org). Additional use will be made of the libraries NumPy 1.9.2 (http://www.numpy.org) and matplotlib 1.4.3 (http://www.matplotlib.org) for visualization. These libraries can be

*Tutorials in Chemoinformatics*, First Edition. Edited by Alexandre Varnek.
© 2017 John Wiley & Sons Ltd. Published 2017 by John Wiley & Sons Ltd.
Companion website: www.wiley.com/go/varnek/chemoinformatics

installed from the Python Package Index (http://pypi.python.org) using the package manager pip. The second part will make heavy use of RDKit (release 2015.03.1) [9], an open-source chemoinformatics toolkit written in C++ with a Python interface. RDKit can be downloaded and installed from the webpage http://www.rdkit.org.

## Similarity Searching Using Data Fusion Techniques

*Goal:* Development of a virtual screening program utilizing fingerprint-based similarity searching.
  *Software:* Python 2.7., matplotlib 1.4.3, RDKit (release 2015.03.1)
  *Code:*

- `generateFingerprints.py`. Calculation of different fingerprints from SD or SMILES files.
- `vs.py`. Similarity searching with multiple reference compounds using fingerprints.
- `vsBenchmark.py`. Performance evaluation of the virtual screening program.
- `analyseResults.py`. Visual analysis of benchmarking results.

  *Data:* Two sets, `zinc-subset10000.smi` and `zinc-subset100000.smi` of 10000 and 100000 random compounds were extracted from the "all clean" subset of the ZINC12[10] database to be used as screening libraries. From ChEMBL20[11] three data sets with compounds having annotated activity for human serotonin 5-HT$_{1A}$ (5-HT1a) receptor (`5ht1a.smi`), dopamine D4 receptor (`d4.smi`), and norepinephrine transporter (`net.smi`) were extracted. Only high confidence data with annotated K$_i$ binding affinities were considered. From the resulting data sets of 1532, 563, and 1040 compounds, respectively, subsets of 100 compounds were selected using a diverse subset selection protocol of MOE.[12]

## Introduction to Virtual Screening

The aim of virtual screening (VS) is to identify subsets of compounds from a database that are enriched with molecules active against a particular target, that is, a receptor protein or an enzyme, using computational techniques. The enrichment is accomplished by scoring (or classifying) compounds stored in a large database and taking a subset (e.g., 1% of the database) of top-ranked compounds. A virtual screening exercise is considered successful if active compounds are significantly enriched in high rank positions compared to random selection. Virtual screening methods can be categorized as either target-based or ligand-based.[13] In target-based methods knowledge of the protein structure and binding site is required and molecules are scored by how well they fit into the binding site using docking algorithms. Ligand-based virtual screening (LBVS) utilizes a set of reference molecules that are known to be active against a particular target in order to score database compounds. One of the most basic LBVS methods is similarity searching, which ranks compounds according to their similarity to known ligands.

The idea of similarity searching can be summarized considering the well-known similarity-property principle[14] stating that structurally similar molecules tend to have similar biological properties. In LBVS, similarity is assessed on the basis of molecular representations. For the virtual screening of large databases and the fast evaluation of molecular similarity, molecules are frequently represented by fingerprints rather than by their structural graphs.

## The Three Pillars of Virtual Screening

LBVS methods can be separated into three separate parts:

### Molecular Representation

Fingerprints are common representations of molecules for virtual screening. They enumerate specific features of a molecule. The design of fingerprints varies considerably, ranging from small fixed-length fingerprints to very large representations enumerating combinatorial features. Fingerprints can roughly be classified as substructure-based fingerprints reflecting a pre-defined dictionary of substructure fragments, pharmacophore fingerprints (enumerating pharmacophore patterns), or combinatorial fingerprints (systematically enumerating bond paths or atom environments). However, there is generally no preferred or "optimal" fingerprint for LBVS applications. For different compound classes, the performance of alternative fingerprints generally varies and it is essentially impossible to predict which fingerprint will perform best in a virtual screen for compounds with a specific activity.[15]

### Similarity Function

A similarity function assigns a similarity score to a pair of molecular fingerprints. It accounts for the distance in the chemical space representation defined by the fingerprint. The most popular one is the Tanimoto coefficient (Tc), which can be understood as the percentage of feature identity of two molecules. However, other similarity metrics based on algebraic concepts such as the Euclidean distance or cosine coefficient are also used for similarity searching.[16]

### Search Strategy (Data Fusion)

Search strategies become important if more than one reference compound is available (which is typically the case). Then there will be a number of similarity scores for each database compound. These scores need to be transformed into a single one, for instance, by determining an average or by considering only the highest score.[17]

## Fingerprints

Fingerprints are bit string representations of chemical structures and properties. A fingerprint typically records the presence or absence of features in a molecule. As already mentioned above, alternative fingerprint designs are, for example, based upon a fixed set of well-defined substructural elements like MACCS,[18] an exhaustive encoding of atom environments (or paths) like extended connectivity fingerprints (ECFP),[19] or a combinatorial enumeration of pharmacophore features.[20] As a bit string, a fingerprint consists of a sequence of ones and zeros where a one or zero in a specific position indicates the presence or absence of a particular feature, respectively. Figure 24.1 illustrates different fingerprint concepts.

### Count Fingerprints

A natural extension of binary fingerprints, especially those encoding structural or pharmacophore features, is the use of feature count fingerprints. A count fingerprint not

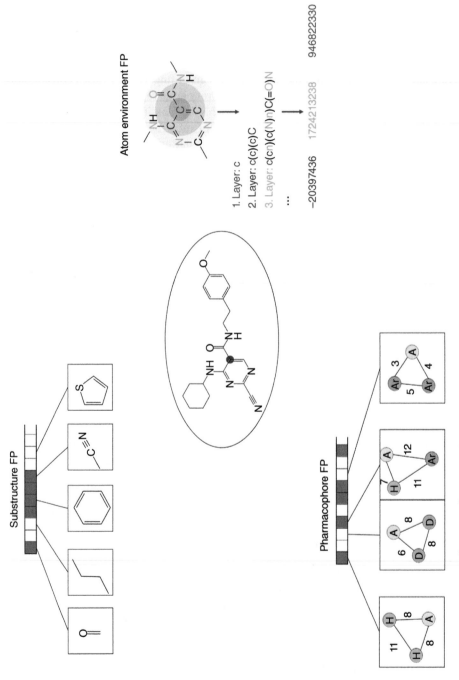

Figure 24.1 Concepts of common fingerprint designs.

only encodes presence or absence of a feature but also records how often a feature occurs. Consequently, a count fingerprint is not thought of as a bit string but rather as a high-dimensional vector of integer numbers.

### Fingerprint Representations

Depending on the type of fingerprint, there are two popular ways how binary fingerprints are represented in files, databases, or computer memory:

### Bit Strings

Fingerprints are represented as sequences of ones and zeros. This is sensible if fingerprints have a well-defined and fixed size like MACCS keys and where a relatively large portion of features is expected to be present.

### Feature Lists

A more general approach is to store fingerprints as feature lists. That is, only the positions of those features that are non-zero are stored. For example, instead of a string 1101001000100001000001 the list 1,2,4,7,11,16,22 is stored. If only a few features are present feature lists are more efficient than bit strings. Furthermore, feature lists are the only feasible alternative for large fingerprints like ECFP,[19] which rely on hashing to generate a theoretical total of more than 4 billion different features. Such fingerprints do not have a fixed format but generate molecule-specific feature sets of varying size. An additional advantage is that this approach can be easily extended to count fingerprints by storing a feature multiple times if it occurs more than once in the molecule. Alternatively, dictionaries or maps can be used for count fingerprints that associate each feature (key) with the number of times (value) it occurs.

For the purpose of our VS program, feature lists are more flexible and will be used for similarity searching, rendering the code applicable to both binary and count fingerprint representations.

### Generation of Fingerprints

*Code:* generateFingerprints.py

In order to perform similarity searching, fingerprint representations of the molecules have to be generated first. Chemoinformatics programs and toolkits usually provide the ability to generate a variety of different fingerprints. The following script can be used to generate text files containing fingerprints necessary for similarity searching from SD files or files containing SMILES. Here the RDKit toolkit has been used.

```
import numpy as np
from rdkit.Chem import MACCSkeys
from rdkit.Chem import AllChem as Chem
import sys

assert len(sys.argv)==5 and sys.argv[1][0] in "BCbc", \
        "Usage: python generateFingerprints.py" \
        "binary|count morgan|ap|maccs < inputFile ><outputFile>"
```

The exemplary call

```
python generateFingerprints.py count morgan input.sdf output.txt
```

will calculate atom environment count fingerprints for the molecules of input.sdf and save the result to output.txt.

```python
def maccs(mol):
    maccs = MACCSkeys.GenMACCSKeys(mol)
    maccsFL = sorted(list(maccs.GetOnBits()))
    return maccsFL
def morgan(mol):
    fp = Chem.GetMorganFingerprint(mol, 2)
    return fp.GetNonzeroElements()
def atom_pair(mol):
    fp = Chem.GetAtomPairFingerprint(mol, 2)
    return fp.GetNonzeroElements()
fp_fct = {'MACCS': maccs, 'MORGAN': morgan, 'AP': atom_pair }
fp_name = sys.argv[2].upper()
fingerprint = fp_fct[fp_name]
```

Functions for generating MACCS, atom environment fingerprints with radius 2, and atom pair fingerprints with minimum bond distance 2 are defined. The functions either return a list of features (MACCS) or a dictionary where keys are the features and values are the counts. fingerprint will refer to the function as specified in the command line.

```python
def to_count_list(fp_dict):
    fp_list = []
    for key,count in fp_dict.items():
        fp_list.extend([key]*count)
    return sorted(fp_list)

def to_binary_list(fp):
    return sorted(fp)

if sys.argv[1][0] in "Cc":
    to_list = to_count_list
    fp_name += "_COUNT"
else:
    to_list = to_binary_list
```

to_list will convert the fingerprint either to a feature list with unique features or with repeated features depending on whether binary or count fingerprints are generated. Count fingerprints cannot be generated for MACCS.

```python
inputFile = sys.argv[3]
outputFile = sys.argv[4]
if inputFile.lower().endswith('sdf'):
    suppl = Chem.SDMolSupplier(inputFile)
else:
    suppl = Chem.SmilesMolSupplier(inputFile)
```

The input file is either an SD file or a text file containing SMILES.

```python
data = []
for mol in suppl:
    if mol is None: continue
    cid = mol.GetProp('_Name')
    fp_data = fingerprint(mol)
    fp_list = to_list(fp_data)
    fp = ",".join(map(str, fp_list))
    data.append([cid, fp])
```

The fingerprints are generated by iterating over all molecules in the input file and are saved to a list of tuples.

```
data = np.asarray(data)
header = ["NAME", fp_name]
np.savetxt(outputFile, data, fmt = "%s", comments = "",
header = "\t".join(header), delimiter = "\t")
```

NumPy provides functions for the formatted output of the data.

For the task of similarity searching, the molecular structures are not required but only the calculated fingerprints. Thus, the fingerprints are stored in a separate file together with an identifier for each molecule. The file contains a header line and tab separated values:

```
NAME    SAMPLE_FP
mol1    1,2,3,4,5
mol2    4,6,7,8,14
mol3    1,6,12,16,74,124
```

The first line is a header. It should be skipped while reading the file. Each following line contains an identifier and the fingerprint of one molecule. The identifier is separated from the fingerprint by a tabulation character. The fingerprint is represented as a sequence of integer numbers that are separated by commas.

The virtual screening program will need to read fingerprint files. To this end the functions `iterate_fp_from_file` and `read_fp_from_file` are defined. An auxiliary function `read_fp_line` converts a line from a fingerprint file into a tuple containing the name and the fingerprint as a list of integers:

```
def read_fp_line(line, idcol = 0, fpcol = 1):
    columns = line.strip("\r\n").split('\t')
    fp = columns[fpcol]
    fp_list = map(int, fp.split(',')) if len(fp) > 0 else []
    return columns[idcol], fp_list
```

Read a single line from a fingerprint file. Return a tuple containing the identifier and list of integer fingerprint features.

```
def iterate_fp_from_file(filename):
    with open(filename) as fin:
        fin.readline() # skip first line
        for line in fin:
            yield read_fp_line(line)
```

`iterate_fp_from_file` is a so-called generator function. It contains a `yield` statement that will suspend the execution of the function and return the supplied value when the function is used as an iterable object. Typical usage of the function is demonstrated in `read_fp_from_file` in the context of a list comprehension.

```
def read_fp_from_file(filename):
    return [fp for fp in iterate_fp_from_file(filename)]
```

All fingerprints in the file `filename` are returned in a list of tuples containing an identifier and a feature list.

Note, that the function `read_fp_from_file` reads in the file of fingerprints at once and returns its whole contents as a list. Depending on the size of the database file this can have a large impact on the memory requirements. However, holding the fingerprints in memory will be useful for the reference fingerprints because for each database fingerprint its similarity to all reference compounds will have to be determined. The function `iterate_fp_from_file`, however, only supplies a single fingerprint at a time and is able to handle arbitrarily large files. Given that database files for screening might be very large and each fingerprint is only needed once to determine its score, using generators is much more efficient.

## Similarity Metrics

In LBVS, the similarity of molecules is used as an indicator as to whether molecules have similar biological activities. For fingerprints, similarity is quantified using a similarity coefficient. There are many possible ways of defining similarity for fingerprints. However, the different coefficients can very often be defined using just three quantities: The number of features present in the first and the second fingerprint and the number of features these fingerprints have in common. Formally, binary and count fingerprints can be defined by a vector whose dimension corresponds to the size of the fingerprint and whose coefficients reflect either the presence or absence of a feature, that is, either are 1 or 0, or whose coefficients reflect the count of each feature, respectively.

Assume two fingerprints are given as vectors: $\mathbf{a} = (a_1, a_2, \ldots, a_n)$ and $\mathbf{b} = (b_1, b_2, \ldots, b_n)$ with $a_i, b_i \in \{0,1\}$ for binary fingerprints and $a_i, b_i \in \mathbb{N}_0$ for count fingerprints, respectively.

Then the following values can be defined:

$$a = \sum_{i=1}^{n} a_i$$

$$b = \sum_{i=1}^{n} b_i$$

$$c = \sum_{i=1}^{n} \min(a_i, b_i)$$

Note, that for binary values $a$ can be interpreted simply as the number of features present in the first molecule, $b$ as the number of features in the second, and $c$ the number of features present in both.

For chemical similarity searching, the Tanimoto coefficient (Tc) is the most popular similarity measure.[16] The Tc coefficient can be defined as the ratio of the number of features two fingerprints have in common to the number of features occurring in total in one or the other fingerprint. It can be intuitively rationalized as the percentage of features that two fingerprints share. This still holds true if the formalism is extended to count fingerprints because $\min(a_i, b_i)$ will account for the common multiplicity of a feature in both molecules accordingly.

Mathematically, the binary form of the Tanimoto coefficient Tc($\mathbf{a}, \mathbf{b}$) can be expressed as:

$$\mathrm{Tc}(\mathbf{a,b}) = \frac{c}{a+b-c}$$

Besides the Tc, other coefficients like the Dice coefficient or the Cosine coefficient[17] can also be defined easily in terms of $a$, $b$, $c$:

$$\mathrm{Dice}(\mathbf{a,b}) = \frac{2c}{a+b}$$

$$\mathrm{Cosine}(\mathbf{a,b}) = \frac{c}{\sqrt{ab}}$$

This tutorial will exclusively use Tc, the gold standard in the chemoinformatics field. However, the programs are easily adapted for the use of other coefficients.

Given that fingerprints are represented in Python as lists of features, the quantities $a$ and $b$ are simply given by the length of the lists. Thus, the central task in determining a similarity coefficient is to determine the number $c$, that is, the number of features two fingerprints have in common. For binary features it would be possible to use Python's set operations:

$$c = \mathrm{len}(\mathrm{set}(fp\_a).\mathrm{intersection}(fp\_b))$$

for two fingerprints fp_a and fp_b. However, this approach will not work for count fingerprints where a single feature can occur multiple times in a list. The following function for determining $c$ makes use of the fact that the features are stored in numerically sorted order.

```
def common_features(fp_a, fp_b):
    i = 0
    j = 0
    common = 0
    while i < len(fp_a) and j < len(fp_b):
        if fp_a[i] < fp_b[j]:
            i += 1
        elif fp_a[i] > fp_b[j]:
            j += 1
        else:
            common += 1
            i += 1
            j += 1
    return common
```

Two indices i and j are used to iterate over the features of fp_a and fp_b. When looking at the features at positions i and j there are two possibilities. Either one of the feature numbers is less than the other or the features are the same. In the first case, the smaller feature number only occurs in one fingerprint and its index is incremented. In the second case, both indices are incremented and the number of common features is increased.

*Note:* This method usually presents a very fast way for determining the number of common features. However, in Python using set operations is usually much faster due to the fact the Python is an interpreted language and explicit loop structures like while or for are usually slow compared to single operations that operate on a whole collection of data objects like set intersection.

Using this function, the determination of the Tanimoto coefficient (and other similarity metrics) is straightforward.

```
def tanimoto_coefficient(fp_a, fp_b):
    a = len(fp_a)
    b = len(fp_b)
    c = common_features(fp_a,fp_b)
    if c > 0:
        return float(c)/(a+b-c)
    else:
        return 0.0
```

*Note*: The function will return 0 in the case when there are no features present in either fingerprint. Numerically, the Tc value is not defined in this case and the choice to return 0 is arbitrary.

## Search Strategy

The search strategy defines how a database compound is scored given multiple reference compounds. Frequently, in LBVS more than one active reference compound is known and the similarities of database compounds to all of these should be taken into account. If there is only a single reference compound known there is not much of a search strategy once one has settled for a molecular representation and a similarity function: the database compounds are ranked according to their similarity to the single reference compound. If, instead, there are multiple reference compounds different strategies become available. Assuming there are $n$ reference compounds, first, the similarities of a database compound to each of these reference compounds can be calculated yielding $n$ different scores. These scores need to be converted into a single score from which to determine the final ranking. In principle, two different strategies might be considered. One would be to first rank the database compounds with respect to each reference compound and then combine the $n$ rankings obtained into a single ranking. As an alternative, the scores themselves can be combined to a single score, which is then used to produce a ranking of the database compounds. Here, we will focus on the second strategy. The process of combining different scores or rankings into a single ranking is known as data fusion. For further information about data fusion methods see references.[17]

To obtain a single similarity score one might, for instance, either take the highest similarity score or determine the average of the scores. The first strategy is known as 1-NN, that is, one nearest-neighbor, and the latter is known as $n$-NN. An intermediate strategy is to take the $k$ highest scores, for a fixed $k$, say 3 or 5, and take the average of these scores. $k = 1$ corresponds to 1-NN and $k = n$ to $n$-NN. In our screening program, the following three search strategies will be implemented:

- **1-NN.** The database compound score is the maximum of the similarity coefficients between the reference compounds and the database compound.
- **$k$-NN.** The database compound score is the average of the top $k$ similarity coefficients between the reference compounds and the database compound.
- **$n$-NN.** The score is the average of all similarity coefficients between reference compounds and the database compound.

These strategies can be implemented in a straightforward manner in a single Python function.

```
def knn(ref_fp_list, fp_b,k):
    tcs = []
    for id,fp_a in ref_fp_list:
        tcs.append(tanimoto_coefficient(fp_a, fp_b))
    tcs.sort(reverse = True)
    if k > 0:
        tcs = tcs[:k]
    return sum(tcs)/len(tcs)
```

Parameter `ref_fp_list` contains a list of reference fingerprints given as a tuple containing the name and the fingerprint as a list of integer features. `fp_b` is the fingerprint for which the score is to be determined and k determines the the $k$-NN fusion strategy. If $k = 0$ $n$-NN will be applied.

## Completed Virtual Screening Program

*Code:* vs.py

The VS program has the following command line parameters:

1) A file containing a reference set of fingerprints from active compounds,
2) A database file containing the fingerprints of the screening library,
3) An integer determining the fusion strategy (1, 1-NN; >1, $k$-NN; 0, average),
4) An integer determining the number of top-ranked compounds to select for the output,
5) Name of the output file.

In combination with the previously defined functions the following code completes the virtual screening program.

```
def screen_db(ref_fps, db_fps, k):
    top_ranks = []
    for name, fp in db_fps:
        sim = knn(ref_fps, fp, k)
        top_ranks.append((sim, name))
    top_ranks.sort(reverse = True)
    return top_ranks
```

The main function for screening takes a list of reference compounds (`ref_fps`) and an iterative object (`db_fps`) for retrieving database fingerprints as parameters. The last parameter k determines the fusion strategy. The function returns a sorted list containing the scores and names of the database compounds.

```
import sys
assert len(sys.argv) ==6, \
    "Usage: python vs.py<reference><database><strategy><keep><result>"
```

The command

```
python vs.py ref-maccs.txt  db-maccs.txt 5 100 result.txt
```

performs a 5-NN search of a database using MACCS fingerprints and saves the top 100 scoring compounds in result.txt.

```
reference = sys.argv[1]
database = sys.argv[2]
k = int(sys.argv[3])
keep = int(sys.argv[4])
result_file = sys.argv[5]
ref_fps = read_fp_from_file(reference)
db_fps = iterate_fp_from_file(database)
top_ranks = screen_db(ref_fps, db_fps, k)
top_ranks = top_ranks[:keep]
with open(result_file, 'w') as rf:
    for sim,id in top_ranks:
        rf.write("%s\t%f\n"%(id,sim))
```

Retrieve the parameters, read fingerprint files, call the main function `screen_db`, and save the results.

Now a similarity search of the given ZINC database can be performed:

```
python generateFingerprints.py count morgan zinc-100000.smi \
zinc-100000-morgan-count.txt
python generateFingerprints.py count morgan d4.smi \
d4-morgan-count.txt
python vs.py d4-morgan-count.txt \
zinc-100000-morgan-count.txt 5 1000 d4-result.txt
```

After generation of the fingerprints the virtual screening program will perform a 5-NN similarity search of the ZINC database for dopamine D4 receptor ligands using atom environment count fingerprints. The file d4-result.txt shows the names and scores of the top-ranked compounds:

```
ZINC19797722 0.529548
ZINC19798550 0.517859
ZINC04364593 0.517044
ZINC23282055 0.514594
ZINC18222227 0.510974
...
```

## Benchmarking VS Performance

Given the variety of different fingerprints and the variety of virtual screening methods, it is important to assess how different fingerprints and search strategies perform when screening for compounds active against a given target. This can be done in a retroactive manner using benchmarking. This section is devoted to develop a small benchmarking program for the evaluation of the performance of different fingerprint representations for a given target. The basic idea is to retain a fraction of the known active compounds and add these to the database as "known" hits. Similarity searching is then performed with the remaining compounds as references and the ranks of the active compounds with respect to the database are determined. The higher the ranks of known active compounds are, the better the VS performance. This process should be repeated multiple times using randomization in order to obtain meaningful results.

In benchmarking, the different VS approaches are scored based on their ability to enrich active compounds. Benchmarking can be used to compare different scoring methods (like $k$-NN compared to the 1-NN approach), different fingerprint representations, or the performance with respect to different activity classes in order to assess the ability of the VS approach to detect active compounds for specific targets. In a benchmarking setting, known active compounds are "hidden" in the database and the success in retrieving them is assessed in detail:

1) Randomly divide a reference set of known active molecules into two sets. One set is added to the database as a set of known active database compounds (ADC). The other set will be used as reference and is termed the "bait" set.
2) Perform a virtual screen of the database. All original database compounds are assumed to be inactive for the target. For most large compound databases this is in fact not known and indeed a small fraction of the database compounds might be active.
3) Perform a virtual screen of the ADC set.
4) Determine the rank of each of ADC compound with respect to the database. More precisely: For each ADC compound determine the number of database compounds ranked higher than it.

## How to Score

One way to score the performance of a VS trial is to record the number of actives that achieve at least a certain rank within the database. For a given percentage $x$ let us consider the set $X$ of top-ranked database compounds (assumed to be inactive). For example, for $x = 0.1\%$ and a database size of $n = 100,000$ this corresponds to the $m = xn = 100$ top-ranked compounds. Now determine the set $Y$ of ADC compounds that have a score at least as high as the lowest-ranked compound of set $X$. This corresponds to a fraction $y$ of all the active compounds. This number is known as the *recall*. The ratio of $y$ to $x$ is called the *enrichment factor*. If it is larger than 1 the method has a positive enrichment in that the number of active compounds found in a small selection is larger than is expected by chance.

The recall can be calculated depending on the fraction $x$ of the database under consideration:

$$y(x) = RR_{xn} = \frac{\left(\text{\# of actives ranked within top } m = xn \text{ database cmpds}\right)}{\left(\text{\# of actives}\right)}$$

The curve $y(x)$ for $x$ ranging from 0 to 1 is known as the *receiver-operating characteristic* (ROC). In statistical terminology, $X$ are the *false positives* and $Y$ are the *true positives*. The ROC is a monotonously increasing function with $y(0) = 0$ and $y(1) = 1$. It can be compared to the main diagonal $y = x$, which corresponds to random selection, that is, the ratio of retrieved actives is identical to the fraction of the whole database selected, which corresponds to an enrichment factor of 1. If the ROC lies above the main diagonal active compounds are prioritized with an enrichment factor greater than 1. From the ROC the area under the curve can be determined. This value is known as AuROC (area under ROC), and is a measure of the overall success of a method. A value of about 0.5 indicates no preferential selection of actives and values close to 1 indicate strong prioritization of active compounds. The ROC is explained in detail in a number of statistical texts, see references[21] for further information.

### Multiple Runs and Reproducibility

In order to obtain statistically meaningful results, individual virtual screening runs should be repeated multiple times (e.g., 10-25 times) each time choosing a different set of bait compounds and ADCs. Randomization of the data can be achieved using built-in pseudo random number generators. However, benchmarking should be reproducible. For instance, when different fingerprints or search strategies are used, it should be possible to generate identical partitions of the data into bait set and ADC set. This might be achieved in different ways. For instance, a set of random partitions might be generated once and saved to a file. In this way, the same partitions can be used in different contexts. Here, we take a slightly different approach. A random seed value is provided that is used to initialize the random number generator so that identical partitions will be generated for repeated runs. In order to retain the information about each partition it will also be written to the output file encoded in a string containing "a"s and "b"s depending on whether the fingerprint in the corresponding position was used as ADC or bait.

## Adjusting the VS Program for Benchmarking

*Code:* vsBenchmark.py

The benchmarking program uses the same functions as the VS program. However, the main part needs to be modified and a few functions are added to handle the random partitioning and the ranking of ADC compounds relative to database compounds. The output will also be different. For each trial the program will output a line that contains the "ranks" for each ADC. Here, ranks refer to the number of database compounds ranked higher than an ADC. This is different from a combined ranking of database and ADCs. In a combined ranking, the rank of an ADC would also be determined by ADCs with a higher score. Although this may seem unintuitive at first, our approach has the advantage that the ranks are not influenced by the number of ADCs in the data set and they correspond directly to the false positive rate giving them a precise statistical meaning.

The command line parameters of the program specify:

1) A reference set of fingerprints,
2) A database set of fingerprints,
3) An integer *k* specifying the *k*-NN strategy,
4) The number of benchmark trials to be performed,
5) The number of or percentage of reference compounds,
6) Name of the output file,
7) A seed value for the random number generator (optional).

The following code contains the added functions and the modified main part of our VS script.

```python
import random,sys
assert len(sys.argv)>=7, "Usage: python vs-benchmark.py" \
    "<reference><database><strategy><num_trials><num_
baits><result>[seed]"
reference=sys.argv[1]
database=sys.argv[2]
k=int(sys.argv[3])
num_trials=int(sys.argv[4])
num_baits=float(sys.argv[5])
result_file=sys.argv[6]
rng=random.Random()
if len(sys.argv)>7:
    rng.seed(sys.argv[7])
ref_fps=read_fp_from_file(reference)

if num_baits<1:
    num_baits=int(len(ref_fps)*num_baits+0.5)
else:
    num_baits=int(num_baits)

db_size=0
for _ in iterate_fp_from_file(database):
    db_size+=1
```

The command

```
python vs-benchmark.py d4-morgan-count.txt zinc10000-morgan-count.
txt 1 25 0.5 bm-result.txt parrot
```

performs 25 benchmark trials with 50% of the fingerprints in d4-morgan-count.txt as baits using 1-NN and stores the result in bm-result.txt. The random number generator is initialized with the seed string "parrot". The size of the database is stored in db_size.

```python
def split_fingerprints(fps,num_baits):
    partition=['b']*num_baits+['a']*(len(fps)-num_baits)
    rng.shuffle(partition)
    baits=[]
    adcs=[]
    for c,fp in zip(partition, fps):
        if c=='b':
            baits.append(fp)
        else:
            adcs.append(fp)
    return baits,adcs, ' '.join(partition)
```

The function takes a list of fingerprints (fps) and the number of baits to be chosen (num_baits) as parameters. A list of characters b and a of length corresponding to the number of fingerprints is generated. The number of b corresponds to num_baits. The list is shuffled and the respective fingerprints corresponding to either b or a are assigned to the bait or ADC set. The sets and the partition string are returned.

```python
from bisect import *
def rank_score(score,ranking):
    left=bisect_left(ranking, score)
    right=bisect_right(ranking, score)
    return (left+right)/2.0
```

```
with open(result_file, 'w') as of:
    of.write("%d\n"%db_size)
    for i in range(num_trials):
        bait_fps,adc_fps,partition=split_fingerprints(ref_fps,
num_baits)
        db_fps=iterate_fp_from_file(database)
        db_scores=screen_db(bait_fps, db_fps,k)
        db_scores=[-x for x,_ in db_scores]
        adc_scores=screen_db(bait_fps,adc_fps,k)
        adc_scores=[-x for x,_ in adc_scores]
        adc_ranks=[str(rank_score(x,db_scores)) for x in adc_scores]
        of.write("# %s\n" % partition)
        of.write("%s\n" % " ".join(adc_ranks))
```

The main loop will perform num_trials trials. First the reference set is randomly split into bait and ADC set. Then the database and ADC set are screened with respect to the bait set. For benchmarking, we are only interested in the scores and not the individual compounds so the compound names are removed from the results. The function rank_scores is used to rank the value score with respect to the scores provided by ranking. If scores are tied, an average rank is returned.

## Analyzing Benchmark Results

*Code:* analyzeResults.py

An exemplary output of the benchmark program will look like this

| | |
|---|---|
| 10000 | Number of database compounds |
| # abbaabaabb | 1st trial partitioning of reference data |
| 0 1 4 45 678 | Ranks of ADCs in 1st trial |
| # baaabbbaba | 2nd trial partitioning of reference data |
| 2 13 76 786 1238... | Ranks of ADCs in 2nd trial... |

The result file can be analyzed in order to statistically evaluate the performance of the virtual screen. As mentioned above, of typical interest are:

- The recall achieved by limiting the set selection to a certain percentage of the database and how the recall varies from trial to trial,
- the ROC, and
- the area under the ROC.

To this end the following script can be used that prints the average recall performances for certain fractions of the database, that is, 0.01%, 0.1%, 1%, and 10% and generates boxplots of the recall for the different fractions. Furthermore, an ROC plot is generated and the AuROC is determined. Usually, the early enrichment performance of VS methods is of special interest because often only a very small fraction of a large database can be experimentally evaluated. The standard ROC plot is not able to resolve early enrichment performance very well. This problem can be resolved by using a logarithmically scaled x-axis.

```
import sys
import matplotlib.pyplot as plt
from collections import defaultdict
import numpy as np
assert len(sys.argv)==2, "Usage: python analyze-results.py<resultFile>"
logarithmic_roc = False
resultFile = sys.argv[1]
percentages = [0.0001, 0.001, 0.01, 0.1]
```

The script only takes the result file as a parameter. The fractions for which recall box plots are to be generated and whether to generate a logarithmic ROC plot can be specified by modifying the script accordingly.

```
recall = defaultdict(list)
all_data = []
with open(resultFile) as f:
    db_size = int(f.readline())
    for line in f:
        if line[0]=='#': continue
        raw=map(float, line.split())
        fractions = [x/db_size for x in raw]
        n=1.0*len(fractions)
        for p in percentages:
            recall[p].append(len(filter(lambda x: x<=p, fractions))/n)
        all_data.extend(fractions)

print "File:",resultFile
for p in percentages:
    print "Recall at %f%%: average: %f%% std.dev.: %f%%" \
        % ((p*100),np.average(recall[p])*100,np.std(recall[p])*100)
```

The recall for each trial is stored as a list in a dictionary for different selection set sizes. Average recall and standard deviation for the recall are printed to the console for different set sizes.

```
fig=plt.figure(figsize=(11,5))
ax=fig.add_subplot(121)
labels=["%g%%"%(p*100) for p in percentages]
data=[recall[p] for p in percentages]
plt.boxplot(data,labels=labels)
plt.ylim([-0.01, 1.01])
plt.xlabel("% of database")
plt.ylabel("Recall")
plt.title('Recall')
```

Boxplots for different fractions of the database are shown.

```
ax=fig.add_subplot(122)
all_data.sort()
x=[0.0, 0.1/db_size]
y=[0.0]
n=len(all_data)
for i in range(n):
    if all_data[i]>x[-1]:
        y.append(1.0*i/n)
        x.append(all_data[i])
y.append(1.0)
x.append(1.0)
y.append(1.0)
```

The ROC curve is determined by accumulating data from all trials. The initial $x$ value of `0.1/db_size` was chosen to avoid problems when plotting the data on a logarithmic scale.

```
area=0.0
for i in range(1,len(x)):
    dx=x[i]-x[i-1]
    area+= dx*(y[i]+y[i-1])/2
```

The area under the ROC is determined.

```
if logarithmic_roc:
    plt.plot(x[1:],y[1:], '-')
    plt.xlim((x[1],1.0))
    ax.set_xscale('log')
else:
    plt.plot(x,y,'-')
    plt.xlim([-0.01,1.01])
plt.ylim([-0.01,1.01])
plt.title('AuROC=%.3f'%area)
plt.xlabel("False positives")
plt.ylabel("True positives")
plt.show()
```

The ROC is plotted using either a linear or logarithmic x-scale.

---

For the three activity classes, serotonin 5HT$_{1A}$ receptor ligands, dopamine D4 receptor ligands, and norepinephrine transporter ligands, 25 benchmark trials using 5-NN have been carried out for binary atom pair and Morgan (ECFP-like) fingerprints. The plots in Figure 24.2 shows the recall boxplots and ROC curves as generated by `analyze-results.py`. Clear differences in the performance are apparent. Class D4 performs best, followed by 5HT1A, and class NET performs worst. Most of the time

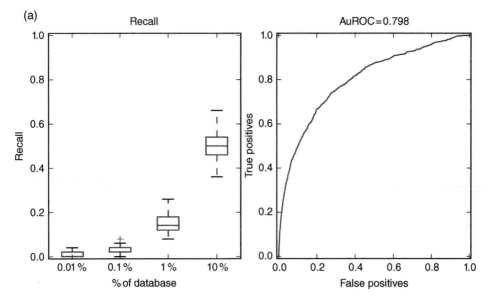

Figure 24.2 **Results of benchmark trials.** Boxplots for recall at 0.01%, 0.1%, 1%, and 10% are shown in the left plots. The ROC and calculated AuROC is shown in the right plots for 25 trials using 5-NN. (a), (c), (e) show results for the atom pair fingerprints while (b), (d), (f) show results for the Morgan fingerprint. (a), (b) serotonin 5HT1A receptor ligands; (c), (d) dopamine D4 ligands; (e), (f) norepinephrine transporter ligands.

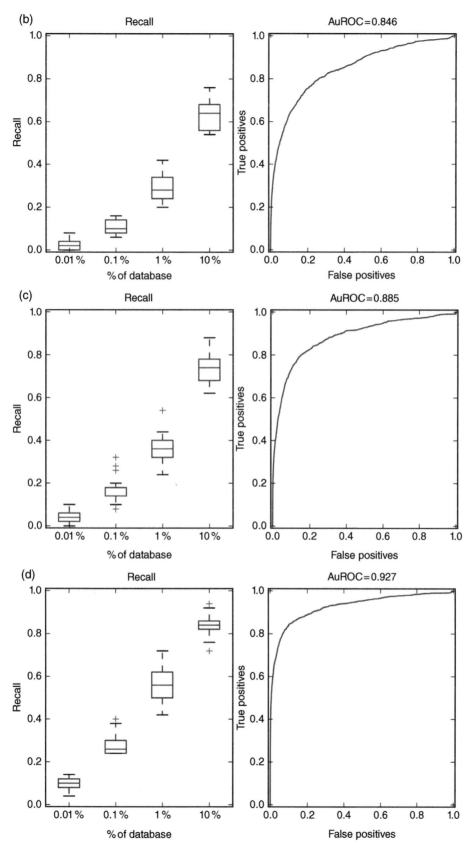

Figure 24.2 (Continued)

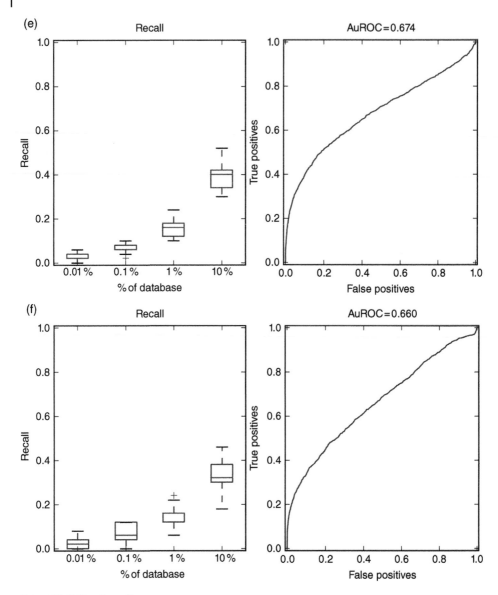

Figure 24.2 (Continued)

the Morgan fingerprints outperform the atom pair fingerprints with the exception of NET. Note that any conclusions drawn from the plots should be substantiated by a statistical significance analysis, which is beyond the scope of this tutorial.

## Conclusion

In this tutorial a simple virtual screening program based on similarity searching using fingerprint representations has been developed in Python. It can be used to screen data-bases of moderate size given a set of compounds with known activity. A benchmark program was developed to assess the performance of different fingerprints and fusion strategies for different activity classes. Finally, a simple analytical tool was developed for graphically displaying the benchmark results. The code focuses on ease of exposition and clarity rather than performance or comprehensiveness, but might form the basis for many modifications and extensions. The calculation of the Tc is a serious bottleneck of the Python implementation that limits its utility. Nevertheless, the code can serve as a template for further enhanced implementations in languages like Java or C++.

## Introduction to Chemoinformatics Toolkits

*Goal:* Introduction to handling and manipulating molecules using chemoinformatics toolkits.
  *Software:* Python 2.7, RDKit (release 2015.03.1)
  *Code:* hill.py

## Theoretical Background

The handling and manipulation of molecular structures on the computer is an essential part of algorithmic chemoinformatics. Central to this task is the representation of the molecule as a structural graph. In its basic form, a computer representation as a graph encodes the individual atoms of a molecule and how they are bonded to each other. Information initially missing in such a representation includes, for instance, three-dimensional coordinates of atoms, stereochemistry, or charge information. Furthermore, much like molecular depictions, hydrogens are often not treated as individual atoms but are instead inferred from atom and charge information of heavy atoms. However, these additional layers of information can be stored as attributes to the individual atoms and bonds.

  The concept of *data structures* as an efficient way of organizing data is one of the fundamental concepts of high-level programming languages like Python or Java. For example, in addition to primitive data types like strings, floats, or integers Python possesses more complex built-in data structures like lists, tuples, and dictionaries for holding collections of data. These general data structures are defined by the way they organize the data and how this data can be manipulated. In addition to built-in data structures, in object-oriented programming languages it is possible to generate user-defined data types by creating appropriate class definitions. A class definition entails how a certain kind of data is organized internally and how it can be accessed and manipulated by methods. The class serves as an abstraction hiding the details how data is stored by defining methods to manipulate the data on a conceptual level. As an example, in chemoinformatic toolkits molecules are implemented using one or more class definitions. They represent the abstraction layer between the way the structure of molecules is organized in computer memory in terms of built-in data structures and primitive types and the way molecules are used and manipulated. The class definition provides an interface that specifies how objects of the class, that is, individual molecules, can be created and manipulated. To this end it is not necessary to know how the information is organized. For instance, molecule classes provide methods for accessing atoms of a class or determining the neighbors of atoms instead of directly dealing with the implementation details. The methods will return atom and bond objects with properties like atom type, bond order, or aromaticity.

Compared to built-in data structures molecules are more complex because they do not have a linear structure that can be easily represented by one of Python's collection of data types. Instead they can be envisioned as mathematical graphs consisting of two different types of objects, namely vertices (or nodes), that is, atoms, and edges, that is, bonds, connecting two vertices. In addition, these objects have properties like the atomic number for atoms or the bond order for bonds.

As mentioned in the introduction, a number of chemoinformatic toolkits are available. For the following tutorials we will use RDKit.[9] Although the interfaces of the toolkits differ quite significantly, they are usually built on similar design principles and it is possible to translate the code presented herein to other languages and toolkits with relatively little effort.

## A Note on Graph Theory

The central mathematical concept used in representing the structure of a molecule is that of a graph. A graph $G = (V,E)$ is a very basic structure consisting of a set of vertices $V = (v_1,...,v_n)$ and a set of edges $E = (e_1 = (u_1,w_1),...,e_m = (u_m,w_m))$ consisting of pairs of vertices $(u_i, w_i)$ indicating which pairs of vertices are connected. A graph with $n$ vertices and $m$ edges is said to have *order n* and *size m*. For a vertex $v$ and edge $(v,w)$ or $(w,v)$, $w$ is said to be a *neighbor* of $v$. The number of all neighbors of $v$ is called the *degree* of $v$.

The structure of molecules can be naturally described by graphs where atoms are represented by vertices and bonds by edges. In order to capture the complete chemical nature of a molecule, vertices and edges of a graph representation need to be associated with properties. That is, each vertex needs to be assigned the element it represents and, in cases where hydrogen atoms are not treated as separate vertices, the number of hydrogen atoms attached to a certain atom or the formal charge of the atom. Attributes associated with edges are typically the bond order and aromaticity. Furthermore, spatial coordinates and/or stereochemical information can be associated with vertices and/or edges.

The description of molecules as graphs reflects the correspondence of basic chemoinformatic tasks and algorithmic problems of graph theory. Two of these problems will be treated in more detail in the following sections.

The first problem is the generation of a unique canonical representation of a molecule using the linear SMILES notation. By generating a canonical string for each molecule it becomes very easy to decide whether two given chemical structures represent the same molecule or not. Mathematically, this is known as the graph isomorphism problem asking whether two graphs $G_1 = (V_1,E_1)$ and $G_2 = (V_2,E_2)$ are isomorphic. Informally, two graphs are said to be isomorphic if they are identical with the exception to the way the vertices are ordered. More formally, two graphs are isomorphic if there exists a one-to-one correspondence between the vertices of both graphs so that an edge between two vertices of one graph corresponds exactly to an edge between the corresponding vertices of the other graph.

The second problem concerns the question of whether a molecule contains a certain substructure or not. In mathematical terms, this is known as the subgraph isomorphism problem asking whether a graph $H = (V_H,E_H)$ is isomorphic to a subgraph of $G = (V_G,E_G)$.

For none of these problems efficient algorithms are available. The subgraph isomorphism problem belongs to the class of NP-complete problems and it is assumed that algorithms for solving this problem must essentially explore every possible way vertices of *H* and a subset of vertices of *G* can be matched. In practice, this means that the running time will increase exponentially with the size of the graphs. Still, algorithms can be found that work well in practice, especially for molecules as these usually have a fairly simple topological structure, at least from a graph theoretic point of view.

## Basic Usage: Creating and Manipulating Molecules in RDKit

For the purpose of the tutorials only elementary functions of the toolkit are required.

### Creation of Molecule Objects

Two principle ways of creating molecule objects are either by parsing a string in SMILES notation or by reading molecule descriptions from a file.

```python
from rdkit import Chem
mol1 = Chem.MolFromSmiles('c1ccccc1O')
mol2 = Chem.MolFromSmiles('c1[nH]c2c(ncnc2n1)N')
```

Two molecules mol1 and mol2 are generated containing the structures of phenol and adenine, respectively.

```python
mol1 = Chem.MolFromMolFile('phenol.mol')
```

A molecule is read from a Mol file.

```python
suppl = Chem.SDMolSupplier('data.sdf')
for mol in suppl:
    if mol is None: continue
    print mol.GetNumAtoms()
```

Multiple molecules are read from an SD file. When a malformed entry is detected in the SD file None is returned.

```python
suppl = Chem.SmilesMolSupplier('data.smi')
for mol in suppl:
    if mol is None: continue
    print mol.GetProp('_Name'), 'has', mol.GetNumBonds(), 'bonds'
```

Molecules can also be read in from a file containing SMILES. For the file data.smi:

```
Smiles Name
c1ccccc1O phenol
c1[nH]c2c(ncnc2n1)N adenine
```

The code snippet will print:

```
phenol has 7 bonds
adenine has 11 bonds
```

Corresponding functions exist for writing molecules to files and generating (canonical) SMILES.

RDKit provides a vast number of methods and functions for accessing and manipulating molecules. Here only the most basic ones needed for implementations of the algorithms are discussed.

```
mol.GetNumAtoms()
mol.GetNumBonds()
```

Return number of atoms/bonds.

```
mol.GetAtoms()
mol.GetBonds()
```

Return a sequence of atoms/bonds of the molecule.

```
mol.GetBondBetweenAtoms(atomIdx,atomIdx)
```

Return the bond between two atoms as given by their indices (see below). If there is no bond between the atoms None is returned.

```
mol.GetProp('_Name')
mol.GetProp(prop)
```

Return the name of the molecule/return value of the property prop. Retrieve additional properties as found in SD files, for instance.

```
mol = Chem.AddHs(mol)
mol = Chem.RemoveHs(mol)
```

Hydrogen atoms may or may not be stored as separate atom objects in a molecule data structure. Explicit hydrogen objects are useful, for instance, for conformer calculations. However, frequently an implicit representation is preferred when the number of attached hydrogens can be regarded as an implicit property of the heavy atoms. Chem.AddHs makes all hydrogens separate objects, while Chem.RemoveHs removes all explicit hydrogen objects.

```
atom.GetAtomicNum()
atom.GetSymbol()
atom.GetDegree()
atom.GetFormalCharge()
atom.GetTotalValence()
```

Return the atomic number/element symbol/degree/formal charge/total valence (sum of all bond orders).

```
atom.GetTotalNumHs()
```

Return the number of hydrogens attached to the atom. This does not include hydrogens that are represented as separate atom objects.

```
atom.GetNeighbors()
```

Return a sequence of neighbor atoms.

```
atom.GetBonds()
```

Return a sequence of the atom's bonds

```
atom.GetIdx()
mol.GetAtomWithIdx(idx)
```

> `atom.GetIdx()` returns the index of the atom. The index corresponds to the position in the internal ordering of the atoms in the molecule data structure. Each atom has a unique index in the range 0 to `mol.GetNumAtoms()-1`. The index can be used to identify the atom.

`mol.GetAtomWithIdx(idx)` returns the atom object corresponding to `idx`.

### Bond Methods

```
bond.GetBondTypeAsDouble()
```

> Returns the bond order, that is, 1, 2, or 3 for single, double, or triple bonds, respectively. 1.5 is returned for aromatic bonds.

```
bond.GetBeginAtom()
bond.GetBeginAtomIdx()
bond.GetEndAtom()
bond.GetEndAtomIdx()
```

> Return the begin/end atom/atom index of a bond. There is no rule as to which atom is considered the begin atom and which the end atom of a bond.

```
bond.GetOtherAtom(atom)
bond.GetOtherAtomIdx(atomIdx)
```

> Given one atom, return the other atom of a bond.

## An Example: Hill Notation for Molecules

*Code:* `hill.py`

As a simple example for the use of the toolkit, the following function will generate the Hill notation for a molecule. The Hill notation is a system for writing chemical formulas, where carbons and hydrogens are listed first for carbon containing molecules and the remaining elements are ordered alphabetically. For molecules that do not contain carbons, all elements are listed alphabetically.

```
from rdkit import *
from collections import defaultdict
def hill_count_to_str(symbol,count):
    if count==0:
        return ""
    elif count == 1:
        return symbol
    else:
        return symbol+str(count)
```

> `hill_count_to_str` takes an atomic symbol and its frequency (`count`) in the molecule as a parameter and returns the string as it should appear in the final formula. Thus, if the count is 0, an empty string is returned, if it is 1 only the symbol is returned, otherwise the symbol is followed by the count.

```
def hill(mol):
    counts = defaultdict(lambda: 0)
    for atom in mol.GetAtoms():
        key = atom.GetSymbol()
        counts[key] += 1
        counts['H'] += atom.GetTotalNumHs()
    formula = []
    if counts['C'] > 0:
        formula.append(hill_count_to_str('C', counts['C']))
        formula.append(hill_count_to_str('H', counts['H']))
        del counts['H']
    del counts['C']
    keys = sorted(counts.keys())
    for s in keys:
        formula.append(hill_count_to_str(s, counts[s]))
    return ''.join(formula)
```

First, the frequency of each element is counted using a dictionary. The dictionary has a default value of 0 so that initially the count of each element is 0. Hydrogens are special because they can appear as attributes of heavy atoms and for each atom the implicit hydrogen count needs to be added.

Once the counts have been determined the formula is constructed, first giving the carbon and hydrogen elements (if present) followed by the elements in alphabetical order.

For instance, the function call

```
hill(Chem.MolFromSmiles('Brc1[nH]c2c(nc(O)nc2n1)NO'))
```

will return the string C5H4BrN5O2 .

## Canonical SMILES: The CANON Algorithm

*Goal:* Implementation of the CANON algorithm for the generation of canonical SMILES.

*Software:* Python 2.7, RDKit (release 2015.03.1)

*Code:* canonicalSmiles.py

## Theoretical Background

### Recap of SMILES Notation

SMILES (Simplified Molecular Input Line Entry System) is a linear notation for representing molecules.[22] A SMILES string encodes the chemical structure of a molecule, that is, its atoms and how they are connected by bonds. SMILES are able to encode, among other things, information about atom type, charges, attached hydrogens, bond types, aromaticity, and stereochemistry. The information encoded typically corresponds to a two-dimensional depiction of a molecule and there are usually many ways in which a molecule can be represented by SMILES. Detailed description of SMILES can be found in the Daylight theory manual[23] or on the OpenSMILES web page (http://www.opensmiles.org).

For the sake of simplicity and clarity, the implementation herein will generate a simplified version of the SMILES outlined below.

**Atoms** are always represented only by their atomic symbol. In order to keep the code simple and to focus on the algorithmic details, the square bracket notation is not used. In standard SMILES the non-bracket notation is only used for the "organic subset" B, C, N, O, P, S, F, Cl, Br, and I. In this case, the number of hydrogens attached to each atom is inferred using the attached bonds and the standard valences of an atom. For example, both "[OH2]" and "O" represent water. Aromatic atoms are indicated by lowercase letters.

**Bonds** are represented by the symbols "-", "=", "#", and ":" for single, double, triple, and aromatic bonds, respectively. Single and aromatic bond symbols will be omitted.

**Branches** are specified by parentheses. An atom is bonded to the first atom of each of the parenthesized SMILES following it, as well as to the first atom following the parentheses. For example, aminoethanol can be represented by the SMILES "C(N)(C)O." The first carbon is attached to the parenthesized nitrogen and carbon, as well as to the final oxygen.

**Cyclic structures** are transformed into acyclic structures by "breaking" one or more ring bonds. In SMILES, the atoms corresponding to a broken bond are designated with a digit to indicate non-adjacent pairs of atoms that are bonded. The first time a digit occurs in a SMILES, a ring is said to be opened. The second time the digit occurs, the ring is closed. After ring closure, a digit can be reused, for example, "c1ccccc1-c1ccccc1" for biphenyl. Note: Technically, the single bond symbol is required in this case because the bond connecting the two phenyl rings is not aromatic. In order to limit the complexity of the code, the implementation of this section will not insert a single bond symbol in this case. Furthermore, stereochemical information will be ignored in the generation of SMILES.

## Canonical SMILES

SMILES representations are not unique. In principle, a SMILES representation can start with any atom in the molecule and at each branching point the order in which the branches appear in the SMILES is arbitrary.

The idea of canonicalization methods is to assign priorities to the atoms of a molecule depending on their location in the molecular graph. The goal is to assign priorities in such a way that when isomorphic representations of a molecule are given, that is, representations that only differ in the way atoms are ordered, corresponding atoms will always get the same priority. Thus the priorities should only depend on the atom and bond properties and the connectivity information of the structural graph. Not all atoms of a molecule can be distinguished this way. For instance, in the molecule shown in Figure 24.3 two pairs of carbon atoms of the phenyl ring are symmetric as well as the two oxygens of the sulfonyl group.

Once priorities have been assigned, a canonical SMILES string can be generated by starting with the atom having the highest priority and, at each branching point, always choosing atoms with the highest priority.

The generation of canonical SMILES can be broken down into two separate steps:

1) Assigning unique priority values to each atom of a molecule based on the graph,
2) Generating a SMILES string based on these priorities.

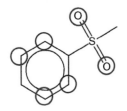

**Figure 24.3 Symmetric atoms.** Some molecules contain atoms that cannot be distinguished by prioritization methods. Pairs of symmetric atoms of this molecule are circled in different colors.

For prioritization, the CANON algorithm proposed by Weininger *et al.*[24] will be implemented. First, however, a function is developed that is able to generate a SMILES notation for a molecular structure.

## Building a SMILES String

The generation of SMILES from a molecular structure corresponds to a depth-first traversal of a graph. Depth-first traversal starts at an (arbitrary) vertex and proceeds to visit vertices along a single path. At branch points one vertex is chosen and all vertices that can be reached from it are traversed before backtracking occurs and alternative paths are explored. The vertices can be either chosen arbitrarily or according to some predefined priorities.

The following pseudocode contains the basic function for depth-first traversal for a molecule *mol*. The additional parameter *atomRanks* contains a list of priorities. The ranks will be calculated by the CANON algorithm explained in detail below.

**procedure** TRAVERSEATOMS(*mol*, *atomRanks*)
    unmark all *atoms* of *mol* as *visited*
    select highest ranked atom as *root* atom
    DFS(*root*, none)
**procedure** DFS(*atom*, *parent*)
    mark *atom* as *visited*
    **for all** neighbors of *atom* except *parent* in order of *atomRanks* **do**
        **if** *neighbor* has not been *visited* **then**
            DFS(*neighbor*, *atom*)

When a SMILES is generated using depth-first traversal a path of bonds is indicated by consecutive atomic symbols and branches are indicated by parentheses. In depth-first traversal, cycles are detected if a neighbor is encountered that is on the currently explored path. These are indicated by ring opening and ring closure numbers in SMILES. In order to account for these, a two-pass approach is used when generating SMILES. In the first pass, all pairs of atoms that are used for ring opening and closure are collected. In the second pass, when the SMILES string is generated, this information is used to apply the correct ring opening and ring closure numbers to the atomic symbol in SMILES generation.

**function** GETSIMPLESMILES(*mol*, *atomRanks*)
    unmark all *atoms* of *mol* as *visited*
    unmark all *atoms* of *mol* as *ancestor*
    select highest ranked atom as *root* atom
    initialize *closureList* as empty list
    GETCLOSURES(*root*, none)
    unmark all *atoms* as *visited*

```
        return BUILDSMILES (root, none)
procedure GETCLOSURES (atom, parent)
      mark atom as visited
      mark atom as ancestor
      for all neighbors of atom except parent in order of
      atomRanks do
            if neighbor is ancestor then
                  closureList←closureList+[(neighbor,atom)]
            else if neighbor has not been visited then
                  GETCLOSURES (neighbor, atom)
      unmark atom as ancestor
```

Note, that the procedure GETCLOSURES requires an additional flag *ancestor* to annotate whether a visited atom is on the currently explored path and a ring closure is required.

After the ring closures have been determined the SMILES can be constructed:

```
function BUILDSMILES (atom, parent)
      mark atom as visited
      seq←BONDSYMBOL (atom, parent) + SYMBOLOF (atom)
      if atom in closureList then
            seq←seq+RINGCLOSURES (atom)
            seq←seq+RINGOPENINGS (atom)
      for all neighbors of atom except parent in order of
      atomRanks do
            if neighor is not visited then
                  if neighbor is last neighbor then
                        seq←seq+BUILDSMILES (neighbor, atom)
            else
                  seq←seq+'('+BUILDSMILES (neighbor, atom) +')'
      return seq
```

A function call BUILDSMILES(*atom, parent*) will create a partial SMILES using the variable *seq* corresponding to the traversal of the molecule from *atom* onwards. These partial strings will be branches if called from within the BUILDSMILES function itself and are concatenated there by putting all but the last branch in parentheses.

The pseudocode can now be translated into Python filling in the missing detail of how ring opening and closure symbols are generated:

```python
from rdkit import Chem

from collections import defaultdict
from heapq import heapify, heappush, heappop
def getBondSymbol(mol, a, b):
    bond = mol.GetBondBetweenAtoms(a, b)
    b = bond.GetBondTypeAsDouble()
    if b == 1.0: return ""
    if b == 1.5: return ""
    if b == 2.0: return "="
    if b == 3.0: return "#"
```

The bond symbol between two atoms is returned. Note that atoms are represented by their integer indices as given by the `GetIdx` method.

```python
def getSimpleSmiles(mol, ranks = defaultdict(int)):
    global atomRanks
    atomRanks = ranks
    atoms = sorted([a.GetIdx() for a in mol.GetAtoms()],
        key = lambda a: atomRanks[a])
    root = atoms[0]
    global visited
    visited = set()
    global ancestor
    ancestor = set()
    global openingClosures
    openingClosures = defaultdict(list)
    getClosures(mol, root, None)
    global closingClosures
    closingClosures = defaultdict(list)
    global digits
    digits = [str(x) for x in range(1,10)]
    visited = set()  # reset visited atoms
    return buildSmiles(mol, root, None)
```

`getSimpleSmiles` takes a molecule and, optionally, a ranking of atoms as input. If `ranks` is not supplied, all atoms have the same rank of 0 and the atoms are not prioritized. The sets `visited` and `ancestor` keep track of which atoms have already been visited and which are on the current path by storing the respective atom indices. `openingClosures` is a dictionary where the key corresponds to the opening atom and the value to the closing atom. It is initialized by calling `getClosures`.

For building the SMILES, two additional data structures are needed. `closingClosures` keeps track of the ring closing digits and bond symbols. `digits` is used as a heap,[25] from which digits are taken when a ring is opened and put pack again for reuse when a ring is closed. By using a heap it is possible to always retrieve the smallest available digit.

```python
def getClosures(mol, atom, parent):
    ancestor.add(atom)
    visited.add(atom)
    atomObj = mol.GetAtomWithIdx(atom)
    nbors = [a.GetIdx() for a in atomObj.GetNeighbors() if
a.GetIdx() != parent]
    nbors.sort(key = lambda a: atomRanks[a])
    for n in nbors:
        if n in ancestor:
            openingClosures[n].append(atom)
        elif n not in visited:
            getClosures(mol, n, atom)
    ancestor.remove(atom)
```

This is a literal translation of the pseudocode, except that closures are stored in a dictionary for efficiency, rather than a list. The opening atom of the ring closure will be the neighbor n and the closing atom the current atom. Thus n is the key and atom the value of the entry added to `openingClosures`.

```
def buildSmiles(mol,atom,parent):
    visited.add(atom)
    seq=""
    if parent is not None:
        seq+= getBondSymbol(mol,parent,atom)
    atomObj=mol.GetAtomWithIdx(atom)
    symbol=atomObj.GetSymbol()
    if atomObj.GetIsAromatic():
        symbol=symbol.lower()
    s+= symbol
    for d in closingClosures[atom]:
        seq+= d
        # The digit is freed and can be used again
        heappush(digits,d[-1])
    for a in openingClosures[atom]:
        # a new digit is taken from digits
        d=getBondSymbol(mol,atom,a)+str(heappop(digits))
        seq+= d
        closingClosures[a].append(d)
    nbors=[a.GetIdx() for a in atomObj.GetNeighbors() if
    a.GetIdx()!=parent]

    nbors.sort(key=lambda a: atomRanks[a])
    branches=[] # Smiles for all branches
    for n in nbors:
        if n not in visited:
            branches.append(buildSmiles(mol,n,atom))
    for branch in branches[:-1]:
        seq+= "("+branch+")"
    if len(branches)>0:
        seq+= branches[-1]
    return seq
```

Partial smiles are constructed in seq. If an opening closure is detected a digit is taken from digits to be used for ring opening/closing using heappop. The ring bond symbol and digit are appended to the opening atom and are added to closingClosures so the correct ring closure symbol can be added to the closing atom. If a ring is closed the digit is added again to the heap of digits using heappush.

While building SMILES branches are first stored in a list and are combined afterward making sure the final branch is not parenthesized.

Note: A heap[25] is an important data structure in computer science that imposes some ordering constraints on how elements are placed in an array so that the retrieval and removal of its smallest element as well as the addition of new elements are efficient. In our implementation the number of ring symbols is limited to single digits to keep the implementation simple. However, after a ring is closed the same symbol can be used again for a new ring.

## Canonicalization of SMILES

Canonical SMILES can be generated once the atoms of a molecule are prioritized in such a way that only depends on the atom and bond types and how they are connected in the structural graph. This can be done in a number of different ways and numerous algorithms and implementations exist for this purpose. It should be noted that implementations are usually different, at least to some extent, even if they are based on the same algorithms, and it should be expected that canonical SMILES generated with different tools or even different versions of the same tool may be different. We note that the InChi representation[26] is a platform-independent identifier for molecules. Generation of InChIs has been standardized and software tools usually rely on the same reference implementation provided by the InChI trust.

The basic idea underlying algorithms for atom prioritization is that of a *vertex invariant*. A vertex invariant is a property of a vertex that is only dependent on the graph topology and not a specific representation, that is, ordering of the vertices. For instance the atom type, number of attached hydrogens, or the degree of an atom are vertex invariants of a molecular graph. Such invariants are usually not enough to distinguish all atoms of a molecule because they do not take connectivity information into account. In order to include the topological information of the graph, an iterative scheme is used. First, some initial (numerical) invariants are calculated and then new invariants are determined by repeatedly combining the invariants of neighbor vertices into new invariants.

One of the basic algorithms making use of this principle is the Morgan algorithm.[27] First each atom is assigned the constant invariant 1. New invariants are then calculated simply by adding the invariants of the neighbors for each atom as illustrated in Figure 24.4.

```
procedure MORGAN (mol)
        for all atom in mol do
            label(0,atom) ← 1
        i ← 0
        repeat
            i ← i+1
            for all atom in mol do
                label(i, atom) ← Σ_{nbor} label(i-1, nbor)
        until number of different values for label is maximal
```

For instance, after the first iteration the invariants will correspond to the degrees of the atoms. In general, the Morgan algorithm calculates for each atom the number of possible walks of increasing lengths along bonds starting at that atom, allowing the same bond to be used multiple times. The Morgan algorithm is not very efficient at distinguishing

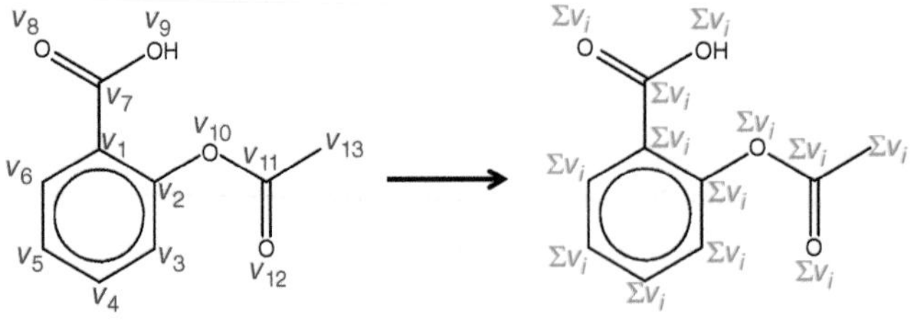

**Figure 24.4 Iteration step of the Morgan algorithm.** New invariants are calculated by summing up the old invariants of the neighbors for each atom.

non-symmetric atoms of a molecule. As a follow-up to the original SMILES specifica-tion,[22] Weininger *et al.*[24] published a more advanced canonicalization algorithm termed CANON that is able to distinguish all non-symmetric atoms for most molecules. It should be noted that algorithms based on the iteration scheme outlined above are not sufficient to discriminate atoms for all possible molecules due to the inherent difficulty of the graph isomorphism problem. To avoid these cases, algorithms are usually augmented by an additional step, where SMILES strings are generated for all possible ways ambigui-ties can be resolved and only the first SMILES string in a lexicographic order is retained.

The CANON algorithm relies on the general principle outlined above; however, it uses more complex initial invariants encoding the properties of individual atoms and utilizes a more sophisticated way to combine neighbor invariants into a new invariant that ensures that if two atoms have a different set of neighbor invariants their new invariants will also differ. Simply adding invariants as in the Morgan algorithm can lead to poorly determined new invariants, for example, atoms with neighbor invariants 1,2,3 and 2,2,2 will both have a new invariant of 6 in the Morgan algorithm although the atoms could have been distinguished based on their neighbor invariants.

The CANON algorithm also provides rules for "breaking ties," that is, resolving the case when atoms cannot be distinguished by the iteration scheme and are, most likely, symmetric. In this case, one tied atom is arbitrarily assigned a new value, and the itera-tion scheme is continued using the updated value until all atoms have received different priorities.

### The Initial Invariant

The initial invariant of each atom encodes atom type information using an eight digit number defined as follows:

1) Number of connections (1 digit),
2) Sum of non-hydrogen bond orders (2 digits),
3) Atomic number (2 digits),
4) Sign of charge (0 for no or positive charge, 1 for negative charge) (1 digit),
5) Absolute charge (1 digit),
6) Number of attached hydrogens (1 digit).

This definition translates directly into the following Python function definition.

```
def invariants(mol):
    inv = dict()
    for atom in mol.GetAtoms():
        i1 = atom.GetDegree()
        i2 = atom.GetTotalValence() - atom.GetTotalNumHs()
        i3 = atom.GetAtomicNum()
        i4 = 1 if atom.GetFormalCharge() < 0 else 0
        i5 = abs(atom.GetFormalCharge())
        i6 = atom.GetTotalNumHs()
        inv[atom.GetIdx()] = int("%1d%02d%02d%1d%1d%1d" % (i1, i2, i3, i4, i5
            , i6))
    return inv
```

The initial invariants of the CANON algorithm are generated and stored in a dictionary.

## The Iteration Step

The CANON algorithm is based on the idea to establish a canonical ranking of the atoms. Thus, the absolute values of invariants are of no interest; instead, it is only important how the atoms are ranked relative to each other. The ranking of atoms is performed after each iteration step and the following step proceeds with the rank positions as new invariants. A ranking of the atoms takes two values into account:

1) The previous rank of the atom,
2) The new invariant calculated in the iteration step.

Atoms are sorted, first, with respect to their previous rank and, second, if ranks are tied with respect to the invariant. New ranks are assigned using consecutive increasing numbers starting at 1. Tied positions receive identical ranks.

```python
def invToRanks(inv,oldRanks):
    sortedAtoms = sorted(inv.keys(),
       key = lambda atom: (oldRanks[atom],inv[atom]))
    rk = 0
    prev = -1
    ranks = dict()
    for a in sortedAtoms:
        if inv[a] != prev:
            rk += 1
            prev = inv[a]
        ranks[a] = rk
    return ranks,rk
```

The sort criterion will sort atoms according to the tuple (oldRanks[atom], inv[atom]), which ensures so that the relative order of previously disambiguated atoms does not change. The function returns the new ranks as a dictionary and the value of the largest assigned rank, which corresponds to the number of different values.

In the iteration step, the generation of new invariants from the ranks of invariants of the neighbors makes use of the number-theoretical fact that each integer has a unique decomposition into prime numbers. Each rank $i$ is associated with $i^{th}$ prime number and a new invariant is calculated by multiplying the prime numbers corresponding to the ranks of neighboring atoms. This ensures that, as long as two atoms have a different set of neighbor invariants, the new invariant will also be different. Figure 24.5 illustrates the iteration step of the CANON algorithm.

The following pseudocode describes the main iteration loop of the CANON algorithm taking a set of (initial) invariants as parameter:

**procedure** CANONITER(*mol, invariant*)
     *ranks(0, atom)* ←rank of *invariant(atom)* **for all** atoms
     $i \leftarrow 0$
    **repeat**
        $i \leftarrow i+1$
        *invariant(atom)* ←$\Pi_{nbor}$ *prime(rank(i-1, nbor))* **for
        all** *atoms*

```
            sort tuples (rank(i-1, atom), invariant(atom))
            rank(i, atom)←updated rank according to sorted
            tuples for all atoms
        until number of different ranks has not increased
```

The implementation of the main iteration loop can now be completed.

```
def newRanks(mol,ranks):
    inv=dict()
    for a in mol.GetAtoms():
        prod=1
        for n in a.GetNeighbors():
            prod *=primes[ranks[n.GetIdx()]]
        inv[a.GetIdx()]=prod
    return invToRanks(inv,ranks)
```

This function performs the iteration step by determining the new invariants and updated ranks.

```
def CanonIter(mol,ranks,rankCount):
    oldRankCount = 0
    while oldRankCount < rankCount:
        oldRankCount = rankCount
        ranks,rankCount = newRanks(mol,ranks)
    return ranks, rankCount
```

This function takes an (initial) ranking as a parameter. rankCount contains the number of different ranks. It performs the iteration until the number of different values does not increase.

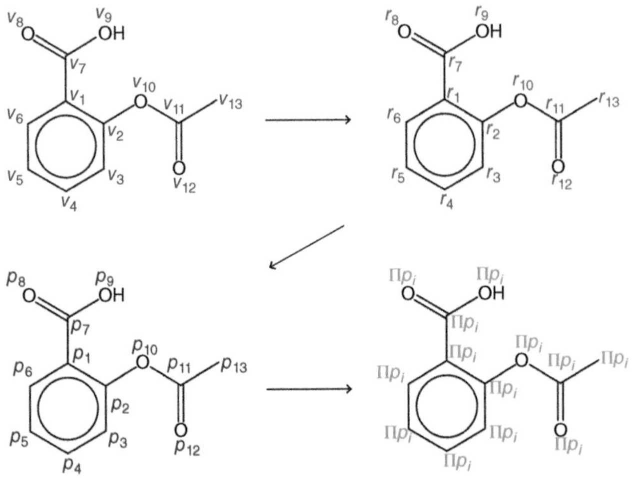

Figure 24.5 **Iteration step of the Canon algorithm.** First, ranks are assigned to the atoms based on previous ranks and the invariants. A prime is assigned to each atom according to its rank. A new invariant is calculated by multiplying the primes of all neighbors for each atom.

The function relies on a given list of primes, which can be determined as follows:

```python
import math
def checkPrime(v):
    "check whether v is prime"
    limit = int(math.sqrt(v)) + 1
    for d in range(3, limit, 2):
        if v%d==0:
            return False
    return True

def getPrimeList(numPrimes):
    "Retrieve a list of numPrimes primes"
    p1 = [2]
    n = 3
    while len(p1) < numPrimes:
        if checkPrime(n):
            p1.append(n)
        n += 2
    return p1
# Get a list of as many primes as atoms in
# the largest molecule the algorithm is expected to handle.
primes = getPrimeList(100)
```

The basic iteration loop will continue until the number of different ranks does not increase. In this case a special step is introduced whose purpose it is to break ties. To this end, the atoms with the smallest tied ranks are selected. These are assumed to be symmetrical atoms of the molecule. This will not always be the case and counter-examples exist representing uncommon molecular structures.[28] One of the tied atoms is arbitrarily assigned a higher rank. Thus, one atom will receive a higher priority in SMILES generation. After a single tie is resolved, iteration can be repeated possibly resolving other ties until either all atoms have received different ranks or another tie must be broken.

The following code completes the implementation of the CANON algorithm in Python:

```python
def breakTies(ranks):
    # count how often each rank occurs
    rankCount = [0]*len(ranks)
    for r in ranks.values():
        rankCount[r] += 1
    # find smallest duplicate rank (rj)
    rk = 0
    while rankCount[rk] <= 1:
        rk += 1
    inv = dict() # new invariants
    for a in ranks.keys():
        inv[a] = 2*ranks[a]
    # change invariant of one of the smallest duplicate ranks
    for a in ranks.keys():
        if ranks[a] == rk:
            inv[a] -= 1
            break
    return invToRanks(inv, ranks)
```

breakTies first counts how often each rank occurs and the selects the smallest rank occurring multiple times. New invariants are assigned by duplicating the value of the old rank. Finally, the first atom encountered corresponding to the smallest duplicate rank is arbitrarily assigned a smaller invariant to break the tie. From the changed invariants updated ranks are determined and returned.

```
def Canon(mol):
    inv = invariants(mol)
    ranks, rankCount = invToRanks(inv, [1]*len(inv))
    while rankCount < len(ranks):
        ranks, rankCount = CanonIter(mol, ranks, rankCount)
        if rankCount < len(ranks):
            ranks, rankCount = breakTies(ranks)
    return ranks
```

The Canon function implements the main loop, which combines iteration and breaking ties in a loop until all atoms of a molecule are disambiguated.

---

Finally, canonical SMILES are generated by first calling CANON and then calling the build SMILES routine:

---

```
def Cansmi(molecule):
    "Return canonicalized Smiles"
    mol = Chem.RemoveHs(molecule)
    ranks = Canon(molecule)
    return getSimpleSmiles(mol, ranks)
```

---

The code has been integrated in the script `canonicalSmiles.py` that takes a number of SMILES as command line argument and converts them to canonical SMILES:

```
$ python canonicalSmiles.py\
"Clc1ccccc1C1=NC(O)C(=O)Nc2c1cc(Cl)cc2"\
"O=C1Nc2ccc(Cl)cc2C(c2ccccc2Cl)=NC1O"\
"c1ccc(c(c1)C2=NC(C(=O)Nc3c2cc(cc3)Cl)O)Cl"

OC1N=C(c2ccccc2Cl)c2cc(Cl)ccc2NC1=O
    <- Clc1ccccc1C1=NC(O)C(=O)Nc2c1cc(Cl)cc2
OC1N=C(c2ccccc2Cl)c2cc(Cl)ccc2NC1=O
    <- O=C1Nc2ccc(Cl)cc2C(c2ccccc2Cl)=NC1O
OC1N=C(c2ccccc2Cl)c2cc(Cl)ccc2NC1=O
    <- c1ccc(c(c1)C2=NC(C(=O)Nc3c2cc(cc3)Cl)O)Cl
```

## Summary

The generation of canonical SMILES corresponds to solving the graph isomorphism problem, for which no efficient algorithm is known. It can be broken down into two steps, 1) the construction of the SMILES string and 2) the prioritization of atoms. SMILES are constructed using depth-first traversal of the molecule. Here, two passes are required to identify bonds to be "broken" and to assign corresponding indices to the broken bonds.

Atoms are prioritized using the CANON algorithm that iteratively constructs new invariants from old ones by considering neighboring atoms. Ties are broken if symmetric atoms are detected. Symmetry detection is not perfect and the algorithm is not guaranteed to correctly handle all molecules. Nevertheless, it works well in practice.

## Substructure Searching: The Ullmann Algorithm

*Goal:* Implementation of algorithms for substructure searching.
*Software:* Python 2.7, RDKit (release 2015.03.1)
*Code:* `subsearchBacktracking.py, subsearchUllmann.py`

## Theoretical Background

Identifying substructures in molecules is one of the fundamental tasks of chemoinformatics. For instance, fingerprints like MACCS[18] are mainly defined by a set of substructures and similarity between molecules is assessed by how many substructures two molecules share. Substructure searching is also important for filtering databases to retrieve subsets of molecules that share certain moieties or structural elements.

In graph theoretical terms, substructure searching corresponds to the subgraph isomorphism problem. Given two graphs $G = (V_G, E_G)$ and $H = (V_H, E_H)$, $H$ is a subgraph of $G$ if there exists a mapping $f$ between vertex sets $V_H$ and $V_G$ that maps distinct vertices of $V_H$ onto distinct vertices of $V_G$, that is, $f$ is a one-to-one mapping, so that there is an edge between $f(v)$ and $f(w)$ in $G$ if there is an edge between $v$ and $w$ in $H$, that is, edges are preserved. A variation of the problem is the induced subgraph isomorphism problem. It has the additional requirement that there must not be an edge between $f(v)$ and $f(w)$ if there is no edge between $v$ and $w$, that is, there are edges between vertices of $H$ if and only if there are edges between the corresponding vertices of $G$. For instance, hexane is not considered an induced subgraph of cyclohexanol because the two terminal carbons of hexane are matched to two bonded carbons in cyclohexanol violating the additional requirement. The algorithms and implementations described here will focus on the original subgraph isomorphism problem; however, they can be easily modified for the induced subgraph problem.

As described, the problems refer to unlabeled graphs. However, when applied to molecules, it is important that atom types and bond orders of the substructure are also matched correctly. Substructure searching algorithms can also be easily generalized to enable pattern matching where a substructure is described using patterns or wildcards for atoms and bonds, for example, where a substructure atom might be described as any heavy non-carbon atom or a bond as either single or double. The most well-known language in which such molecular patterns are described is SMARTS.[23] SMARTS is an extension of SMILES, which allows the definition of atom and bond patterns and most chemoinformatics toolkits enable substructure searching using SMARTS patterns. For the purpose of our implementation, substructures will simply be molecules and a mapping between substructure and molecule will only check for correct atom type and bond order.

Substructure searching is a hard problem and algorithms in principle have to check every possible way substructure atoms might be matched to molecule atoms. This leads to methods whose runtime increases exponentially with substructure size. However, it is not necessary to evaluate all possible $m^n$ mappings between a substructure of $n$ atoms and a molecule of $m$ atoms if matches between substructure and a structure are created incrementally. This is the principle idea behind *backtracking* approaches. Starting with a partial mapping between substructure atoms and molecule atoms backtracking systematically tries to expand such mappings atom by atom. This approach will either lead to a successful match of all substructure atoms or to a dead end if it is not possible to find a mapping for an additional substructure atom. In this case backtracking occurs, that is, one or more matched atoms from the partial mapping constructed so far are removed and alternative mappings are explored. This process continues until either all atoms have been successfully mapped or until all possibilities have been exhausted. The basic backtracking strategy, which is explained in detail in the following, will only check whether a currently assigned partial mapping can be extended. More advanced algorithms like the Ullmann algorithm, which will be discussed in the following, essentially keep track of all possible assignments for all remaining substructure atoms to be matched and are able to detect infeasible partial mappings much sooner.

## Backtracking

*Code:* `subsearchBacktracking.py`

Backtracking constructs partial mappings of increasing size between substructure atoms and molecule atoms. To this end, the atoms of the substructure can be given in any order, and partial mappings are constructed by starting with the first atom, and then augmenting the mapping by considering the next atom and so on. If one pictures the substructure atoms lined up from left to right according to their index (see Figure 24.6) the algorithm starts at the left, that is, at index 0, and tries to match the first atom of the substructure with every possible atom of the molecule.

The only requirement being that the two atoms possess the same atom type. If no match can be found obviously the whole substructure cannot be matched to the molecule and the search fails. Once two matching atoms are found the algorithm proceeds one step to the right to the next substructure atom. This atom can only be mapped to a previously unmatched atom of the same atom type. If a matching atom is found the algorithm needs to check whether the matching so far is a valid substructure match. Given the definition of a subgraph the following criterion has to be fulfilled:

"If there is a bond between the first and second substructure atom, a bond of the same type exists between the matched atoms of the molecule".

At the top of Figure 24.6 the first two atoms have been matched and the bond between the first two substructure atoms is mapped onto a bond of the molecule. If the criterion is met the method proceeds to the next substructure atom to the right. If not, all other molecule atoms are systematically checked for a match. If none of them fulfill the criteria the backtracking step is performed by going to the atom to the left and trying a different match for that atom. Thus, at each step, either a match is successful and the procedure proceeds to the right, or it is unsuccessful and returns to the left. For successfully matching an atom the following criteria have to be met:

- The atom types match,
- For every bond of the substructure atom to a previously matched substructure atom a corresponding bond in the molecule exists.

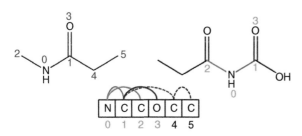

**Figure 24.6 Substructure backtracking.** The top shows the status of backtracking after a successful match of the first two atoms. Corresponding atoms are numbered accordingly and matched bonds have the same color. The bottom shows the status after four atoms have been successfully matched. The fifth atom cannot be matched given the current mapping and the algorithm has to backtrack.

The bottom of Figure 24.6 illustrates the situation after the first four atoms have been matched. The fifth atom cannot be matched to any molecule atom so that its bond to the second atom is mapped to a molecule bond. At this point a dead end is reached and the algorithm backtracks.

The method proceeds until one of two things happen:

1) It reaches the end on the right if all atoms of the substructure have been successfully matched and the molecule contains the substructure,
2) Or it terminates on the left past the first atom, in which case all options have been exhausted and the molecule does not contain the substructure.

The procedure is formalized in the following pseudocode description:

```
function SUBSTRUCTURESEARCH (sub, mol)
        mol has atoms m₁, …, mₖ
        sub has atoms S₁, …, Sₙ
        matchSoFar←empty
        return SEARCH(1)
function SEARCH(i)  # S₁, …, Sᵢ₋₁ have already been matched
        if i>n then
              return true # reached 'right end', success
        else
              for j=1, …, k do
                    if mⱼ has not been matched and EXTENDSMATCHING (i, j) then
                          matchSoFar(s_i)←m_j
                          if SEARCH(i+1) then
                                return true
```

```
            else
                    remove (sᵢ, mⱼ) from matchSoFar
        return false # current matching cannot be extended
        backtrack
function EXTENDSMATCHING(i, j)
      if atom types of sᵢ and mⱼ do not match then
            return false
      for r=1, …, i-1 do
            if sub has bond (sᵣ, sᵢ) and does not match with
            bond (matchSoFar(r), mⱼ) then
                  return false
      return true
```

The recursive SEARCH function performs the basic backtracking approach. The crucial function is EXTENDSMATCHING, which checks whether edges between a newly added vertex and previously added vertices of the partial matching correspond to edges of the matched atoms.

The pseudocode is implemented in Python as a class definition.

```python
from rdkit import Chem
class SubstructureBacktracking:
    def __init__(self,sub):
        self.sub = Chem.RemoveHs(sub)
        self.n = self.sub.GetNumAtoms()
        self.matchSubToMol = dict()
        self.matchMolToSub = dict()
    def find(self,mol):
        self.mol = Chem.RemoveHs(mol)
        self.matchSubToMol.clear()
        self.matchMolToSub.clear()
        return self.search(0)
```

For a substructure molecule sub a class object can be constructed using

```python
ssbt = SubstructureBacktracking(sub)
```

Subsequent calls to the method find perform substructure searches for molecules. The method returns True if the substructure was found in mol and False otherwise. In case of success the dictionaries matchSubToMol and matchMolToSub contain the mapping between the atoms of sub and the atoms of mol and vice versa, respectively.

```python
def search(self,subIdx):
    if subIdx >=self.n:
        return True
    for atom in self.mol.GetAtoms():
        molIdx = atom.GetIdx()
        if (molIdx not in self.matchMolToSub and
                self.extendsMatching(subIdx,molIdx)):
            self.matchSubToMol[subIdx] = molIdx
            self.matchMolToSub[molIdx] = subIdx
            if self.search(subIdx+1):
                return True
            else:
                del self.matchSubToMol[subIdx]
                del self.matchMolToSub[molIdx]
    return False
```

The recursive search method has been directly translated from the pseudocode.

```
def extendsMatching(self,subIdx,molIdx):
    if (self.sub.GetAtomWithIdx(subIdx).GetAtomicNum() !=
        self.mol.GetAtomWithIdx(molIdx).GetAtomicNum()):
        return False
    for i in range(subIdx):
        subBond=self.sub.GetBondBetweenAtoms(i,subIdx)
        if subBond:
            j=self.matchSubToMol[i]
            molBond=self.mol.GetBondBetweenAtoms(j,molIdx)
            if (not molBond or
                        subBond.GetBondTypeAsDouble()!=
                            molBond.GetBondTypeAsDouble()):
                return False
    return True
```

extendsMatching is a direct translation of the pseudocode.

---

This code has been implemented in the program subsearchBacktracking.py.

## A Note on Atom Order

The order in which substructure atoms are matched does not matter to ensure correctness of the algorithm. However, the order can greatly influence the efficiency of the algorithm. The underlying idea is that if a certain mapping is invalid it should be detected as early as possible. For instance, starting the substructure search with a heteroatom will limit the number of possible matches of the first atom to atoms of the same type in the molecule, which are usually rare compared to carbon atoms. Furthermore, subsequent substructure atoms should contain bonds to previously matched atoms restricting the number of possible matches due to the bond criterion of EXTENDSMATCHING.

## The Ullmann Algorithm

*Code:* subsearchUllmann.py

The Ullmann algorithm[29] provides a significant improvement over the basic backtracking approach by keeping track of a list of all possible matches for all substructure atoms. This allows for a much better pruning of the search tree. Compared to the basic approach above backtracking will not only occur if the current substructure atom cannot be matched. Instead, at each step, the potential matches for all remaining atoms are considered. If a single atom is detected for which no potential matches remain the current partial mapping is unsuccessful and backtracking occurs.

To record which matches are possible for each substructure atom the algorithm keeps track of a mapping $f$ from the set of substructure atoms to the power set of molecule atoms. That is, for each atom $v$ of $H f(v)$ is a subset of atoms of $G$ containing all possible matches for $v$. In the original description of the algorithm this map was described by a matrix data structure $M = (m_{ij})$, which we term feasibility matrix here, where each row corresponds to an atom of $H$ and each column corresponds to an atom of $G$. An entry of 1 indicates a possible match and an entry of 0 an impossible match. In our implementation the potential matches will be stored using a Python dictionary where the keys are substructure atoms and the values are Python sets of molecule atoms. For convenience, in the explanation of the method, the matrix notation will be used and the class implementation of the data structure will be appropriately called `FeasibiliyMatrix`.

The feasibility matrix is first set up during an initialization phase and is later updated during the search phase by a refinement procedure that seeks to eliminate infeasible matches based on the current partial mapping.

Initially, any substructure atom can be matched to any molecule atom as long as two conditions are met:

1) The atom types match,
2) The degree of the molecule atom is at least as large as that of the substructure atom.

The second condition enforces the requirement that all bonds attached to any substructure atom must also be matched to bonds of the corresponding molecule atom. Thus, a matching molecule atom must contain at least as many neighbors as the substructure atom. Figure 24.7 illustrates the initialization of the matrix for a sample substructure and molecule. For instance, carbon atom 3 having three neighbors has to be matched to either atom a or c because these are the only carbon atoms of the molecule with at least three neighbors.

After initialization, at least one possibility for each substructure atom must remain, that is, each of the rows of the feasibility matrix must contain at least one 1. In this case, we call the matrix *alive*. If at any point during the algorithm a complete row becomes 0 the matrix cannot lead to a successful substructure match and the matrix is said to be *dead*.

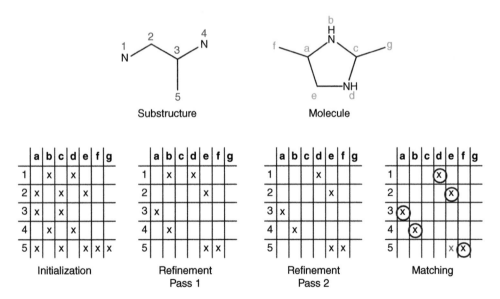

**Figure 24.7  The feasibility matrix.** The rows of the matrix correspond to substructure atoms and the columns to molecule atoms. A possible match between substructure atom and molecule atom is indicated by an "x." After initialization, the matrix is refined repeatedly followed by a backtracking search. In the example shown only one potential matching remains after refinement.

The recursive backtracking search of the Ullmann algorithm is similar to the basic algorithm of the previous section. The Ullmann algorithm explores partial mappings systematically row by row by considering all potential matches. A partial match between a substructure atom and a molecule atom is realized by modifying the matrix in the following way. Let $m_{ij} = 1$; in order to match substructure atom $v_i$ and molecule atom $w_j$ set $m_{ik} = 0$ for all $k \neq j$ and set $m_{kj} = 0$ for all $k \neq i$. Using the modified matrix, the algorithm then proceeds to the next row to systematically try all potential matches for the corresponding substructure atom.

The crucial step, however, is that each time the matrix is modified (once after initialization and then each time a substructure atom is matched with a molecule atom) all positive entries in the matrix are examined to find out whether or not they still are viable matches. This step is called the *refinement procedure*. For every entry $m_{ij}$ with $m_{ij} = 1$ the refinement procedure tests whether a match between substructure atom $i$ and a molecule atom $j$ is possible by verifying the following criterion:

"For each neighbor $v_k$ of $v_i$ there must exist a neighbor $w_\ell$ of $w_j$ so that $m_{k\ell} = 1$, that is, the match between $v_k$ and $w_\ell$ is possible, and that corresponding bonds match".

If the criterion is not fulfilled $m_{ij}$ is set to 0. This check is performed repeatedly for every positive entry of the matrix until either the matrix does not change anymore or becomes dead. The refinement step is illustrated in Figure 24.7. In the first pass, two options for atom 2 can be eliminated, only option e remains because only atom e has a neighbor, in this case atom c, that is a potential match for the neighbor atom 3 of atom 2. In a similar way, infeasible options for atoms 3, 4, and 5 are eliminated. A second pass eliminates more potential matches. In this case, after initial refinement, a single potential match for the substructure atoms remain, effectively reducing the search tree to a single branch.

The algorithm is implemented in Python using two classes. Class `Ullmann` contains the main recursive search method, and class `FeasibilityMatrix` contains the functionality for initializing and refining the feasibility matrix.

```python
from copy import deepcopy
class FeasibilityMatrix:
    def __init__(self,sub,mol):
        "determine initial feasibility matrix"
        self.sub = sub
        self.mol = mol
        f = dict()
        self.isAlive = True
        for i in range(self.sub.GetNumAtoms()):
            sub_atom = self.sub.GetAtomWithIdx(i)
            f[i] = set()
            for j in range(self.mol.GetNumAtoms()):
                mol_atom = self.mol.GetAtomWithIdx(j)
                if (sub_atom.GetAtomicNum() == mol_atom.GetAtomicNum()
and sub_atom.GetDegree() <= mol_atom.GetDegree()):
                    f[i].add(j)
            if len(f[i]) == 0:
                self.isAlive = False
        self.f = f
```

The feasibility matrix `self.f` is implemented as a dictionary of atom/set of atoms pairs. For each substructure atom the potential molecule atoms are recorded in the set according to the above criterion for matrix initialization. The flag `self.isAlive` is used to hold the status of the matrix.

```
def possibleMatches(self,i):
    return self.f[i]
def fixMapping(self,i,j):
    copy = deepcopy(self)
    copy.f[i] = set([j])
    for k in range(i+1,copy.sub.GetNumAtoms()):
        copy.f[k].discard(j)
    return copy
```

fixMapping(i,j) returns a copy of the feasibility matrix that has been modified to reflect the matching between substructure atom i and molecule atom j. This means j becomes the only element of self.f[i] and j is removed from all possible matches of the remaining rows.

```
def refine(self):
    change = True
    while change:
        change = False
        for i in range(self.sub.GetNumAtoms()):
            change |= self.refineRow(i)
            if not self.isAlive:
                return False
    return self.isAlive
```

The main loop of the refinement procedure updates every row of the matrix repeatedly until the matrix remains stable for one pass over all rows. The method returns True if the matrix is alive after refinement and False, otherwise.

```
def refineRow(self,i):
    change = False
    subNbors = self.sub.GetAtomWithIdx(i).GetNeighbors()
    for j in list(self.f[i]):
        molNbors = self.mol.GetAtomWithIdx(j).GetNeighbors()
        if not self.isEntryGood(i,j,subNbors,molNbors):
            self.f[i].remove(j)
            change = True
            if len(self.f[i]) == 0:
                self.isAlive = False
                return change
    return change
```

All possible matches for substructure atom i are checked by calling the method isEntryGood, which requires a list of the neighbor atoms of both substructure and molecule atom. If the test fails the appropriate molecule atom is removed from the set of potential matches. If the number of potential matches reaches 0 the respective atom cannot be matched and the matrix becomes dead.

```
def isEntryGood(self,i,j,subNbors,molNbors):
    for sn in subNbors:
        snIdx = sn.GetIdx()
        foundPossibleMatch = False
        for mn in molNbors:
            mnIdx = mn.GetIdx()
            if (mnIdx in self.f[snIdx] and
                self.sub.GetBondBetweenAtoms(i,snIdx).GetBondTypeAsDouble() ==
                self.mol.GetBondBetweenAtoms(j,mnIdx).GetBondTypeAsDouble()):
                foundPossibleMatch = True
                break
        if not foundPossibleMatch:
            return False
    return True
```

isEntryGood implements the refinement criterion for substructure atom i and molecule atom j as explained above.

```
class Ullmann:
    def __init__(self,sub):
        self.sub = Chem.RemoveHs(sub)
        self.n = self.sub.GetNumAtoms()
        self.matchSubToMol = dict()

    def find(self,mol):
        self.mol = Chem.RemoveHs(mol)
        self.m = self.mol.GetNumAtoms()
        self.matchSubToMol = dict()
        feas = FeasibilityMatrix(self.sub,self.mol)
        if not feas.isAlive:
            return False
        else:
            return self.search(0,feas)
```

The class Ullmann is similar to the SubstructureBacktracking class. The feasibility matrix is an additional parameter to the recursive search method and contains information about the current partial match as well as all potential matches for yet unmatched substructure atoms.

```
def search(self,i,feas):
    if i >= self.n:
        self.retrieveMatching(feas)
        return True
    if feas.refine():
        for j in feas.possibleMatches(i):
            f = feas.fixMapping(i,j)
            if self.search(i+1,f):
                return True
    return False

def retrieveMatching(self,feas):
    for i in range(self.n):
        self.matchSubToMol[i] = feas.possibleMatches(i).pop()
```

Using the feasibility matrix class the recursive search function can be implemented in a straightforward manner. Note that the recursive call self.search(i+1, f) uses a copy of the feasibility matrix as argument to avoid corrupting the matrix feas for subsequent calls. On a successful match a matching between substructure atoms and molecule atoms is retrieved from the feasibility matrix and stored in self.matchSubToMol.

## Sample Runs

This code has been implemented in the program subsearch-ullmann.py. The programs subsearchBacktracking.py and subsearchUllmann.py require a molecule file as input. Substructures are given as additional parameters. The programs search for the provided substructures in all molecules and print a fingerprint output for each molecule.

```
$ python subsearchUllmann.py 5ht1a.smi NCCN c1ccccc1 \
O=S=O "FC(F)(F) " "CC(C)C" "c1ccccc1CCc1cc(OC)ccc1"
CHEMBL377697 111010
CHEMBL215010 010000
CHEMBL2206397 010000
CHEMBL384538 000010
```

```
CHEMBL1548 010000
CHEMBL481839 010001
...
```

The results of both programs are identical. However, the Ullmann algorithm is considerably faster than the simple backtracking approach.

## Summary

Substructure searching corresponds to subgraph isomorphism, an NP-hard problem. Although no efficient algorithms are known, backtracking algorithms provide a feasible approach for identifying substructures in molecules. In this tutorial, two backtracking approaches were implemented, a straightforward simple backtracking approach and the Ullmann algorithm that requires more "housekeeping" to keep track of potential matches for substructure atoms, thereby increasing the complexity of the implementation. Although the Ullmann algorithm carries out more operations during each step than the simple backtracking approach, it is much more efficient as it enables the pruning of large portions of the search tree.

## Atom Environment Fingerprints

*Goal:* Exemplary implementation of atom environment fingerprints.
*Software:* Python 2.7, RDKit (release 2015.03.1)
*Code:* ecfp.py

## Theoretical Background

Fingerprints encode a well-defined set of features of a molecule. Most commonly, fingerprints indicate the presence or absence of features but variants exist that also account for their frequency. The former are called binary fingerprints and the latter count fingerprints. Fingerprints encoding substructure features of molecules are among the most popular ones. The features encoded in a fingerprint might either be represented by a predefined dictionary of structural features like, for instance, the 166 pre-defined features of the MACCS keys[18] or might be obtained by systematically generating features from the molecular structure. Popular variants of the latter type include path-fingerprints such as the Daylight fingerprints[23] or atom environment fingerprints, for example, the extended connectivity fingerprints (ECFP),[19] which are the topic of the exemplary implementation of this section. The basic idea of these fingerprints is to comprehensively enumerate specific substructure features of a molecule such as paths up to a certain length or atom environments up to a certain bond distance. Each of the generated substructures, in principle, can represent a different feature and the fingerprint of a molecule is given by the set of all features present in the molecule where multiple occurrences of a substructure are either ignored, giving rise to a binary fingerprint, or are accounted for according to their multiplicity, giving rise to count fingerprints.

Here, we follow closely the definition of extended connectivity fingerprints (ECFP) given in the references.[19] It is noted that different implementations of a fingerprint typically differ in some details from toolkit to toolkit much like canonicalization algorithms for SMILES, as mentioned previously. The reasons are manifold ranging from the way in which aromaticity is accounted for by different software packages over slight modifications of implementation details to the way in which features are represented as numbers. The latter issue is especially relevant for ECFP fingerprints that rely on a hashing function converting the representation of an atom environment into an integer value. In extended connectivity fingerprints, the structural features comprise the atom environments of increasing radius around each atom. Figure 24.8 illustrates the concept. The features can be organized in levels. The first level consists of the atom itself, the second of the atom and its neighbors, the third of the atom, its neighbors, and the neighbors of the neighbors, and so on. Common usages of ECFP fingerprints use the first two, three, or four levels and are named ECFP2, ECFP4, and ECFP6, respectively. Here, the number refers to the maximum diameter of the encoded substructures.

In ECFP, the generation of representations for substructures is similar to the way invariants are generated in canonicalization algorithms. First, an initial invariant is generated, that only encodes information about the atom and its bonds much like the invariant of the CANON algorithm. In each step, a new invariant is generated that combines the original invariant with the invariants of its immediate neighbors, effectively encoding atom environments of bond diameter 2. In the next iteration, the invariants of the previous iteration are combined thus encoding substructures with a diameter of 4. During each iteration, the atoms of increasing distance to the central atom influence the final invariant. Invariants might be generated to uniquely distinguish between all possible environments, for instance, by using string representations similar to SMILES. However, for efficiency and

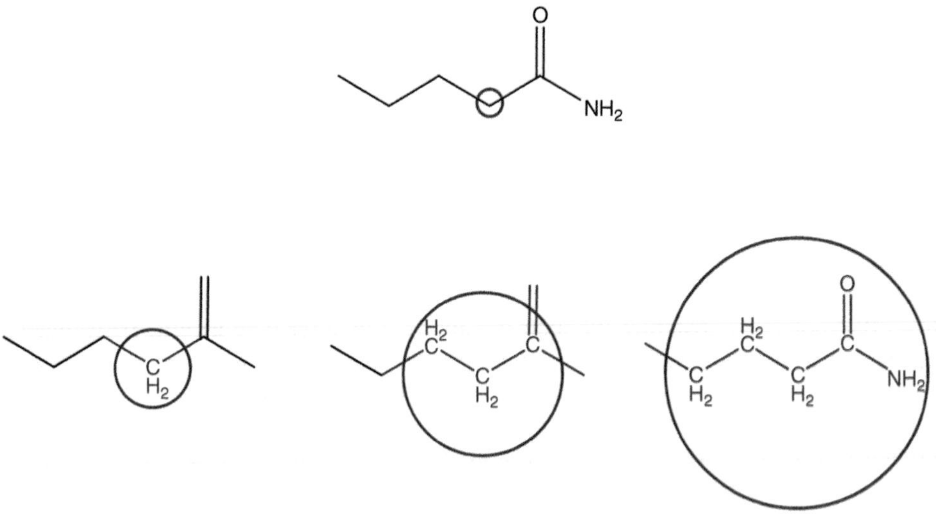

Figure 24.8 **Atom environments.** For an atom of an exemplary molecule atom environments of increasing radius are depicted. The atom environment of radius 0 (left) encodes information about atom type, hydrogen atoms, and attached bonds. The radius 1 environment (middle) includes immediate neighbors and the radius 2 environment (right) covers all atoms within a bond distance of 2 to the central atom.

practical purposes ECFP features are encoded by (32 bit) integer values. Thus, data structures describing any given atom environment need to be transformed into integer values. The process of transforming any data structure to an integer value of fixed length is called hashing. Hash functions play an important role for data structures of programming languages; dictionaries and sets in Python make use of hashing to efficiently store objects for searching and identification. Hash functions can transform any data object into an integer. However, the set of all objects such strings or, as in our case, encoded atom environments, is usually much larger than the set of fixed-length integers and thus it is not possible to associate a unique integer with each object. To alleviate this problem, effective hash functions produce seemingly random (but well-defined) numbers for different objects so that the occurrence of collisions, that is, two distinct objects having identical hash values, is probabilistically restricted. For instance, for 32-bit integers, which represent ECFP environments, one can estimate that for a set of about 77,000 different atom environments, the probability of having no collision is about 0.5.

During the generation of ECFP fingerprints, first, a unique representation for an atom environment of a given radius is generated. Then this environment is transformed into an integer using a hash function. In the next iteration, these integers are used to represent the atom environment and form the basis for the generation of representations of atom environments of increased radius.

## Implementation

### The Hashing Function

Python has a predefined hash function that generates hash values for built-in primitive data types like strings, integers, and floats, but also for immutable data types like tuples. However, there is no guarantee that hash values are reproducible across different versions, platforms, or implementations. In fact, recent versions of Python randomize the hashing function from run to run for security reasons. Thus, specifying our own hash function guarantees that our fingerprint implementation will produce identical fingerprints on different systems and versions of Python.

The following implementation defines a hash function for integer, string, and tuple objects, which will be used to encode atom environments. The hashing algorithm follows a simplified deterministic version of the Python hash function. It distinguishes between strings, integers, and tuples. Recursion is used to generate hashes for nested tuple structures.

```python
from numpy import int64, int32, integer

def myHashLong(obj):
    if not obj:
        return 0
    if isinstance(obj, basestring):
        mult = int64(1000003)
        value = int64(ord(obj[0]) << 7)
        for char in obj:
            value = (mult * value) ^ int64(ord(char))
        value = value ^ len(obj)
    elif isinstance(obj, int) or isinstance(obj, integer) or
```

```
   isinstance(obj,long):
        value = int64(obj&0xffffffff)
        obj = int(obj) >>32
        while obj!=0 and obj !=-1:
            value = value ^ (obj & 0xffffffff)
            obj = obj >>32
    elif isinstance(obj,tuple):
        mult = int64(1000003)
        n = int64(len(obj))
        value = int64(0x345678)
        for item in obj:
            value = mult * (value ^ myHashLong(item))
            mult += int64(82520)+n+n
        value += int64(97531)
    else:
        raise RuntimeError("Hash not implemented for %s"%type(obj))
    return value

def my_hash(obj):
    return int32(my_hash_long(obj) & 0xffffffff)
```

### The Initial Atom Invariant

The initial atom environment of ECFPs encodes the:

1) Number of non-hydrogen bonds,
2) Sum of the non-hydrogen bond orders,
3) Atomic number,
4) Formal charge,
5) Number of attached hydrogen atoms, and
6) Flag indicating whether the atom is part of a ring.

These atom properties are almost the same as used in the CANON algorithm with the addition of the flag indicating whether an atom is part of a ring. It is straightforward to implement the invariant using a string representation.

```
def invariant(atom):
    "calculate initial invariant"
    return "%d%02d%02d%+d%d%d"%(atom.GetDegree(),
        atom.GetTotalValence()-atom.GetTotalNumHs(),
        atom.GetAtomicNum(),
        atom.GetFormalCharge(),
        atom.GetTotalNumHs(),
        1 if atom.IsInRing() else 0,
        )
```

### The Algorithm

The ECFP fingerprints are generated in levels of increasing radius around each atom. For the first level representing radius 0 the initial invariants for each atom are calculated and are stored in a *feature* dictionary. In addition to the invariant, the algorithm also keeps track of the substructure represented by each environment using a *bondSet* dictionary encoding the bonds of the substructure. The bonds are not defined at first because the initial invariants

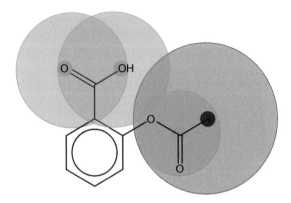

Figure 24.9 **Common atom environments.** Specific environments around different atoms can encode the same substructure of the molecule. The two environments of radius 2 of the carboxyl group are identical and the substructures of the carbon atoms of the acetyl group of radius 1 and 2, respectively, are also identical.

only encode the central atom information. This is important because substructures defined by an atom environment around a central atom can be identical for different central atoms and different radii as the example in Figure 24.9 shows. For each substructure (identified by its bond set) only a single canonical feature value is calculated. This is accomplished by associating a single invariant with a specific bond set, again using a dictionary data structure (`bondSetFeatures`). The dictionary records all substructures encountered thus far. The same substructure may be encountered multiple times during the algorithmic procedure yielding different invariants. In order to resolve ambiguities two rules are applied for generating canonical invariants for identical substructures:

1) Always select the invariant generated from a smaller radius,
2) If invariants of identical substructures are generated for atom environments of the same radius, retain the one with the smaller value.

During an iteration, a new invariant for an atom representing an environment of increased radius is calculated from the old one by encoding:

1) The current level (radius),
2) The previous invariant of the atom,
3) A sorted tuple of invariants of all neighbors and their bonds for which a hash value is determined.

```
from rdkit import Chem
import numpy as np
from collections import defaultdict
def ecfcFingerprint(mol, radius):
    graph = dict()
    for atom in mol.GetAtoms():
        nbors = []
        for bond in atom.GetBonds():
            nbr = bond.GetOtherAtom(atom)
            nbors.append((nbr.GetIdx(),
                        bond.GetIdx(),
                        int(2*bond.GetBondTypeAsDouble())))
        graph[atom.GetIdx()] = nbors
```

For convenience, a graph structure is initialized where keys are the atoms and the values represent information about the bonds: the attached atom index, bond index, and bond order.

```
features0 = dict(((atom.GetIdx(),my_hash(invariant(atom)))
      for atom in mol.GetAtoms())
bondSets0 = defaultdict(frozenset)
atomIndices = features0.keys()
```

The first level contains the initial atom invariants and encode substructures with no bonds.

```
bondSetFeatures = dict()
oldBondSets = bondSets0
oldFeatures = features0
for lvl in range(radius):
    newBondSets = dict()
    for a in atomIndices:
    # Bond set of next level consists of old bond set + ...
    bondSet = set(oldBondSets[a])
    nb = []
    for nbor,bond,order in graph[a]:
        # ...+ old bondsets of neighbors
        bondSet.update(oldBondSets[nbor])
        # ...+ bonds to neighbors
        bondSet.add(bond)
        nb.append((order,oldFeatures[nbor]))
    nb.sort()
    bondSet = frozenset(bondSet)
    newInv = my_hash((lvl,oldFeatures[a], tuple(nb)))
    newBondSets[a] = bondSet
```

When determining new invariants in the main loop, for each atom, bondSet represents the atom environment of increased radius and nb combines the old invariants of the neighbors and their respective bonds in a list. The new invariant newInv is determined as a hash value combining level, previous invariant, and neighbor information.

```
if (bondSet not in bondSetFeatures or
    (bondSetFeatures[bondSet][0] == lvl and
     bondSetFeatures[bondSet][1] > newInv)):
    bondSetFeatures[bondSet] = (lvl,newInv)
```

The new invariant is added to the dictionary of bond set features, or if an invariant for the encoded substructure already exists, the dictionary is updated according to the rules outlined above.

```
    # Update invariant values for current layer from bondSetFeatures
    newFeatures = dict()
    for a,bs in newBondSets.items():
        newFeatures[a] = bondSetFeatures[bs][1]

    oldBondSets = newBondSets
    oldFeatures = newFeatures
```

The features for the current level are updated from the bond set features dictionary for the next iteration.

```
# get all features from bondSetFeatures and the initial invariants
features = sorted([inv for _,inv in bondSetFeatures.values()] +
                  features0.values())

return features
```

The final fingerprint combines the initial features with the combined features for the additional levels. The list contains duplicates, if encoded substructures occur multiple times in the molecule.

```
def uniq(featureList):
    "remove duplicate features in sorted list"
    uniqList = featureList[:1]
    for f in featureList:
        if f != uniqList[-1]:
            uniqList.append(f)
    return uniqList

def ecfpFingerprint(mol, radius):
    features = ecfcFingerprint(mol, radius, invariant)
    return uniq(features)
```

The binary ECFP version of the fingerprint is obtained from the count version by removing duplicates.

The code has been integrated in the script `ecfp.py` and can be used to generate binary and count versions of atom environment fingerprints of different radius similar to the `generateFingerprints.py` program of the first section of this chapter.

## Summary

Atom environment fingerprints encode substructures of predefined radii around a central atom using a hashing function. The substructures are encoded in a data structure that is hashed to yield an integer value representing the substructures. Features are generated in levels of increasing radius. In each level the invariants of the neighbors of the central atom are combined analogously to the invariants of canonicalization algorithms. Because identical substructures can be represented by different central atoms care must be taken to ensure that a canonical version of the invariant is chosen as a feature.

## References

1 Molecular Operating Environment (MOE). http://www.chemcomp.com, Chemical Computing Group: Montreal, Canada. (accessed 09/16/2015)

2 Pipeline Pilot. http://accelrys.com/products/collaborative-science/biovia-pipeline-pilot, BIOVIA, San Diego, CA, USA. (accessed 09/16/2015)

3 Berthold, M.R.; Cebron, N.; Dill, F.; Gabriel, T. R.; Kötter, T.; Meinl, T.; Ohl, P.; Sieb, C.; Thiel, K.; Wiswedel, B. KNIME: The Konstanz Information Miner. In *Studies in Classification, Data Analysis, and Knowledge Organization (GfKL 2007)*, Springer: Heidelberg, Germany, **2007**.

4 Steinbeck, C.; Han, Y.Q.; Kuhn, S.; Horlacher, O.; Luttmann, E.; Willighagen, E.L. *J Chem Inform Comput Sci* **2003**, *43*, 493–500

5 Daylight Toolkit. http://www.daylight.com/products/toolkit.html, Daylight Chemical Information Systems, Inc., Laguna Niguel, CA, USA. (accessed 09/16/2015)

6 JChem. http://www.chemaxon.com, ChemAxon, Budapest, Hungary. (accessed 09/16/2015)

7 O'Boyle, N.; Banck, M.; James, C.; Morley, C.; Vandermeersch, T.; Hutchison, G. *Journal of Cheminformatics* **2011**, *3*, 33.

8 OEChem TK. https://www.eyesopen.com, OpenEye Scientic Software, Santa Fe, NM, USA.

9 RDKit: Open-source cheminformatics. http://www.rdkit.org.

10 Irwin, J.J.; Shoichet, B.K. *Journal of Chemical Information and Modeling* **2005**, *45*, 177–182

11 Bento, A.P.; Gaulton, A.; Hersey, A.; Bellis, L.J.; Chambers, J.; Davies, M.; Krüger, F.A.; Light, Y.; Mak, L.; McGlinchey, S.; Nowotka, M.; Papadatos, G.; Santos, R.; Overington, J.P. *Nucleic Acids Research* **2014**, *42*, 1083–1090

12 Molecular Operating Environment (MOE). http://www.chemcomp.com, Chemical Computing Group: Montreal, Canada. (accessed 09/16/2015)

13 Bajorath, J. *Nature Reviews Drug Discovery* **2002**, *1*, 882–894

14 Johnson, M.A.; Maggiora; G.M. *Concepts and Applications of Molecular Similarity*, John Wiley & Sons: New York **1990**.

15 Stumpfe, D.; Bajorath, J. Similarity searching. *Wiley Interdisciplinary Reviews: Computational Molecular Science* **2011**, *1*, 260–282

16 Willett, P. *J Chem Inform Comput Sci* **1998**, *38*, 983–996

17 Willett, P. *Journal of Chemical Information and Modeling* **2013**, *53*, 1–10.

18 MACCS Structural keys; Accelrys Inc.: San Diego, CA, USA.

19 Rogers, D.; Hahn, M. *Journal of Chemical Information and Modeling* **2010**, *50*, 742–754.

20 McGregor, M.J.; Muskal, S.M. *J Chem Inform Comput Sci* **1999**, *39*, 569–574

21 Brown, C.D.; Davis, H.T. *Chemometrics and intelligent laboratory systems* **2006**, *80*, 24–38

22 Weininger, D. *J Chem Inform Comput Sci* **1988**, *28*, 31–36

23 Daylight Theory manual. http://www.daylight.com/dayhtml/doc/theory/, Daylight Chemical Information Systems Inc., Laguna Niguel, CA, USA. (accessed 09/16/2015)

24 Weininger, D.; Weininger, A.; Weininger, J.L. *J Chem Inform Comput Sci* **1989**, *29*, 97–101.

25 Cormen, T.H.; Leiserson,C.E.; Rivest, R.L., Stein, C. *Introduction to Algorithms*, 2nd edition. MIT Press: Cambridge, MA, USA, **2001**.

26 IUPAC International Chemical Identifier (InChI). http://www.iupac.org/home/publications/e-resources/inchi.html. (accessed 09/16/2015)

27 Morgan, H.L. *Journal of Chemical Documentation* **1965**, *5*, 107–111

28 Carhart, R.E. *J Chem Inform Comput Sci* **1978**, *18*,108–110

29 Ullmann, J.R. *Journal of the ACM* **1976**, *23*, 31–42.

# Index

*Tutorials in Chemoinformatics*, First Edition. Edited by Alexandre Varnek.
© 2017 John Wiley & Sons Ltd. Published 2017 by John Wiley & Sons Ltd.
Companion website: www.wiley.com/go/varnek/chemoinformatics

Printed and bound by CPI Group (UK) Ltd, Croydon, CR0 4YY

27/10/2024

14580304-0002